UNDERSTANDING SOCIAL ANIMALS

By

Dr. Amita Sarkar
Dept. of Zoology
Agra College
Agra (U.P.)
(India)

DISCOVERY PUBLISHING HOUSE PVT. LTD.
NEW DELHI-110 002

First Published-2010

ISBN 978-81-8356-509-7

Published by:

DISCOVERY PUBLISHING HOUSE PVT. LTD.
4831/24, Ansari Road, Prahlad Street,
Darya Ganj, New Delhi-110002 (India)
Phone: 23279245 • Fax: 91-11-23253475
E-mail: parul.wasan@gmail.com
info@discoverypublishinggroup.com
Website: www.discoverypublishinggroup.com

Printed at:

Sachin Printers
Delhi

Preface

The present title "Understanding Social Animals" has been written for those students interested in careers in diverse fields of biological sciences. It provides a structured approach to learning by covering all the important topics in a uniform, systematic format. The book has been comprehensively designed incorporating recent advances in this fast moving field. It also provides accessible information on social animals in compact form for undergraduate students in biology and related life sciences. It is intelligible to the educated layman, though it deals with some complex ideas. It is an adequate text for all the requirements of students in this area. In addition, busy lecturers who require a quick reference compendium will find it useful, particularly for tutional planning. Simple, yet hopefully clear figures and tables are provided throughout the book.

The over-riding goal of this book, and indeed of the whole *Understanding series*, is to present the essential information concering social animals in a compact, readily accessible form which leads itself to student learning and revision. The convergence of various approaches has generated a rich panorama of detail, the significance of which we are still attempting to unraval. The present text has been written as an introduction to this rapidly growing field.

To make the work more comprehensive and informative, the author has consulted many authoritative books, research journals, abstracts, monographs etc., so there can be no claim to originality except in the manner of treatment.

The author expresses his thanks to his friends and colleagues whose continue inspirations have initiated him to bring out this book.

The author expresses his gratitude to Mr. Wasan and staff of M/s Discovery Publishing House Pvt. Ltd. for their whole hearted co-operation in the publication of this book.

In the mean time, the author will remain sincerely responsible for any shortcomings of the book and be grateful to the readers for their suggestions and constructive criticism for the continuous betterment of the book. He takes this opportunity to appeal to the readers to send their suggestions straightaway to his Publisher.

Author

Contents

1

INTRODUCTION

Animals forms groups of enhance their foraging success and gain better protection against predators; but group living also exposes them to diseases, competition, and social interference. The costs and benefits of social living vary with the age, sex, experience, and physical condition of individuals. Social groups are maintained by chemical, visual, auditory, and tactile signals. Signals benefit both receivers and senders, but animals do not always communicate their intentions clearly and accurately. Complex social systems have evolved from offspring remaining with parents to help rear future broods and from cooperation among adults of the same generation.

This chapter deals with costs and benefits of group living, roles of the sexes, types of communication signals and their evolution, choice of associates, and the evolution of animal societies.

Even solitary organisms must interact to exchange gametes. Among many sessile marine organisms, the gametes do the associating by themselves in the water, but most mobile animals get together to copulate, even if they part immediately afterward. From these brief and simple interactions, social behaviour had developed into more and more elaborate systems, culminating in the complex societies of social insects and vertebrates. How these systems have evolved, why it was advantageous for animals to associate with one another, and how social groups are maintained are the subjects of this chapter.

Costs and Benefits of Social Living

Humans tend to think that the highly complex societies of some vertebrates, ourselves included, represent a pinnacle of evolution. This self-congratulatory view implies that "highly evolved" creatures, like ourselves, represent a form of life superior to the less highly social species. To adopt this attitude is to ignore the broad spectrum of potential disadvantages, as well as possible advantages that arises from social living. One of the more thorough explorations of the costs and benefits of living in a group is the study of bank swallow behaviour by *John Hoogland* and *Paul Sherman*. Bank swallows form nesting colonies of dozens to hundreds of pairs in sand quarries and river banks. *Hoogland* and *Sherman* showed that by nesting together the swallows are able to reduce the effectiveness of some of their diurnal predators. When a hunting jay or weasel approaches he colony, the birds mob the predator, much as gulls do and sometimes succeed in driving their enemy away. In addition, pairs seek to acquire a centrally located nest burrow and so to "use" the other members of the colony as a shield against nest-robbing predators that tend to attack the peripheral members of a group first. Moreover, because the birds link their reproduction with the pairs nesting around them, they avoid producing eggs or immature young when other birds are not reproducing. If the breeding period were extended, a predator could harvest the offspring of the swallows one by one, as they were generated. By synchronizing egg laying, a pair of swallows adds its progeny to a large pool of

Table 1.1. Some major advantages and disadvantages of sociality*

Advantages

Reduction in predator pressure by improved detection or repulsion of enemies

Improved foraging efficiency for large game or clumped ephemeral food resources.

Improved defense of limited resources (space, food) against other groups of conspecific intruders.

Improve care of offspring through communal feeding and protection.

Disadvantages

Increased competition within the group for food, mates, nest sites, nest materials, or other limited resources.

Increased risk of infection by contagious diseases and parasites

Increased risk of exploitation of parental care by conspecifics

Increased risk that conspecifics will kill one's progeny.

* For solitary species certain of the disadvantages outweigh any benefits of social living; for social species the costs are more than matched by certain of the advantages of sociality.

vulnerable offspring that exists for only a short period. Because predators can consume only a few nestling per day, synchronized breeders may have a better chance of getting their clutch through the dangerous period than a synchronous reproducers.

Some workers have suggested that an additional benefit of colonial nesting in this species might be improved foraging success. Birds that had failed to find much food might be able to follow other individuals to ephemeral concentrations of aerial insects these luckier or superior hunters had found. Although *Hoogland* and *Sherman* found little evidence to support this possibility in the colonies they observed, it might apply in other years or in other areas.

Offspring the benefits of colonial nesting in bank swallows is a number of documented costs. Nestings in large colonies suggests that the presence of many swallows in an area depletes food resources and actually makes it harder, not easier, for birds to find insects. Moreover, the swallows fight for nest sites and nesting materials; a solitary pair would not run the risk that a neighbour would make off with the hard-earned feathers it had collected for a nest lining. Nor would the male of a solitary pair run the same risk of cuckoldry, which would cause him to spend his breeding season caring for the progeny of an adulterous mate and another male. "Illicit" matings of this sort do occur in colonies. Finally, Hoogland and Sherman counted the bird fleas in bank swallow burrows, no doubt an edifying task, and found that the probability that a nest was infested was higher in large than in small colonies.

Because bank swallows never nest alone, but always in groups, one presumes that the benefits, primarily improved antipredator defense, outweigh the numerous disadvantages. However, the benefits of sociality vary from individual to individual, depending on the success of groups mobbing, the ability of the birds to claim central nesting sites, and their capacity to synchronize egg laying with other members of the colony. In other social species, the costs and benefits of group living vary so much from place to place or season to season that social units will change dramatically in size or even disband entirely when this is advantageous to individuals.

Animals in groups gain a number of advantages related to defence. While a group of animals is bigger than a single one, and so can be spotted from further off, if the predator only eats one prey at a time it will take longer to find meal if the prey are clumped than if they are evenly spread out. Furthermore, there are more pairs of eyes in a group, so the predator is less likely to creep up undetected. To some extent this is offset by the fact that animals in groups are able to be less vigilant than solitary individuals. Lone geese, for instance, raise their heads much more frequently while cropping the grass than they do when feeding in groups. Because there are more pairs of eyes in a group, so the predator is less likely to creep up undetected. To some extent this is offset by the fact that animals in groups are able to be less vigilant than solitary individuals. Lone geese, for instance, raise their heads much more frequently while cropping the grass than they can afford to keep their heads down and feed for longer periods with the security that others will raise a warning if danger threatens.

Another reason for grouping was pointed out by William Hamilton in a paper called 'Geometry for the selfish herd.' He made the simple point that, if predators just take one prey at a time, the best prey strategy is to keep another individual between oneself and the predator. Prey animals that do this will tend to form groups simply because being in a group minimises the area of danger to each animal. Being in a group can also confuse a predator because it has difficulty fixating upon and chasing a single individual. Some animals do cooperate in defence, however, giving yet another advantage to group living Musk oxen set upon by a pack of wolves will form a circle facing outwards so that wolves are confronted by a solid wall of horns and cannot reach any of their vulnerable flanks.

A final example of grouping geared to predator defence is a very curious one and this is in the nesting behaviour of ostriches. Several female ostriches lay eggs in one nest, but only the first to lay incubates: she accumulates a much larger number of eggs in this way than she could lay herself. After the eggs hatch the female may have around 25 chicks following her as she sets off across the savannah, but later she may collect even more as a result of chasing other hens away from their chicks. Brain Bertram, who studied this extraordinary system, even recorded one group in which two adults were accompanied by 105 chicks.

What advantage can there be to a female ostrich in caring for the eggs and chicks of others? The answer seems to be that chicks survive better in larger groups so that her own chicks benefit by a 'dilution effect.' Predators such as golden jackals will eat ostrich eggs and chicks, but the more of these that a female has with her the less likely is the one that is taken to be one of her own offspring. This cannot be the whole story, however, otherwise a female would gain whether she cared for her young herself or let another female do so. At the stage of incubation it seems that there are advantages to the hen that sits, for she can recognise her own eggs and she gives preference to them so that they remain in the centre of the nest and are more likely to hatch. She is not therefore begin a generous nanny helping out others, but is running the creche for her own advantage. It remains to be seen whether females with chicks also gain from giving care themselves rather than leaving them to be looked after by others.

Finding Food

A further range of benefits to group living is related to the exploitation of food supplies. If food is clumped, as is very often the case, animals that feed on it tend to become so too. Seeing another individual feeding may be a good indication of where food is abundant, so groups may form for this reason alone without there being many advantage in being a member of a group. Indeed the animals may interfere with each other's feeding efforts but, if they are conspicuous, there is no way in which they can stop newcomers arriving to share in the spoils.

Not all feeding groups, however, are passive clumps. While many predators, such as frogs, snakes and domestic cats, are solitary hunters, using stealth to catch small prey, some others, like lions, spotted hyaenas and wild dogs, cooperate in the hunt. This may increase the chances of success, because individuals can take turns in chasing and because they can spread out and so limit escape routes. It also enables them to tackle for larger prey than could an individual on its own : a pack of hyaenas can pull down a zebra or a wildebeest, or some of them can distract a rhino mother while others attach her calf.

A final advantage of grouping may also be important in animals that are not carnivorous. This is that groups may be able to make use of resources in a more systematic way than can isolated individuals. The bumble-bee flying from flower to flower may well be following close behind another one and so getting less reward than it would otherwise do. If individuals form flocks and move round their range together, as is often true of small birds outside the breeding season, they can be sure that they are all arriving in a fresh area where the food has had a chance to replenish itself since last they came that way.

Many monkey groups occupy quite large territories and move over them in a single partly searching for the fruit or buds on which they live. They too may gain from the systematic foraging that group living allows, but their groups are more integrated and the relationships of individuals within them more varied than in a flock of birds or a herd of deer. The advantages of food finding or predator avoidance may explain why they, like many other animals, come to live in groups, but to little to account for the richness of the interactions between the group members. It is to various facets of the behaviour of individuals within groups towards one another that we shall now turn.

Kinship

Not surprisingly, animals that live in social groups tend to be related to one another. Young ones are born into the group to which their mother and father belong and may often stay in it to become parents themselves. However, if this was always so, animals in the group would become more and more inbred. Too much inbreeding leads to young which are less viable, so it is discouraged by natural selection. What usually happens, therefore, is that, as maturity approaches, young animals of one sex or the other move away from the group in which they were born and so do not mate with close relatives. In birds it is usually the females that disperse to different areas, whereas in mammals the males are more likely to leave, although there are quite a few exceptions to this rule, including species in which both sexes move away.

Which sex disperse may have a profound effect on various aspects of social behaviour. Male white-crowned sparrows sing songs which are almost identical with those of their neighbours and quite different from those of birds a few kilometres away. Song learning usually only occurs in the first three months of life and so it complete long before they set up their territories and start to sing. But they breed only a short distance from where they hatched and thus the song dialects are maintained. Although female birds do not usually sing, their choice of mate is probably influenced by the songs of prospective partners. In white-crowned sparrows, for example, there is evidence that females may prefer males from the same dialect area as themselves, though there is some controversy about this.

By contrast with birds, groups of many mammalian species are matriarchal, females staying with their mothers, while males leave to find other groups which they can join. Lions are a case in point here. As already mentioned that males leave their natal group and go in search of another that they can take over. Females, however stay in the group in which they were born so that the lionesses in a pride are usually closely related to one another. They often give birth at around the same time and it has been found that they will suckle each other's cubs, something one would only expect if they had an interest in them through being related. Furthermore, again in line with what kinship would predict, they suckle their sisters' offspring more than those of their cousins.

Alarm calling is another behaviour where kinship is important. Making a nose when a predator is around is obviously a risky thing to do, and one would expect to find some advantage to offset this. Many birds only call when they have young in the nest, so saving these form being discovered seems to be the benefit. Using a stuffed badger as a 'standard predator', John Hoogland has shown that black-tailed prairie dogs in America produce more alarm calls when they have close relatives in their group than when they do not. Males called a lot in their natal group but ceased to do so after they moved elsewhere, only to start calling again once their own young were born in the new group that they had joined. All these finding fit in well with the idea that animals are more likely to call when relative may benefit.

As far as the relatedness of their members of concerned, the most complex groupings are amongst the social insects. In the honey-bee, only the queen lays eggs and all other females in her hive are sterile workers. The eggs the queen lays are of two sorts: unfertilised ones, which will develop into fertile males (drones), and fertilised one, which usually develop into infertile females (workers) but, if nourished only on royal jelly, will form a new queen. This happens when the old queen dies or when the colony has grown to a point where it must split. A peculiarity of this breeding system is that drones have half the number of genes that queens or workers to and, when they fertilise an egg, all their genes pass to their offspring instead of just half of them. As a result of this system, which is known as haplodiploidy, workers share 3/4 of their genes with their sisters rather than 1/2, as most animals do. It is thought to be for this reason that it benefits them to raise sisters which will become queens rather than having daughters of their own. Through a quirk of their reproductive system, their sisters are more closely related to them than their daughters would be and are thus a better investment in the future of their genes.

The social system of bees is thus extraordinarily intricate, and this is probably because these animals are especially closely related to others in the hive to which they belong. However, having this unusual mode of inheritance is not a prerequisite for a social system of this sort. Termites are not haplodiploid yet they have a colony structure which is just as complex. More remarkable still is the naked mole rat, for this is a mammal which has colonies much more like those of a social insect. These animals live communally in groups of around 40 individuals which share a burrow. For must of the time they huddle together in one underground chamber, and it is here that the only female to breed has her young. She is large, and males approaching her size may mate with her, but most other large animals in the group neither breed nor forage : their main role seems to be to keep the colony warm. Smaller individuals are like the workers of social insects. They build nest and more around through the burrow system foraging for roots and tubers which they bring back for all the animals to feed upon. If there is a disturbance, all the members of the colony join in the task of carrying the young off to safety.

As well as the food that the workers bring back being shared, the young obtain nourishment form adults by eating their faces. Individuals also spread their urine round the colony and groom with it. These habits, unsavoury as they may seem, probably serve to pass information around the group for, if the breeding female dies or is removed, another one comes into reproductive condition very soon. This is just like the social insects. In these, queen substance is spread throughout the colony and stops the workers from rearing other queens. But, if the queen dies, this substance is no longer there, and the workers quickly start to rear a replacement.

Male rats are not very mobile and their burrow systems are isolated from each other. It is likely, therefore, that most of the animals within a colony are relatives and that their close social relationship stems from the common interest that this gives them in the offspring of the breeding female. Unfortunately we do not know just how closely related they are to each other, and this is very often the case in studies of social behaviour. It is easy to speculate that kinship is involved when we observe animals being altruistic to one another, but only in a few cases, such as that of the social insects, do we know in any detail exactly what the relationships are between the individuals involved.

INSECT SOCIETIES

Social Life of Insects

Ethologists honour four groups of insects—ants, bees, wasps and termites—with the description of 'social' insects. The crucial characteristic of these four groups is that, within their nests, they show a reproductive division of labour. Within the nest of a typical ant species there is a single queen who lays nearly all the eggs that are laid in the nest. The rest of the ants are sterile 'workers'. The workers carry out all the duties necessary to keep the colony going except for the laying of eggs.

There are more than 12,000 social insect species, and they show a fascinating range of ways of life. Within the ants, for example, there are 'army' ants with huge colonies of up to 22 million individuals, which bestride the jungle floor eating everything edible in their path. There are 'fungus gardening' species which grow fungus on specially prepared rotting leaves and live on the produce of the fungus. Other ant species live by milking honeydew from herds of little insects called aphids. In yet others, workers from living 'honeypots', as they hang upside down from the roof of their nest, their abdomens hugely distended with honey. Australian aborigines dig up the nests, take the ant's head between their fingers, and bite off the honeyed abdomen. The marvels of the social insects are almost endless, but we shall concentrate on some general properties of their social life that are closely related to their altruistic habits.

We must consider ants, bees and wasps separately from termites. Ants, bees and wasps belong to the insect order Hymenoptera; termites make up the order Isoptera. A typical hymenopteran colony is founded by a single queen after her 'nuptial' flight. If the species is an ant, the foundress bites off her wings—she will never fly again. She excavates a small nest, lays her first brood, feeds and rears the larvae. She will never rear larvae again, because this first generation of workers themselves rear the next lot of eggs. It is an important fact that all the workers, in the first and later generations, are females. They are sterile and therefore in a sense sexless, but they contain the genetic make-up of females. Once the colony has reached a certain size the queen lays eggs which are reared as reproductives. The time when reproductives are first produced varies between species. In *Myrmica rubra*, a common garden ant in Europe, it is not until about 9 years after founding, when the colony has grown to about 1000 workers.

Altruism in Insect Societies

The distinctive property of social insects that cries out for explanation is the sterility of the workers. Why do these workers slave away to their death only to enhance the reproduction of another individual? An important part of the answer may be provided by *Hamilton's theory*. For Hymenoptera have an idiosyncratic genetic system, rather different from the usual Mendelian one. They are 'haplodiploid'. The females, like most animals have two sets of chromosomes (they are 'diploid'); but the males have only one (they are 'haploid'). Male offspring contain no genes from their father; they develop from unfertilized eggs. Females contain genes from both their mother and their father. Now haplodiploidy, as *Hamilton* pointed out, will alter the normal pattern of relatedness. The sister of one family are exceptionally closely related because they all share exactly the same set of genes from their father; their father has only one set of genes to give. Many of the values of relatedness are changed under haplodiploidy.

The relatedness between siblings under diploidy was calculated under the premise that if a father contributes a gene to one sibling there is only a chance of one half that he will give it to another. Under haplodiploidy that chance is one, not a half, and the total relatedness between sisters is therefore (½ × 1) + (½ × ½) = 3/4. Because the relatedness of mother to daughters, and one, other things being equal, 'breed' more efficiently by making sisters than by reproducing daughters. They may be why female hymenopterans have so often evolved sterility. Males, by contrast, have not evolved sterility. But then they are not exceptionally closely related to their siblings.

Table 1.2. Relatedness under haplodiploidy. The relatedness is the probability, given a gene is in one kind of individual, that it is in another kind of individual. They can therefore be asymmetrical, as between mother and son.

Relationship			*relatedness (r)*
*	Mother	daughter	½
*	Mother	son	½
*	Father	daughter	1
*	Father	son	0
*	Daughter	mother	½
*	Son	mother	1
*	Brother	sister	½
*	Brother	brother	0
*	Sister	sister	3/4
*	Sister	brother	¼

A large controversial literature has grown up around Hamilton's explanation of sterility in hymenopterans. It would be inappropriate to consider it here. No ethologist would doubt that kin selection has had some part in the evolution of hymenopteran social behaviour, but this statement leaves plenty of room for disagreement. The same exact theory cannot apply to termites. Termites are diploid, therefore the relatedness of siblings is normally the same as that of parents to offspring. But sterile castes have evolved in termites. As Hamilton's theory predicts, in termites both sexes of offspring become sterile workers; in termites, unlike Hymenoptera, there is no force of kin selection predisposing one sex to evolve sterility. However, although the theory can explain this difference from Hymenoptera, there is no widely accepted explanation of the evolution of sterile castes in termites. Suggestions have been made. The relatedness among siblings can be higher than that of parents to offspring if siblings are produced by highly incestuous matings but the offspring produced by outbreeding. A termite that 'went alone' might outbreed, whereas it could help to rear closely-related, inbred siblings if it stayed on as a helper. However, the idea is theoretical only, and is not well tested. I only mention it to illustrate that the theory of kin selection has been applied to termites as well as Hymenoptera. The theory is more easily applied to Hymenoptera.

Recognizing Relatives

We have not proved beyond doubt that kin selection has caused the evolution of altruism in avian helpers and social insects, but the theory probably contains a large measure of truth in these cases. We might pause, therefore, to ask how animals actually recognize their genetic relatives. For if animals are to act altruistically towards relatives, they must be able to recognize them. However, do they do so? No work has been done on helpers at the nest. They could 'recognize' relatives merely as other young birds in the same nest as they were reared in as they would usually be siblings. In practice, recognition may be more sophisticated, but in the absence of facts speculation is unnecessary, because there is no difficulty of principle here. There has, however, been some work on insects merits mention. Colonies of social insects have been known for many years to possess individual colony odours. Individuals can tell members of their own colony (who are genetic relatives) from members of other colonies (unrelated) by their characteristic smell. The source of colony-specific odours is not completely known; but it probably develops from the diet of the colony, as individuals frequently regurgitate food to one another, and minor differences in the diets of different colonies would give them different characteristic odours. Different colonies perhaps secrete their own pheromones as well, but the importance of diet has been indicated by experiment. Kalmus and Ribbands moved two hives of honeybees from a typical honeybee environment to an isolated moor which had only one species of flower. The level of fighting between the two hives decreased, perhaps because they increasingly came to recognize each other as members of the same hive. Likewise, when parts of a hive were isolated and fed on different diets, the level of fighting within a hive increased.

Each member of the colony has to learn the colony odour. Young ants do not challenge other ants in the colony, regardless of their odour; but older ants challenge any ant that does not have the colonial smell. The learning may even take place during a sensitive phase of imprinting.

What I have said so far would probably have passed until quite recently as a short summary of how social insects recognize nestmates. It was supposed that relatives are recognized by cues of environment origin, by learning. A rush of recent evidence now suggests that story is incomplete. In several species, individuals have been found to be capable of distinguishing relatives even though they had no chance of learning to recognize them: they seem to possess an innate ability to distinguish relatives. The first important experiment was conducted on sweat bees, *Lasioglossum zephyrum*, by *Greenberg* in 1979. By various genetic tricks, he bred colonies in the laboratory that had twelve different kinds of relatedness to one another, varying from completely unrelated colonies ($r \sim 0$) to different colonies produced from inbred lines (r*1). In his experiments he introduced a bee from one colony to the entrance of another colony. At the colony entrance a 'guard' bee generally admits nestmates and challenges foreigners, and *Greenberg* observed the rate at which guard bees admitted introduced bees bearing differing degrees of relatedness to them. His result was strikingly positive; guard bees were more likely to admit bees the more closely related the introduced bee was. The important part of Greenberg's experimental design was that in nearly all cases the guard bee had never had any opportunity at all to learn the introduced bee's smell: in nearly all cases guard and introduced bee had spent all their lives in different colonies. For a minority of cases the bee would have been separated at the larval stage from the same colony, but no learning is thought to take place this early, and, in any case, the positive result still stands even if this minority of experiments is ignored. Sweat bees, it appears, can recognize degrees of relatedness among conspecifics of which they have no experience at all. Since Greenberg's work on sweat bees, moreover, a comparable ability has been demonstrated with varying degrees of certainly in other bees, and other animals, such as ants, mice, quails and tadpoles. In now seems reasonable to guess that genetic abilities to recognize relatives are widespread. Our original understanding must therefore be modified. Animals can recognize their relatives by a mixture of innate and learned information, in ways that probably differ among species and are not, at this early stage of research, will understood.

Primate Societies

The Social Life of Primates

Different species of primates live in different kinds of societies; indeed, the same species may form different kinds of social groups according to the conditions. The Hanuman langur (*Presbytis entellus*) forms both harems, with one male and several females, and multimale groups, with several males and a larger number of adult females. The reason is uncertain, but may be related to population density, for multi-male groups are commoner when the total population density of an area is lower. Before considering the altruistic behaviour of primates, let us consider briefly some of the different kinds of social groups.

The white-handed gibbon (*Hylobates lar*) lives in the trees of South East Asia in monogamous family groups. The male and female are similar in size and appearance, and live as a pair with 0-4 young. The pair defend a territory of about half a square mile. Within the family there is no dominance or aggression: male and female live together peaceably as equals. Other primates too, such as the spider monkeys (*Ateles*), also live in family groups. But how different are the societies of hamadryas baboons! The hamadryas baboon (*Papio hamadryas*) is a terrestrial species, inhabiting the plains of North East Africa and South West Arabia. The hamadryas baboon society has several levels of organization. The lowest level, the breeding unit, is a harem of one to ten females with a single male. The male treats his females aggressively, often attacking them if they stray away. The male is about twice as large as the female, which influences the dominance relations of the sexes, but the large size of males is probably mainly an adaptation for fighting off enemies, especially rival males. At the next level, several of these harems may walk around together in larger bands while feeding, and these bands may act as a unit to defend a food source from a rival band. The hamadryas baboon, however, does not defend a territory. The bands do confine their wandering to a large area of about 12-15 square miles, but different bands may overlap in their use of this area. It is, therefore, called a home range, to distinguish it from a defended area, which would be called a territory. At a higher level, several bands may join into larger groups for sleeping. If suitable shelters for sleeping are difficult to find, as many as 700 hamadryas may sleep together.

Whereas in the hamadryas baboon the mating unit is a single harem aggressively controlled by a single male, in the howler monkey (*Alouatta*), several adult males may live peaceably together in a single group. When Ray Carpenter watched howlers on Barro Colorado Island, in Central America, the group sizes were variable but contained an average of 3 males, 8 females, and 7 young. The male howler monkey is about 30% larger than the female, and has an enlarged voice box covered by a beard. The males roar daily, making the loudest animal noise in the American forests, a noise which carries for over a mile. This howling serves to space out the different groups. Within each group there is little aggression and no obvious dominance hierarchy.

These three species live in entirely different kinds of societies. One is monogamous, another polygamous within single male groups, another polygamous with multi-male group. One shows dominance and agression within groups, the other two do not. Two defend territories, the other does not. Other primate species live in yet other kinds of societies. Ethologists would like to be able to explain why different primate species live in different kinds of societies. Why are some species aggressive, others not? Some territorial, others not? These questions cannot yet be answered satisfactorily. Our understanding of the diversity of primate societies is still limited. One trend which can be explained is as follows. There is a tendency in species in which there are many females per male in the group for the males to be bigger than the females: in monogamous species the male is about the same size as the female; in polygamous species the male is larger. This has presumably arisen because sexual selection has favoured larger males in polygamous species, because larger males are more successful in fights over females.

Altruism in Primates

Many kinds of altruistic behaviour can be seen in primate groups. There is the feedings carrying and defence of young, not only by their mothers; cooperative searching for, hunting and exploitation of food; food-sharing; cooperative group defence against enemies; and most common of all, that favourite pastime of primates, grooming. Most, perhaps all, of these habits should probably be explained by kin selection. We, however, have considered the application of that theory in two examples already, and will therefore, use a primate example to illustrate another reason why altruism may evolve: the theory of reciprocal altruism.

The example concerns 'consorting', which is a habit of males, in many species that live in multi-male groups, whereby a male stays close to a female during the receptive phase of her oestrous cycle, and defends her from the advances of other males. It is an adaptation produced by the male competition component of sexual selection. In the olive baboon (*Papio anubis*), Craig Packer observed that two males may occasionally co-operate to fight off a single male: clearly, two will be stronger than one, which makes the advantage of co-operation clear. The advantage, however, is only for the one male that copulates with the female; he has gained a benefit of as much as one extra offspring: the other male has paid a cost of the risk of injury in a fight, but obtains no benefit. He has behaved altrustically. If what we have seen so far were the end of the matter natural selection should eleminate the altruistic habit. But it is not the end of the matter. The next stage arrives when another female in the troop comes into oestrus. The roles of the same two males may now be reversed. Packer saw ten cases in which a male who had previously been 'solicited' into co-operating to defend a female (but did not copulate with her) himself solicited a male into co-operative defence. In nine of the ten cases the solicited male was the individual whom he had previously helped. It looks, therefore, as if males form co-operating pairs to defend females and take turns in the copulating that is the end of the defence. If so, it would be an example of 'reciprocal altruism', Altruism can evolve under individual selection, without any need for the animals to be related, if the altruist is more than paid back later.

The danger of any such reciprocal arrangement is that they will be cheated on. There is a clear short-term advantage to receiving altruism but then not paying it back; that way the cheat gains the benefit but does not pay the cost. This begin so, reciprocal altruism is expected mainly to evolve in species that form stable groups, with individual recognition. Then a cheat, after gaining a short-term benefit, can be recognized and excluded from future transactions; the cheating will then not pay. Without the opportunity and mechanism of discriminating against cheasts, reciprocal altruism is likely to break down. Because reciprocal altruism requires rather special conditions, it may be rare than kin-selected altruism. It is, however, a theoretical possibility; and in olive baboons it has been realized in fact.

Manipulated Altruism : The Control of Behaviour by Parasites

Natural selection, we have seen normally males animals behave in their own selfish interest. Even when if favour altruistic behaviour, it is only in the interest of some broader form of selfishness. However, conflicts of animals open many opportunities for exploiting other individuals, and animals may not therefore always behave in their own interests. There are probably numerous subtle forms of exploitation within a social group—a possibility we have considered before in relation to animal signals—but the idea of 'manipulation' in animal behaviour is a relatively recent one, and has not been the subject of much research. For c' ·ar examples we are forced to go to those unambiguous situations of conflict, parasite—host relations. Here are many examples in which animals do not behave in their own selfish interests, or the interests of their genetic relatives. Hosts under the influence of parasites do show altruistic behaviour, according to our adopted definition, but only because they are in some sense forced to, against their own interest. I should emphasize that I have picked parasite—host examples only because they unambiguously illustrate a process that may be expected to be of wider importance. Many cases of altruistic behaviour in nature may be due to still subtler forms of manipulation than those we are about to consider.

A young cuckoo (*Cuculus canorus*) being fed by its foster parent, such as a reed bunting (*Emberiza schoeniclus*), is a striking example of behavioural manipulation by a parasite. It is not in the reed bunting's interest to feed the cuckoo: natural selection favours reed buntings that rear their own offspring, not cuckoos. To our eyes, it is very strange that a reed bunting cannot tell when it is feeding a cuckoo rather than its own offspring. For the insatiable demands of the cuckoo are still met by is tireless foster parents even after the cuckoo has grown larger than its foster parent. It should be easy, we think, for a reed bunting to distinguish a great, ugly cuckoo from a reed bunting. Yet the parent does not. It continues to pour worms down the cuckoo's all consuming throat. Something about the continual gaping and begging of the cuckoo compels the reed bunting to continue to provide. The reed bunting is being forced by a parasite to do something against its won best interests. The behaviour of feeding the young, however, is not abnormal, it is just misdirected.

Let us now move on to some stranger examples, in which the parasite actually changes the behaviour of its host. The reasons are to be found in the life-cycles of the parasites. Many parasites grow up in a series of different host species. They may start off in one species, be transferred to an intermediate host, and then to a final host; but there can be a problem in effecting the transfer from an intermediate host to the next one. The normal behaviour of the intermediate host might not take it near the final host. A kind of fluke (flukes are small, flattened, worm-like animals) called *Dicrocoelium dendriticum*, for example, lives an ants as its intermediate host, and sheep as its final host. How is it to get from in ant to a sheep? The trick is to have the ant eaten by a sheep. Ants normally, understandably enough, avoid being eaten by sheep; they stay down in the soil away from grazing sheep. However, an ant harbouring a *Dicrocoelium* changes its behaviour. The infected ant typically contains about fifty *Dicrocoelium* individuals. One of these burrows into the ant's brain and somehow causes the ant to climb a blade of grass and fastens its jaws to the top of the blade; it clings fast until eaten, perhaps by a sheep. The parasite has then reached its goal.

There are many fascinating examples of this kind. Here are two more. Nematomorphs (small work-shaped animals), in live in insects as intermediate hosts, but in water as adults. The parasite has to bring its insect to water, and the nematomorphs do somehow bring this about. In one dramatic record, a bee was flying over a pond when suddenly, as it was about 6 feed above the water, it dived in. As soon as the bee splashed into the water, the worm exploded from the body they had so abused, and swam away, leaving it dead. A final example concerns the parasites of the amphipod *Gammarus lacustris*. *Gammarus* is a small shrimp-like animal which lives in fresh-water. Normally, *Gammarus* avoid the light, sheltering under pebbles on the bottom. However, when infected by the parasitic worm *Polymorphys paradoxus*, their reaction to light is reversed: they now seek the light and swim just below the surface. The parasite's motives is that the next host after *Gammarus* is a duck that feeds by dabbling at the surface; the *Polymorphus* changes the *Gammarus's* behaviour to make ti be eaten by its final host. The duck gets an easy meal, but it infects itself with a parasite when taking it.

Some biologists and psychologists would not include the manipulated altruistic behaviour in these parasitological examples as examples of real 'altruism'. They would prefer to count only kin selected and reciprocal altruism as real altruism, and exclude from the term behaviour carried out in the genetic interests of others. The process of natural selection does differ between the cases of kin selection, reciprocal altruism, and manipulation, and this might seem to provide a reason to separate them. But, following most authorities. I have defined altruism not be the process of natural selection that favours it, but in term of its reproductive consequences. An altruistic act is one that increases the number of offspring left by the recipient, and decreases the number left by the altruist. Manipulated behaviour fits the definition of altruism.

The consequentialist, non-intention definition of altruism is far different from the normal everyday usage in which we thin of an altruist as intending to be kind. And perhaps some are tempted to include kin selection and reciprocal altruism as true altruism because the subjective intention (in the normal meaning of the word) can, by loose and wholly erroneous reasoning, be read into the genetic 'interest' of an individual. In kin selection and reciprocal altruism the individual does behave in its *genetic* self-interest (although not its own self-interest). However, it is not part of the definition of altruism that an animal should be behaving in its won genetic self-interest. The word was defined in terms of the effect on the number of offspring produced, regardless of whose genes those offspring carry.

Cooperation Through Group Advantage

W.C. Allee, in *Cooperation among Animals* (1951), synthesized much of the thinking about animal societies through the first half of this century. The ideas expressed are best understood in relation to contemporary development in the study of community ecology. *Clements* (1936) classified plant communities and started a school of "plant sociology", which emphasized the interrelationships of species and recognized the community as a "superorganism" with attributes transcending those of single-species populations. According to *Clements* living organisms reacted to the environment and so modified it that other species could not east there.

The various assemblages of plants were so interdependent that they were distributed as single units. The idea of communities and ecosystems as superorganisms with emergent properties reached a peak with the writings of *Margalef* (1963), who discussed the qualities of mature and immature ecosystems. He treated the ecosystem as the functional biological unit that organizes itself so as to conserve and manage information. Although not explicitly stated, such reasoning led to the notion that the whole ecosystem was the unit of selection. Among animal behaviourists, the popularity of group selection reached its peak with Wynne-Edward's theory that social behaviour evolved mainly for the function of population regulating.

Allee (1951) recognized the importance of natural selection at the level of the individual and did much work on egoistic, or selfish, behaviours such as competition and dominance. However, he felt that altruistic or cooperative forces were stronger, and he devoted much study to what he called *natural cooperation*, a force supposedly separate from natural selection. Allee believed that the phenomena he studied could be represented as several grades of social behaviour, listed here in generally increasing complexity:

1. *Invertebrate coloniality.* Allee thought the complexity and division of labour shown by such invertebrates as coelenterates in involuntary and, therefore, less advanced than the more voluntary cooperation seen in the vertebrates.
2. *Aggregation.* Organisms may be forced together by the actions of wind, tides, and other phenomena over which they have no control.
3. *Orientation to stimuli.* Animals may be brought together by their response to environmental gradients—for example, insects aggregate around lights. Such organisms must have at least a tolerance for each other.
4. *Locomotion to favourable locations.* When resources are in patches, animals may gather in taking advantage of them—for examples, birds in mulberry tree when the fruit is ripe.
5. *Clumping in the absence of substrate.* Mutual attraction and clumping occur in an attempt to find missing substrate. Brittle stars (Ophiuroidea) are typically dispersed in eel grass, but in aquarium devoid of objects they cling to each other. If small glass rods are planted as artificial substrate, they disperse and attach to the rods.
6. *Sleeping group.* In many species, individuals that are more or less isolated come together to sleep. Such behaviour may provide superior predator protection but may attract predators as well.
7. *Complex social life.* Allee considered highly developed social life an extension of sexual and family relations over a large portion of the life span. He posited an unconscious tendency to cooperate that predisposes such species to develop complex societies.

Although he appreciated the negative effects of crowding, Allee also pointed out the unfavourable consequences of undercrowding. Typical of his experiments was the finding that goldfish (*Cyprinus* species) reared in a toxic colloidal suspension of silver survived longer in groups that they did alone; the slime produced by fish present in the "conditioned" water precipitated the silver to the bottom of the tank, thus protecting them. Other experiments showed that planarian flatworms (*Planaria* spp.) gain protection from ultraviolet rays by grouping and that the effect is not due simply to shading of one worm by another.

2

Social Life

Animals form groups to enhance, their foraging success and gain better protection against predators; but groups living also exposes them to diseases, competition, and social interference. The cost and benefits of social living vary with the age, sex, experience, and physical condition of individuals. Social groups are maintained by chemical, visual, anditory and factile signals. But the animals do not always communicate their intentions clearly and accurately. Complex social systems have evolved from offspring remaining with parents to help near future broods and from co-operation among adults of the same generation. Many persons would regard any form of interactive group behaviour as social - a herd of deer, perhaps, or a winter roost of monarch butterflies. Students of vertebrates often consider parental care - for example, among song birds - as social behaviour. As we saw in the preceding chapter, therc are many examples of parental care among insects that are usually regarded as "solitary." In some cases, the mother even feeds the offspring progressively, resulting in much mother-offspring contact. In instances of communal nesting, several females share a nest, each preparing and provisioning her own cells. But all of these are examples of what entomologists call *presocial behaviour*. The highest level 'of cooperative behaviour, termed *eusocial behaviour* (literally, "truly social") is found in only four groups: termites, ants, and some bees and wasps. Eusocial insects have the following attributes in common:

1. Brood care is cooperative; that is, individuals often feed offspring that are not their own.
2. There is a *caste system* involving a reproductive division of labour, such that many colony members are sterile.

3. There is an overlap of generations some offspring assisting the parental generation in the rearing of further offspring.

The question of how this unique form of behaviour may have evolved is a topic of much lively discussion among entomologists. The diversity of lifestyles among presocial insects suggests that there have been several routes to eusociality. In this chapter we shall be concerned with types of behaviour especially characteristic of eusocial insects, but we shall also take occasion to consider the advantages of presocial and of eusocial behaviour and to ask why (considering the success of ants, termites and the like) not all insects have evolved eusociality.

SOCIAL COMMUNICATION

The behaviour of the many individual in the colonies of social insects must be integrated in such a way that there is cooperation in defense, building, foraging, brood rearing, and the production of reproductive individuals at appropriate times. This can only be accomplished by broadcasting messages throughout the colony. Since visual signals would be difficult to transmit within the darkness and complexities of the nest, social signals are more often acoustic, tactile, gustatory, or especially olfactory. We present here on outline of the types of messages conveyed with a few examples of the mechanisms that have evolved in the various groups of wasps, bees and ants (Hymenoptera) and termites (Dictyoptera). The nests of social insects present a rich source of food for vertebrate predators (such as skunks or bears) and for arthropod predators (such as other social insects). Not only are they often filled with thousands of larvae and pupae, but there may be large amounts of food in storage. All social insects have evolved methods of dealing with predators, either by attacking en masse or by fleeing to a place of safety. But first a message must be transmitted quickly throughout the colony concerning the danger. Such *alarm signals* are often highly volatile pheromones, which fade quickly unless renewed. Rapid movements within the colony, with much tactile stimulation, may also convey alarm. Termites, when alarmed, vibrate their bodies against the substrate, producing a sound audible to humans, and some ants are able to stridulate by rubbing together specialized ridges on parts of the abdomen. Alarm usually mobilizes defense, but in some cases the alarm signals themselves may also serve a defense function.

Attractants also play a major role in social life, especially with respect to the queen. In most social insects the queen rarely if ever leaves the nest and must be fed and groomed by the workers. In a

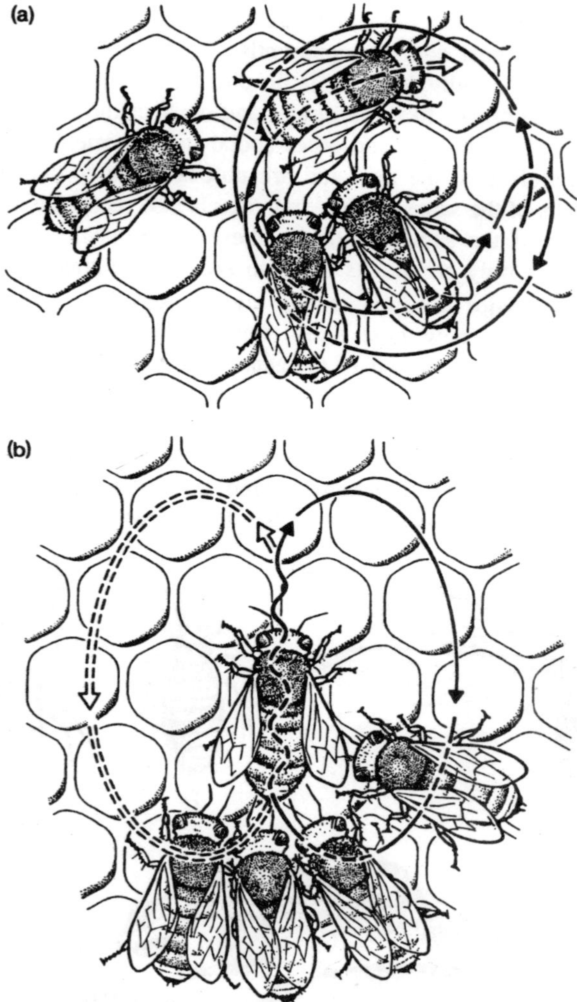

Fig. 2.1. A queen honey bee surrounded by a retinue of workers.

honey bee hive, for example, the queen is usually accompanied by a circle of attending workers. Indeed, the term *queen* was first suggested by the retinue surrounding the mother of the colony, who of course "reigns" only in the sense that she produces chemical signals causing workers to forego reproduction and to attend to the queen. In the honey bee the signal is the continued production of ketodecenoic acid (often called "*queen substance*") from her mandibular glands.

The most prevalent attractants are the subtle combinations of odors usually called "nest odor" - apparently compounded of phero-mones on the cuticle of individuals; volatile pheromones in the colony; and the odor of the brood, food, substrate, and other unknown odor sources. Individual ants, termites, and other social insects orient toward these odors when close to the nest and are able to identify their own nest among others of the same species. Individid:.als also recognize one another as members of the same colony by these same cues, which may be said to be not only attractants but also *recognition signals*.

Another major group of social signals serves in *recruitment* to food sources or new nest sites. One of the simplest forms of recruit-ment is shown by ants of the genus *Leptothorax*. When a foraging worker is successful, she returns to the nest and regurgitates food to nestmates. She then raises her abdomen, extrudes her sting, and discharges a droplet of fluid. This serves to attract other workers, one of whom touches the abdomen or hind legs with her antennae and proceeds to follow the forager to the food source. The leader then

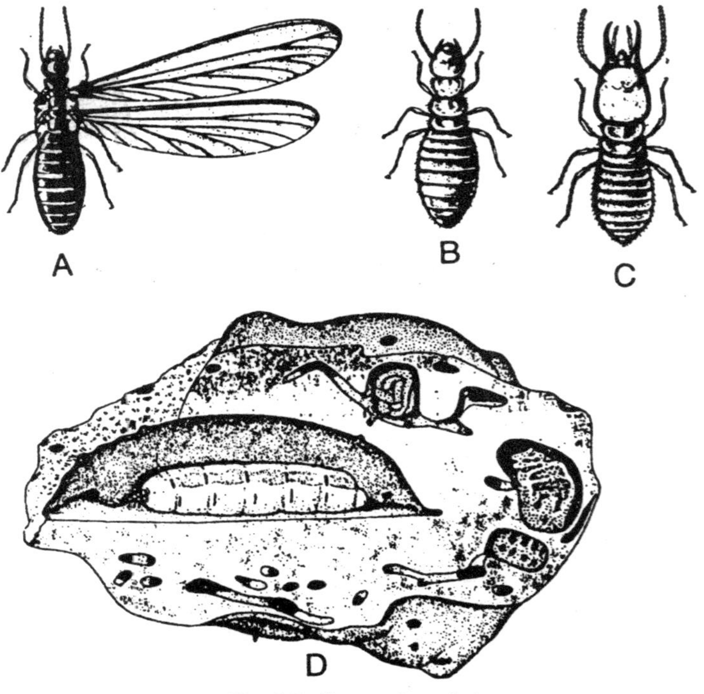

Fig. 2.2. Castes of termites.

lowers her abdomen, but if the follower becomes lost, the leader once again elevates her abdomen and initiates calling behaviour. Tandem running, as this is called, involving only two or at most a very few individuals, appears an inefficient system for recruitment to a food source, but it is interesting as an apparent evolutionary precursor of the more common and efficient method of laying *odor trails*. As E.O. Wilson has shown when a fire ant worker locates a source of food, she returns to her nest with her abdomen lowered, dragging her sting

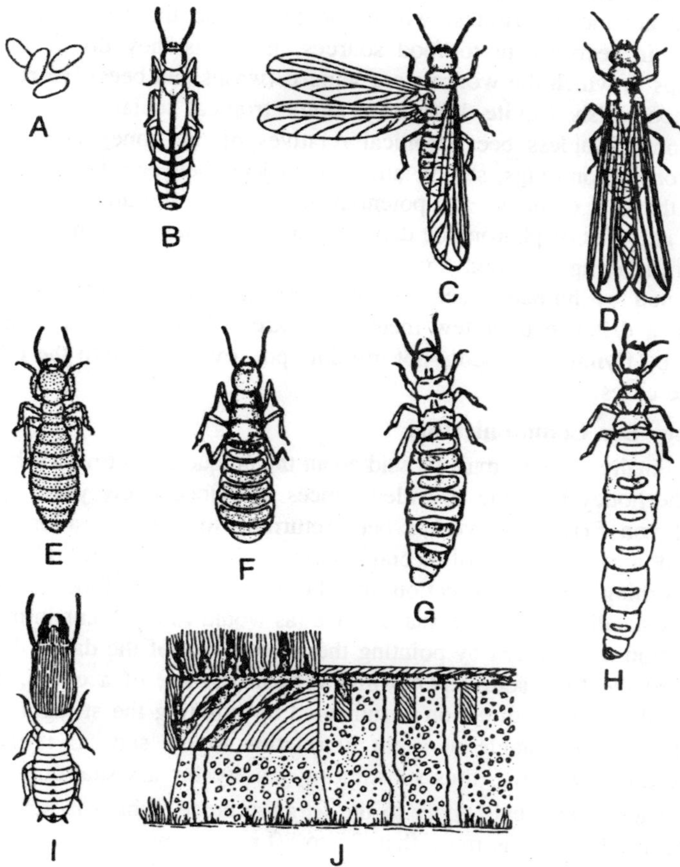

Fig. 2.3. Polymorphism in Reticulitermes. (A) Eggs, (B) Nymph, (C) Female winged form, (D) Male winged form, (E) Workers, (F) King, (G) Secondary reproductive (female), (H) Queen (abdomen distended with enlarged ovgaries), (I) Soldiers, and (J) Earthn tubes from soil across surface of concrete foundation to wooden sturucture.

lightly over the substrate and depositing small quantities of a trail-marking pheromone. When she arrives at the nest and regurgitates to some of the workers, they are able to follow the trail to the food source, reinforcing it on the way back as long as the food lasts.

Termites also lay odor trails from abdominal glands a remark-able case of evolutionary convergence, since ants and termites belong to quite unrelated orders of insects. In more primitive termites, odor trails are used in recruiting workers to breaches in the nest wall. In more advanced termites, which forage outside the nest, odor trails serve in recruitment to food sources, much as they do in ants. In groups in which the workers are winged (wasps and bees), recruitment is accomplished quite differently. Some tropical social wasps as well as some stingless bees (tropical relatives of the honey bee) deposit pheromone on twigs, stones, and other objects between the food source and the nest or between a potential new nesting site and the old nest. The amounts of pheromone deposited are much larger than in the case of trail-laying ants and termites, and in some cases the odor can be detected by human observers. Workers are able to detect the odor from a distance of a few meters and are able to proceed along the trail by flying from odor spot to odor spot, even high into the trees in some cases.

Honey Bee Communication

Finally, a word must be said about the unique recruitment behaviour of the honey bee, the so-called dances described many years ago by Karl von Frisch. A worker bee, returning to the hive after having found a rich source of nectar, conveys information to other bees concerning both the direction and the distance of the food source. If she is standing on a horizontal surface (as would rarely occur in nature), direction is indicated by pointing the straight run of the dance directly toward the food source. Only the vertical surface of a comb, in the dark of the hive, direction is indicated by pointing the straight run to conform to the angle of flight in relation to the sun, as if the sun were directly above, thus substituting gravity for a visual cue.

These statements apply to the *waggle dance*, which is performed if the food source is more than 35 to 80 meters away (different races of the honey bee differ in this regard). In this dance more precise information on distance is indicated by the duration of the straight run: The greater the distance, the greater the duration of the straight run. If the food source is close to the hive, however, a different dance having no information on direction or distance is performed, the so-called *round dance*. This consists of a circle, at the end of which the worker turns around and repeats the circle facing in the other direction. Workers reading this message fly out in all directions near the hive and search for odors like those carried on the body of the dancing bee. Odors of the food source are also important in the waggle dance, which in itself serves only as an approximation of distance and direction. The dances are decoded within the hive by other bees, using tactile, olfactory and acoustic cues. Bees cluster about the dancer; contact her with their antennae; and having read the message, fly to the vicinity of the food source. At times there may be several workers dancing on behalf of several food sources. In this case the most persistent dancers tend to recruit the most workers. These same dances are also used for recruitment to new nesting sites at the time of swarming.

In recent years, there have been a number of criticisms of von Frisch's research possibly he underestimated the importance of odor of the food source clinging to the body of the dancers and the information in the wing sounds they produce; and it is true that different races of honey bees have different "dialects." But recent novel experiments have tended to confirm von Frisch's findings. James Gould, of Princeton University, established two feeding stations equidistant from the hive but in opposite directions, both similarly scented or unscented. In one the concentration of sugar was twice that in the other. Most of the foragers danced to the direction and distance of the station with the highest concentration of sugar, and nearly all recruitment was to that station. Factors such as site-specific odors, wind direction, and so forth were controlled, and it was also shown that in the absence of dancing almost no bees found the food sources. In a later experiment Gould developed a sophisticated design in which some of the bees responded to the information in the dances even though these were performed by bees coming from a different food source than that specified in. the dance. That is, the recruits responded to distance and direction cues in the dances rather than to odors conveyed by the dancers.

The communication signals of social insects are parsimonious in that the same signal is often used in different contexts to transmit different messages. "Queen substance" of the honey bee inhibits worker ovarian development and the initiation of queen cells, stimu-lates grooming and feeding of the queen by the workers, maintains colony cohesion during swarming, and attracts males during the queen's nuptial flight—a single substance conveying different information in different

Fig. 2.4. Waggle dance of the honey bee. On a horizontal surface (a) the straight part of the dance points directly toward a food source, but on the vertical surface of a comb (b) the straight part points to an angle with the sun as if the sun were directly above.

circumstances. This by no means exhausts the subject of social communication. For example, larvae of paper wasps may indicate hunger by hitting their heads against the cell walls, producing an audible sound, queens on the nests of these same wasps may signal their dominance by tail wagging and other visual signals. interchange of food and secretions between adults and between adults and larvae conveys many messages on conditions within the colony. for example, the proper balance between the castes may be maintained by the amount of caste-specific pheromone being circulated in the colony. Studies using radioactive tracers have shown that in *Formica* ants substances imbibed every by a single worker are rapidly spread throughout the colony, and within 27 hours every member of the colony has received some of it. The behaviour of individual social insects is intimately guided by these messages. Since colonies may contain many thousands of individuals, it is difficult to imagine how they would function coherently without such a system of mass communication.

Division of Labour in Social Insects

The most unique feature of social insects is the existence in each species of several castes: at least two (queen and workers), often a third (soldiers), and sometimes several subcastes differing in appearance and function. Wasps and bees lack a distinct soldier caste; in the ants soldiers are simply large and specialized workers; but in the termites the soldiers constitute a caste quite different from the workers. Worker and soldier termites may be either male or female, and workers may be either immatures or adults. In contrast, worker Hymenoptera are always adult females. Workers and soldiers do not mate; they represent the worker force and defense of the colony, and reproduction becomes the sole function of the queen and males. In termites, a male

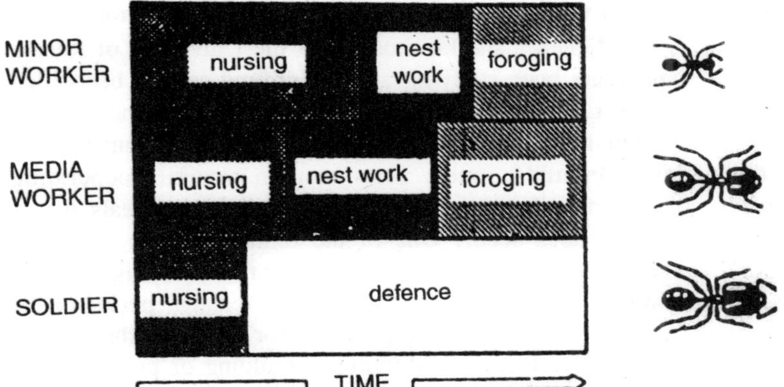

Fig. 2.5. Polymorphism and polyethism in an ant of the genus Pheidok. Smallest workers, called minors, are at first nurses, feeding and tending larvae. Larger work ers, or medias, spend less time as nurses and more in other activities. The largest workers, called majors, may also be termed soldiers, since a major part of their time is spend in colony defense.

reproductive ("*king*") is a permanent attendant of the queen, but in Hymenoptera males die soon after mating.

Members of different castes of one species often differ radically in appearance, a striking case of *polymorphism* that has, for the most part, an environmental rather than a genetic basis. Members of different, castes and subcastes also exhibit very different behaviour, a phenomenon called *polyethism*. A soldier termite is a specialist in defense and often cannot even feed itself. But in other cases polyethism has a temporal element, as in the subcastes of worker ants. In the case of the honey bee, division of labour among the workers is wholly temporal. There is no polymorphism within the worker caste, but workers perform different tasks depending on their age.

How can all these diverse behaviour patterns be built into one individual? And how is one species able to produce several kinds of individuals that differ so much in appearance and in behaviour? If workers and soldiers do not reproduce, how are their genes conveyed to the next generation? If nature selection favours those individuals that produce the most surviving offspring, how does one account for the existence of sterile castes? Obviously the social insects present a host of questions that cannot easily be answered. Perhaps some will seem less intractable if we look briefly at certain species that seem not to be "fully eusoc ial" that is, that appear to be on the verge of acquiring a worker caste.

Determination of Caste in "Primitively Eusocial" Hymenoptera

Charles D. Michener and his students at the University of Kansas have for some years been studying a small ground-nesting bee called *Dialictus zephyrus*. These bees form aggregations in earthen banks, provisioning their nests with pollen and nectar gathered from flowers in the vicinity. In midsummer, nests contain several cooperating females, all superficially alike. But closer inspection reveals that a division of labour does occur. One of the females proves to be the major egg layer, while the others act as guards or foragers. The major egg layer ("queen") nudges the other females periodically, causing them to move into guarding positions, or she leads them down into the burrow where there are stimuli for building or provisioning. When the subordinates lay eggs, as they do occasionally, the queen eats them and replaces them with her own. If the queen is removed, one of the subordinates assumes her role. Ovarian development in the subordinates is believed to be retarded as a result of the queen's behaviour, but all the workers have the potential to mate and become queens (males are present throughout the summer).

A somewhat different situation prevails among bumble bees, which live in small colonies in cavities in the soil. Inseminated females overwinter and start colonies in the spring, and the workers and the new crop of males and queens are all the foundress's offspring. The queen dominates the workers by pheromones transmitted on contact, aided by aggressive behaviour, and the workers come to assume various roles in the colony, the smaller ones mostly as nurses, the larger ones mostly as forgers, although they may change their role as a result of age or the needs of the colony. If the queen is removed, of one of the larger workers will assume her role. Workers tend to be larger late in the season, and eventually a new crop of large females destined to be queens is produced (as well as males). Late in the season there are more workers present in the colony in relation to the number of larvae. The increase in size of individuals may simply be the result of the increasing worker: larva ratio; larvae receive more food and become larger adults.

Determination of Caste in Advanced Eusocial Insects

In the honey bee, behavioural dominance by the queen has been replaced by secretion of a pheromone (queen substance) that prevents development of the ovaries in the workers. Queens are reared in especially large cells and receive not only more food but food enriched by the contents of hypopharyngeal glands in the heads of the workers ("royal jelly"). In most of the more advanced social insects, the role of queen is sustained by pheromones. In colonies containing thousands of individuals, control by behavioural interactions between queen and workers would scarcely be practicable.

By and large, polymorphism in social Hymenoptera has a *trophogenic* basis; that is it is the result of differential feeding of the larvae or of different-sized eggs. Larvae receiving a minimal amount of food develop into small, sterile workers, a phenomenon sometimes called "nutritional castration." Better-fed larvae tend to produce larger individuals. In the ants these may have disproportionately large heads and mandibles, the result of allomevic growth, and may serve as soldiers. Allometric growth is a genetically determined tendency for certain body parts to grow at a more rapid rate than other parts and thus to be larger in the adult stage. Reproductives—the queens of the next generation and their mates are produced from larvae that are abundantly

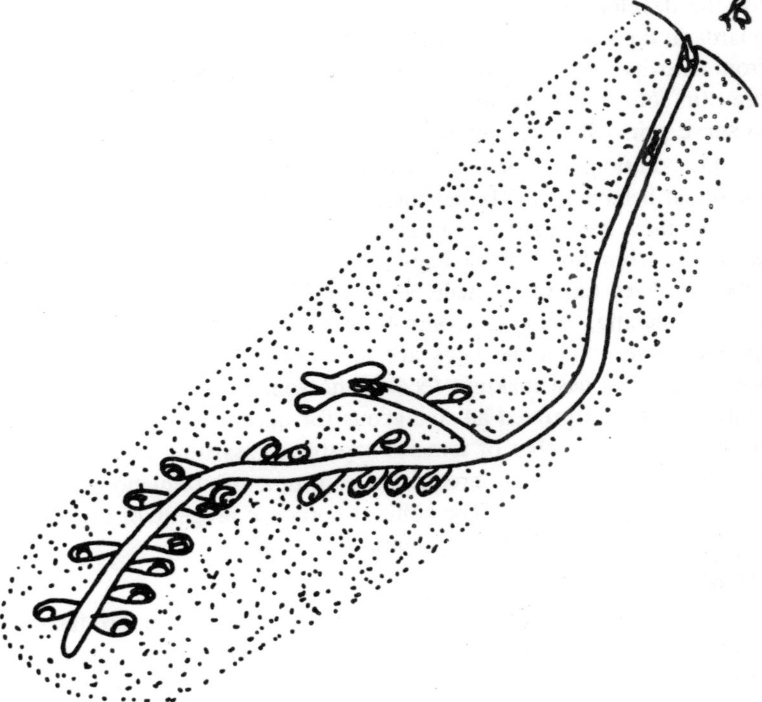

Fig. 2.6. A nest of the sweat bee Dilic us zephynts *in a soil bank, in early summer. Four females are present; the queen is deepest in the nest near a cell being provisioned, the guard is at the entrance, and a forager is returning with pollen and nectar.*

fed an sometime provided with food of different quality. Often they are reared in larger cells, the combination of large cells and a high worker: larva ratio ensuring a crop of reproductives.

Hymenoptera appear especially well suited for the development of social behaviour, as reflected in the fact that eusociality has evolved several times independently in the ants, in the wasps, and several time sin the bees; but only one other group of insects, the termites, has evolved eusociality. The Hymenoptera appear to have several features that together render them likely to become eusocial. These insects must have evolved fail-safe homing abilities as well as the ability to protect the larvae from predation, desiccation, and so forth. Limitation in sites suitable for nesting may cause them to aggregate in certain places and to produce more permanent, many-celled nests. Aggregated nests may improve the opportunities for parasites and predators, who do not have to look far for their hosts-thus, the need for the development of defense mechanisms such as the presence of guards in the nest entrance. Another potent defense, the sting, evolved from an ovipositor that had first evolved into a mechanism for paralyzing prey (in solitary wasps) and then, in the eusocial Hymenoptera, came to serve primarily as a defensive weapon.

On additional factor may have been important in setting the stage for social behaviour in Hymenoptera. Male Hymenoptera are haploid; that is, they have half the usual number of chromosomes and are produced from unfertilized eggs. Females have the capacity to lay either fertilized (diploid, female-producing) eggs or unfertilized (haploid, male-producing) eggs by controlling the flow of sperm from the spermathecae when the eggs are laid. Since the female sex is the one involved in nest making, feeding the young, and so forth, even in solitary forms, it proved advantageous for queens of social species to produce a large number of females capable of assisting in the work of the colony, delaying male production until time of production of a new brood of potential queens. The ability to produce progeny of the desired sex, as needed, is one that we humans may envy.

Termites

Most of these remarks do not apply to termites, which are believed to have evolved from the ancestors of cockroaches, which do not make nests and are diploid. Hence, they seem to lack the preadaptations of Hymenoptera. Nevertheless many do aggregate. The group of cockroach like insects that gave rise to the termites probably lived gregariously in logs and digested cellulose with the aid of intestinal microorganisms as some species of cockroaches still do. Since these symbionts are cast off with each molt and must be reacquuired from others, and newly emerged individuals must obtain them from their elders, isolated individuals would soon starve. Both woodfeeding cockroaches and termites frequently feed at the anus of other individuals, thereby obtaining intestinal symbionts needed for digesting their food. It is apparently this factor that enhanced the development of eusocial behaviour in the Dictyoptera.

Caste determination in termites differs considerably from that in Hymenoptera. Since termites have gradual metamorphosis, the young are not helpless, and in fact by the third instar members of both sexes assist in work of the colony-an example of "child labour" unique in the insects. These immatures molt further and may in some cases develop into soldiers or reproductives, depending on a complex system of pheromones circulating in the colony. Soldier termites are typically wingless adults that differ greatly from the workers and reproductives. Termite colonies also contain "supplementary reproductives," which

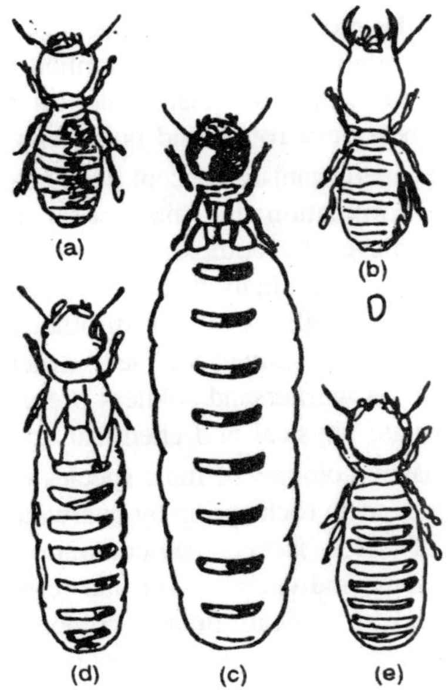

Fig. 2.7. The castes of termites. In these insects, in contrast to Hymenoptera, the "king" (not shown here is a permanent member of the colony. The queen (c) may ultimately become very large. The termites in (d) and (e) are supplementary queens: the one in (a) is a worker, in (b), a soldier. The worker is about 5 mm long in the species figured.

have wing pads and the capacity to become winged reproductives if the queen or king dies. Development of these various castes is mediated by the interaction of pheromones and hormones in ways that are far from fully understood. Presence of supplementary reproductives means that colonies are potentially immortal, and indeed the nests of termites sometimes persist for many years and result in structures of incredible size, considering the size of the builders. The nests of each species of termite are distinctively different, so that taxonomists sometimes find it easier to identify the species by the nest rather than by the termites themselves.

Social Homeostasis

Maintenance of a functional steady state in an organism or in a colony of organisms is termed *homeostasis*. The colonies of social insects have sometimes been compared to organisms; the individuals, to cells, sopte of which are specialized for reproduction, others for nutrition, others for protection. Colonies have a birth, a growth cycle, and an eventual death. As we have said, the nests are species specific, just as individuals can be recognized as members of different species. Thus, the term *superorganism* is sometimes applied to these colonies, though a little reflection will show that analogies between colonies and organisms are at best very rough and not overly instructive.

Nevertheless, the superorganism concept does help one to visualize the vast amount off coordination that must occur within the colony coordination that is achieved by complex systems of communication operating with respoot to individuals that are diverse but limited the their responses and in their abilities to control other colony members. Integrity of the colony is maintained via nest structures that cannot easily be breached and via workersand soldiers prepared to respond to intruders with appropriate physical and chemical attacks.

In temperate climates, colonies of most species of bees and wasps are annual affairs, begun anew each spring by overwintered, inseminated females. Thus a nest of yellow jackets may contain only a few hundred or at most a few thousand individuals. Ant and termite colonies are, however, perennial, and in tropical climates colonies of some species reach enormous size. A single colony of army ants may sometimes contain over a million workers. Queens of certain termites are reported to lay as many as 30,000 eggs per day, or 10 million per year. Homeostatic mechanisms in colonies this size must be complex beyond belief.

Most social insects are able to control temperature and humidity of the nest at least to some extent. In temperate climates, many ants

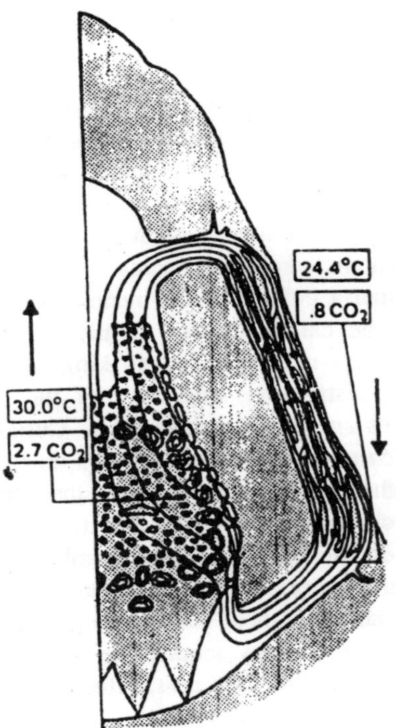

Fig. 2.8. Temperature regulation in the nest of an African termite.

start to breed under stones in early spring, taking advantage.of the capacity of stones to absorb heat, then they move deeper into the soil in summer. Aerial nests of wasps are built in protected places, and those of hornets and yellow jackets are enclosed in protective paper sheaths. Fanning at the nest entrance and the carrying of water to the nest play a role in cooling the nest in hot weather for many species. Honey bees survive the winter by clustering in the hive. Clustering bees consume honey and move about within the cluster, generating enough heat so that the temperature of the cluster does not fall below about 20°C, even though the outside temperature may frequently be below 0°C. Similarly honey bees are able to maintain a hive temperature of about 35°C even on the hottest summer days, so long as water is available to them. Some of the most remarkable instances of "air conditioning" occur among some of the fungus-growing termites of Africa. Warm air generated in the core of the nest rises and passes into a series of small chambers in the nest walls, where it is cooled and where fresh oxygen is obtained; it then flows back into the core of the nest at the bottom, and a slow but continuous circulation is maintained.

Salavery in Ants

Before leaving the social insects, mention should be made of another unique phenomenon: slavery among ants. This behaviour is justly termed unique, since it differs from slavery among humans in usually, occurring between species; that is, members of one species enslave those of another, related species. It also differs in being a biological rather than a cultural phenomenon. Thus one could argue that it should not be called slavery at all, though the superficial resemblances so slavery in humans are striking. Slave-making species of ants are widely distributed, and most of the reported "battles" between ant colonies are actually slave raids. Certain reddish species of the genus *formica* undertake frequent raids on colonies of some of the common and rather docile black species of the same genus. Columns of workers approach and surround the nest to be plundered, causing alarm, attempts to escape, and often some combat among the workers. Eventually the more aggressive slave makers enter the nest, seize larvae and pupae, and carry them back to their own nest. When these develop into adult workers, they acquire the colony odor and perform their normal behaviour in the alien nest, effectively supplementing the worker force of the slave-making species.

Recent research by E.O. Wilson and his colleagues at Harvard University has shown that slave makers of at least some species have enlarged Dufour's glands in their abdomens. When they approach a nest of the slave species, they spray the contents of these glands onto the workers. The secretion consists of a mixture of acetates that causes the workers of the slave species to become alarmed and disorganized. Curiously the effect of the acetates on the slave makers is quite the reverse: They produce attraction and excitement as Wilson, says, "exactly the responses needed to conduct successful slave raids." By and large, colonies of most ant species defend themselves well against incursion by alien colonies, sometimes resorting to physical combat to maintain territorial boundaries. In honeypot ants of the deserts of the southwestern United States, territories are defended by ritualized displays in which workers of opposing colonies "*stilt wall*," at first facing one another, then beginning to circle.and trying to push one another sideways. Such "tournaments," recently described by Bert Holldobler, of Harvard University, may go on for days. If one of the colonies is unable to recruit enough workers to the tournament area,

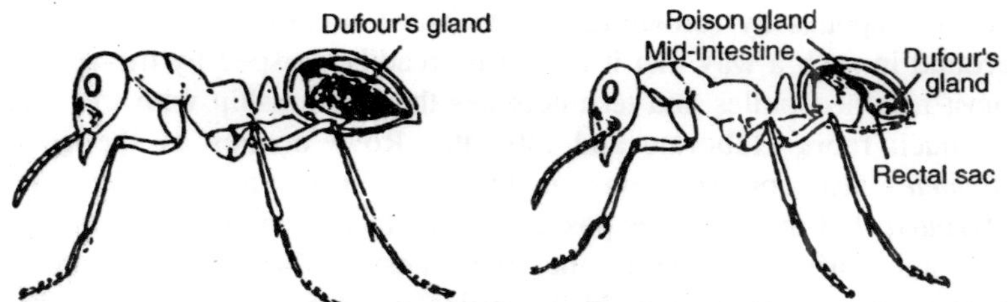

Fig. 2.9. The enlarged Dufour's gland of a slave-making ant (Formica subintegra) (left) as compared with that of a member of a slave species (Fonntca subsereceaa) (right).

it may be overrun by the stronger colony and the brood and other nest contents carried off. The surviving workers then become incorporated into the raider's nest. This is the only known example of slavery occurring within one species of ant.

Symbionts in Social Insects

In spite of the various behavioural and chemical defenses of the colonies of social insects, quite a number of arthropods have evolved ways of breaching these defenses and taking advantage cf the security of the nest and the abundance of food provided. They do this by tapping into the communication codes-that is, by acquiring the ability to "speak the language" of their hosts. Many of them have glands that secrete substances attractive to their hosts, who "adopt" them as part of the colony. Altogether, according to E.O. Wilson, insects of at least 120 different families and 17 different orders have become "guests" of social insects, along with a variety of mites, millipedes, and other arthropods. As the noted student of social insects W. M. Wheeler put it many years ago: Were we to behave in an analogous manner we shouldd live in a truly Alice-in-Wonderland society. We should delight in keeping porcupines, alligators, lobsters, etc., in our homes, insist on their sitting down to table with us and feed them so solicitously with spoon victuals that our children would either perish of neglect or grow up as hopeless rachitics.

Symbionts associated with ants are termed *mynnecophiles* (literally, "*ant lovers*") while those found with termites are called *termitophiles*. Relatively few occur in the more open societies of the social bees and wasps.

One of the simplest associations is shown by the "*highwayman beetle*," which intercepts black wood ants on their trails and solicits food from them by tapping their mouthparts. Frequently the ants soon realize they are being "tricked" and begin to attack the beetle. The beetle responds by retracting its legs and flattening itself against the ground in such a way that it cannot be readily grasped by the ants. In most myrmecophiles and termitophiles the relationship with the hosts is much more elaborate and intimate. Rove beetles of the genus *Atemeles* also approach ants, in this case brown ants of the genus *Myrmica*, but these beetles possess glands that the ants find attractive. The ants carry the beetles into their nests, where they spend the winter among ample food. In the spring the *Atemeles* beetles migrate to colonies of the mound-building wood ant, a species of *Formica*, which provide a richer source of food in the summer. Here they lay their eggs, and the larvae, also provided with special glands, solicit food from their hosts and also consume larvae of the host ants.

Not all symbionts have such elaborate adaptations, but all have some means of escaping, appeasing, or otherwise "duping" their hosts. Termitophiles often have grossly swollen abdomens that exude substances that are licked by the termites. In many cases we do not known precisely how the symbionts. "break the code" of their hosts or what they feed on in the nest. Although the large colonies of army ants and some of the tropical termites contain hundreds of "*guests*" of great variety, many of the symbionts appear quite rare. Evidently this is a precarious existence; as in espionage among humans, a slip may be fatal.

Advantages and Disadvantages

A great many insect species live essentially solitary lives, contacting other individuals chiefly at time of mating-and a few even dispense with this "inconvenience," reproducing parthenogenetically. Clearly a solitary way of life has advantages: Such individuals can more effectively hide from predators and avoid competition with similar lifestyles; they can also more effectively live in small spaces and exploit dispersed food sources. Group living also has obvious disadvantages: There are opportunities for the spread of disease and for the incursion of unwanted "guests" into large societies. Why then live in groups? One advantage might be that escape from predators is actually enhanced if, via communication resulting in alarm and defense behaviour i ha of the group, permits predator is driven off. Another might better exploitation of food sources that are strongly localized or so large and tough that one individual could not well exploit them, as seen in first-instar sawfly larvae feeding on pine needles or ants attacking a beetle. Still another factor might concern limitations in living space - that is, cockroaches may aggregate because only limited areas provide suitable conditions of warmth and moisture; or ground nesting bees or wasps, because soil of suitable consistency is patchily distributed.

Individuals of a species commonly compete severely for food and living space. For group living to succeed, individuals must survive and reproduce more successfully within the group than apart from it. This means that they must evolve a degree of tolerance, if not actual cooperative behaviour, with other members of that species. Clearly it will not pay an individual within a group to devote energies to competitive interactions; but it may very well pay to produce or respond to an alarm signal. Wherever, the advantages of group living outweigh its disadvantages, in terms of individual reproductive success, group living may evolve.

Social groups among insects (and among animals generally) usually consist of related individuals parents and offspring, siblings, or at least members of a local population having many genes in common. Often there are mechanisms for occasional outbreeding, such as dispersal at time of mating. Webs of a tent caterpillar commonly contain individuals from a single egg mass; lady beetles clustering for the winter are commonly the descendaatsof those that clustered there the previous winter. As we have seen, the colonies of eusocial insects are essentially "extended families," even when the queen has mated more than once or when there is more than one queen, there is genetic relatedness among colony members. Individual *Darwinian fitness* (that is, reproductive success) may in such cases be replaced by *inclusive fitness* (defined as net genetic representation in succeeding generations, including other relatives in addition to offspring). That is to say, an individual may devote energies to cooperative behaviour to the detriment of its own reproductive success to the extent that the individual it helps carries genes identical to its own.

The expenditure of energy in cooperative behaviour that might otherwise be spent in individual effort is often spoke of as *altniism*, and natural selection that involves inclusive fitness is often spoken of as *kin selection*. Whether or not unselfish behaviour among humans is a product of kin selection is a controversial topic we need not consider here; but in any case the word *altruism* as applied to insects does have this connotation and does not imply purposefulness on the part of the performers. Even with low degrees of relatedness among members of a colony, kin selection favours the evolution and maintenance of sterile castes if the benefit cost ratio is high enough. In this context it is much easier to understand the behaviour of a worker honey bee, for example, who leaves her sting in the body of a predator and dies thereafter: She is ensuring the survival of many thousands of individuals of very similar genetic constitution.

All of this is helpful in explaining the unique properties of eusociality. Workers and soldiers have lost all individual fitness (or most of it, since workers of Hymenoptera do in some cases lay viable, male-producing eggs). But the colony, as an extended family, may be enormously successful in survival and reproduction and may exploit and even modify the environment in ways no individual could do. A soldier termite guards a colony of thousands of individuals, and at much risk to its own life. But ultimately its genes (which are also those of the reproductive caste) will persist in succeeding generations via inclusive fitness. We have suggested reasons why eusocial behaviour may have evolved in termites and in Hymenoptera but not elsewhere. In Hymenoptera, eusociality has in fact evolved several times independently. In addition to permitting females to control the sex of their offspring, male haploidy, in this order, may result in an unusually close relationship among siblings. Since each male in the result of "virgin birth," every sperm he produces has all his genes, while each egg produced by a female has only half of hers (as in diploid species).

Because anY daughter of a male has a full set of his genes, sisters related through both parents are unusually closely related (by 3:4). On the other hand, daughters are related to their mother only 1 : 2. The assumption is that females are therefore more likely to evolve a tendenCY to cooperate with their sisters to rear more sisters than to start then own nest and produce daughters. How important this factor is in enhancing altruism in colonies of ants, bees and wasps in a moot point, bait one that must be kept in mind.

Eusocial Behaviour

Eusocial Wasps

All of the eusocial wasps belong to the family Vespidae. Among these, the *generapolistes* (paper wasps), *Vespa* (hornets), and *Vespula* (yellow jackets) are the best known. Each of these genera have three castes: *queens* (fertile females), *workers* (nonfertile females), and *drones* (alts). In temperate regions, wasp colonies are established each spry by overwintered queens and last but a single season. When a queen comes out of hibernation, she immediately begins the construction of a nest and rears the first small group of daughters. In the genus *Polistes*, a founding queen may be joined by other overwintered queens, but these auxiliary queens are always subordinate to the foundress, which

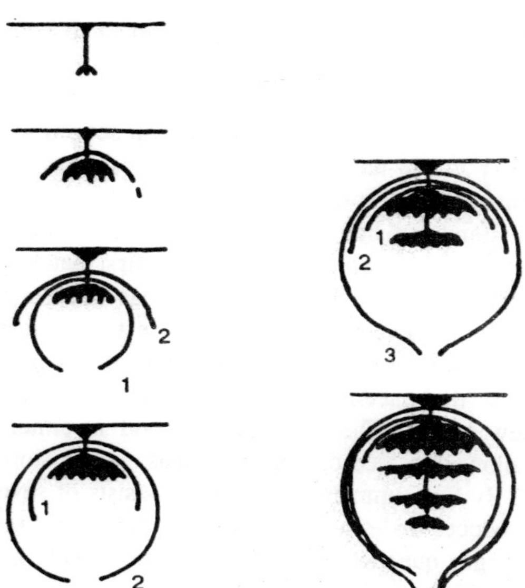

Fig. 2.10. The steps in the construction of a nest by a vespid wasp.

eats any eggs her helpers might lay. Normally, however, wasp colonies are started by a single foundress that constructs an initial tier of hexagonal paper cells, rears the first group of larvae on macerated insects, and keeps the small nest clean. The first group of offspring are all workers, which upon emergence add new cells to the nest, tend the queen, and forage for food. From this time on, the founding queen rarely leaves the nest and spends the rest of her life laying eggs.

The colony grows rapidly throughout the summer as each generation of workers increases the size of the nest and the number of young that can be reared at one time. Toward the end of summer, some larger brood cells are constructed in which a number of males and new queens are reared. When the young queens and drones emerge, they leave the nest to mate and never return. The founding queen dies, and the colony declines to nothing as the remaining workers expire.

Eusocial Bees

The eusocial bees consist of a number of species with a variety of social life styles. Among them the well-known bumblebees and honeybees represent quite different levels of social complexity, whereas the stingless bees of the genera *Trigona* and *Malipona* display an interesting combination of social complexity/and primitive nesting behaviour. The bumblebees are primarily adapted to cooler climates. Like the eusocial wasps, they normally overwinter as queens that survive the cold moths in the solitude of protected hibernations sites. In the spring, the queen emerges and sets out in search of a suitable subterranean nest site in the form of some preexisting cavity. Within the nest the constructs a cuplike cell from wax that is secreted by the intersegmental glands of her abdomen. The cup is then provisioned with a ball of pollen, on which she lays her first batch of eggs, and the cell is sealed. After the initial egg cell has been completed, the queen constructs a honeypot near the entrance to the nest and fills it with nectar. The first group of offspring are all workers that upon emergence assist the queen in producing more larval cells and in feeding the new brood. The subsequent brood is either fed by the workers or is allowed to feed directly on a mass pollen store, depending on the species. By late summer, the colony consists of several hundred workers and produces males and new queens; the colony then abandons its nest. The virgin queens are met by waiting males, copulate, and then enter hibernation.

The tropical, stingless bees construct their wax and resin nests in hollow tree trunks or roots or in subterranean crevices. The nest is usually divided into a brood area and food storage area. The brood combs consist of horizontal tiers of cells that open upward and are constructed one above the other like an upside-down wasp nest. Each tier is begun with a central cell to which rings of peripheral cells are added until each comb reaches its full diameter. Unlike the wasps, stingless bees construct one cell at a time, provide it with an egg, then provision and seal it before the next cell is built. Unlike the honeybees, which use their combs over and over again, the stingless bees tear down their cells after the young emerge. The food storage area consists of a series of wax pots that are filled with honey and pollen. Stingless be colonies expand to thousands of individuals and, unlike those of bumblebees, persist from year to year. As with the honeybees, stingless bees produce maiden queens and drones, from time to time, and the number of colonies in increased and dispersed by swarming. Unlike honeybee swarms, however, stingless bee swarms consist of a cluster of workers and a maiden queen rather than the old queen.

The honeybee originated in the tropics or subtropics but seems to have been preadapted, through its ability to regulate its nest temperature, to expand its range into regions of colder climate and still maintain perennial colonies. Entire colonies, instead of just fecund queens as in the wasps and bumble bees, overwinter. This not only permits the colonies to become much larger but permits and requires other changes as well. Because honeybee queens do not initiate new colonies each year by their own labour, they do not require any of the capabilities of their worker caste. Consequently, they have become much more differentiated from the workers morphologically and serve solely the functions of reproduction. New colonies are founded through the process of colony division by swarming, as in the stingless bees. The swarming process usually begins when a colony is strong and crowded. In response to some unknown stimulus, the queen reduces the level of queen substance the produces, and the workers respond by constructing a small number of queen cells. The larvae that hatch from the eggs"taid in these larger cells are fed exclusively on royal jelly and complete their development in about two thirds of the time it takes workers and drones to develop.

When virgin queens are developing within the colony the resident queen's egg production declines, and she is treated less hospitably until more or less forced to leave with a large retinue of older workers. Swarms containing the old queen are called prime swarms and may be followed a short time later by one or more afterswarms, each containing a new queen produced within the hive and already mated.

Swarms usually settle in a temporary cluster while a permanent nesting site is located. Once scout bees have found an communicated the location of a suitable nesting place, the temporary cluster breaks up and moves to the new site. The workers immediately begin to construct vertical sheets of hexagonal wax cells suspended from the roof of the nest site. The wax combs are constructed a precise distance apart and consist mainly of regular, well-fitted worker cells. The size and shape of the cells determine the kind of egg the queen lays therein and how each cell is provisioned. Worker cells receive diploid eggs and are provisioned with royal jelly for a few days, followed by a mixture of pollen and honey called bee bread. When a queen encounters an irregular, enlarged cell, she will lay a haploid egg, which becomes a drone. When an egg is laid in a cuplike cell, the workers will feed the larvae that develops only on royal jelly and thereby produce maiden queens. Thus the structural details of the wax comb play an integral part in the social organization of the colony. Since the queen that accompanies a primary swarm is mated and the workers can immediately begin constructing a brood comb, the queen can soon proceed with egg laying, and the new colony grows very quickly. Meanwhile, the virgin queen left with the original colony must go on one or more nuptial flights to obtain enough sperm to last the 5 to 7 years that she might live. The original colony also grows rapidly, since the brood left by the departed queen emerges and gradually takes over the hive chores from the again workers that remained after the swarm.

Drones are produced throughout the more favourable part of the year. They are usually most abundant when the colonies are reproducing by swarming and therefore contain virgin queens that must be inseminated. Drones contribute absolutely no labour to the colony but are tolerated by the workers until late in the summer. As the supply of food in the field begins to decline in the fall, the workers become intolerant of the drones, driving them from the hive and not permitting them to reenter.

In respect to brood rearing and overwintering, the ability to regulate nest temperature is highly beneficial. The precision with which honeybee colonies do this is truly remarkable, but it would not be possible if it were not for the type of nest site selected and the way it is modified and utilized. The actual thermoregulatory process, however, is accomplished by the bees themselves rather than physically, as with the termites. During the months when a brood is being reared, the brood chamber is maintained at a temperature of 34.5 to 35.5°C, even though the outside air temperature may reach more than 60°C. In the winter, the cluster temperature ranges from 20 to 30°C and is never permitted to drop below 17°C, regardless of the outside temperature. When the outside temperature drops, the temperature in the hive is maintained by the metabolic heat generated by the workers, whose behaviour changes as the temperature declines. At first, the workers form a loose cluster toward the center of the hive. As the temperature drops, the cluster tightens. The workers at the center consume small amounts of honey and generate heat by vibrating their muscles; those forming the outer layers of the cluster act as an insulating blanket. As time passes, the outer bees move toward the center of the cluster, while those near the center move into the outer layers. During hot weather, the temperature of the brood area is maintained by a circulation of air in the hive created by workers that fan with their wings at the hive entrance. When this activity proves to be inadequate, other workers carry water into the hive and distribute it over the brood cells, which are then cooled as the water evaporates.

Ants

All ants are eusocial. The lack of presocial behaviour of the kinds found among the wasps and bees can be explained by the fact that the ants probably evolved from a group of presocial wasps. The variety of life styles displayed by the ants has long been of fascination to naturalists, perhaps because so many aspects of their social behaviour seemingly parallel our own. All living species of ants have a caste system that generally parallels that of the eusocial wasps and bees and clearly reveals a division of labour. The principal casts consist of males, queens, and workers the latter group being subdivided into different types in most species. Several genera also have castes intermediate between males and workers and between workers and queens, but these are generally less important than the main castes. Typical males are winged reproductives that contribute absolutely nothing to the labour of the colony although they do groom other adults and are groomed in return. Queens are the colony founders and mothers. They are usually winged when they emerge but shed their wings after their nuptial flight. During the establishment of a new nest, the queen performs all of the tasks later performed by the workers; but , once the first group of workers has emerged, the queen's activities are reduced to egg laying and grooming.

Workers are sterile females, and in the broad sense include both the labourers and defenders of the colony. There are often three subcastes of workers, based partially on size. The largest are the *majors* or *soldiers*, which are often considered to be a distinct caste

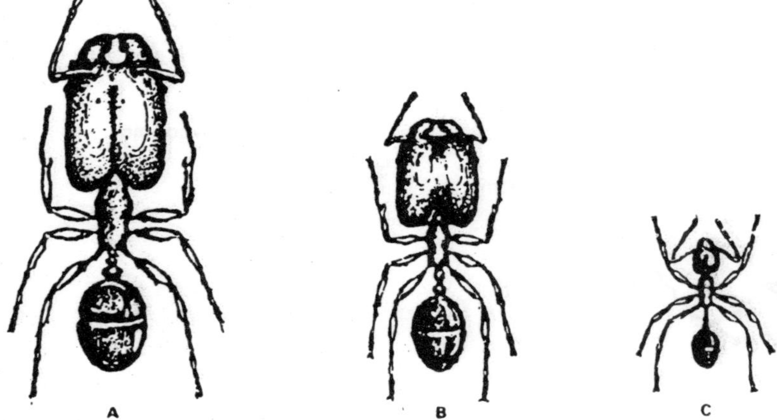

Fig. 2.11. Different forms of the worker cast of ants. (A) Major. (B) Media. (C) Minor.

because of the oversized (allometric) development of their heads and mandibles. The other workers are distinguished as being *media*, if of intermediate size, and *minors*, if small. The degree of workers polymorphism varies greatly from species to species, but, in species in which it is pronounced, a clear division of labour among the subcastes can be observed. The most unusual workers are the repletes of the honey ant *Myrmecocystus*. These individuals serve as casks for the storage of the honeydew brought to the nest by the foragers. When filled, the abdomen of a replete becomes a large swollen sac.

Ants are extremely numerous and ubiquitous. Their food habits are diverse, although most are omnivorous, and their social groups range in size from hundreds to perhaps millions. Certainly, the most spectacular of these groups are the legionary or army ants of the humid tropical forests. These ants do not construct nests but form temporary clusters called bivouacs in the shelter of a fallen tree trunk or other such partially exposed location. The workers form a solid mass up to a meter across, consisting of layer upon layer of individuals linked together by their tarsal claws. The queen, thousands of larvae

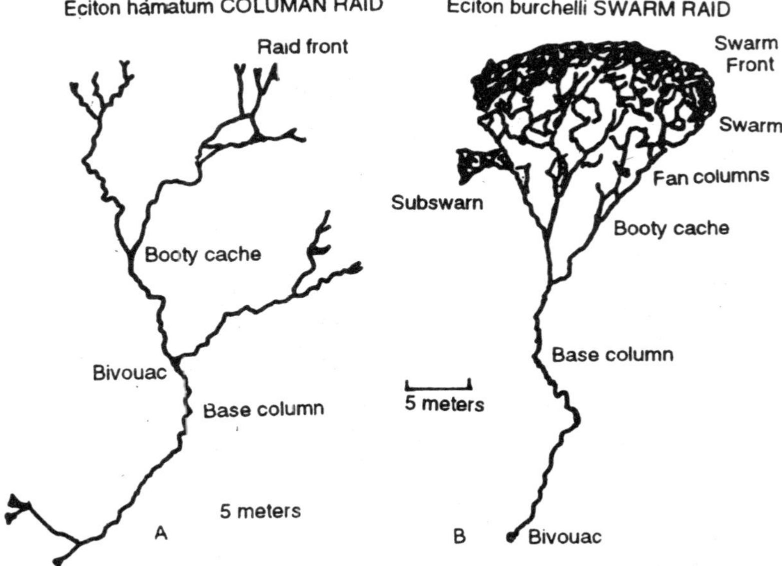

Fig. 2.12. Basic raiding patterns employed by army ants. (A) The column raid of Eaton hamatum *with an advancing front composed of several narrow columns of workers. (B) The swarming raid of* Eciton burchelli *composed of a mass of work ors advancing together.*

and pupae, and, at certain times of the year, up to 1000 males and a few virgin queens, are located near the center of the ball of up to a half million workers.

The night is passed in a tight cluster, but, at first light, the chains of workers become detached and a hoard of individuals begins to move away from the bivouac site in all directions. Soon, one or more raiding columns form and move out in search of food. The workers lay a pheromone trail to guide those that follow, while the soldiers guard the column's flanks. Some species engage in column raids, whereas others employ a swarming tactic. Regardless of their method, however, the foraging ants form a number of changing columns behind the advancing hoard of workers that flushes and captures a variety of mainly orthropod prey. The prey are stung to death dismembered, and transported rearward as food for the developing larvae. At the end of each day of marauding, the protective cluster is reformed.

As they colony rapidly reduces the food supply in the immediate vicinity of its bivouac, it must move to a new location. However, in at least one species, the change in bivouac site has been correlated with the reproductive cycle within the colony, which in turn has an effect on its food requirements. According to Schnierla, the queen periodically undergoes rapid ovarian development, which results in great distension of her abdomen. Over a period of several days, such a queen lays 100,000 to 300,0000 eggs, which hatch about 3 weeks later. during this *statary phase*, the larvae from the previous period of oviposition complete their pupal development and emerge as young workers. The addition of these new workers seems to stimulate the colony to increase the intensity of its daily foraging raids and change the location of its bivouac site each evening. The *nomadic phase* therefore coincides with the period during which a large number of larvae must be fed. Once these larvae complete their development and pupate, the nomadic behaviour declines, and the colony enters another reproductive period, during which a single bivouac site is utilized.

People living in the temperate zone most familiar with the nest building ants, such as members of the genus *Mynnica*, which usually nest in the ground under a large stone, or the genus *Formica*, which construct the familiar mounds of twigs and other plant debris. These ants are characterized by the production of winged reproductives during the summer and early fall. In some species the emergence of the reproductives produces quite spectacular swarms, during which mating occurs. The nuptial flight results in the ubiquitous spread of potential foundresses of new colonies. After a short time, the newly mated queen sheds her wings and excavates the beginning of a new nest into which she seals herself. In the spring, she lays a small group of eggs and tends the larvae until they attain adulthood and become the colony's initial group of workers. These first offspring may take more than a year to complete their development, but the queen never leaves, relying on her internal food reserves and flight muscles for her own sustenance as well as that of her young.

Over the succeeding years, the colony grows very slowly until enough workers are present for it to enter a period of accelerated growth. In areas characterized by cold winters, the colonies enter hibernation for several months, and no eggs are laid until warmer conditions prevail in the spring. In most species, several years pass before winged m^les and females are produced to leave and establish new colonies in their turn. Colonies of many species contain only a few hundred individuals, whereas some contain many thousands; most nests contain only one queen, but some *Formica* nests have hundreds of active queens. In spite of the general similarities displayed by the nest-building ants, as far as the establishment and growth of new colonies is concerned, they show extensive adaptive radiation in other aspects of their behaviour. Most display a high level of variability in their choice of prey. Like the army ant, they forage widely for almost any kind of small animals, especially arthropods, that are available. A single nest of the European red ant, *Formica polyctena*, is said to gather up to a kilogram of such food in the course of single day.

Fig. 2.13. Media worker of the fungus-growing ant, Aaa cephalotes, *cxutting and carrying a portion of leaf while being protected against attack from a phorid fly by a minor worker.*

Chauvin concluded that the estimated population of 300 million red ants, which inhabit the Italian Alps, would be capable of destroying 15,000 tons of insects a year. But not all ants have the catholic tasii of *F. polyctena*. Some feed specifically on certain kinds of arthropods of their eggs. Others, often called "harvester" or "agricultural" ants, subsist entirely on seeds, and thereby become pests in some grain and grass-growing regions. When protein-rich foods are not needed for the rearing of larvae, many ants feed upon the nectar of flowers.

A number of ant species feed exclusively on the anal excretion of aphids, scale insects, and other homopterans. The so-called "honey-dew" is rich in sugars and free amino acids and forms an inexhaustible source of nutrient throughout the time when the producing insects are actively feeding. Some species of ants can be seen to stroke the cornicles of aphids to induce the flow of honeydew from the anus. Ants, in their tending of homopterans, can seriously interfere with biological cantrol programmes directed against these insects by actively repelling parasites and predators. In the fall, some ants actually transport aphids or their eggs from their host plant to underground "barns" in which they are cared for over the winter. In the spring, the ants carry the aphids back to their host plant once again to "milk" them of their honeydew.

Members of the New World tribe Attini are sophisticated fungus culturers. Many species in the group gather small pieces of green leaves or the petals of flowers to form underground beds on which they culture a specific variety of fungus. Members of the genus *Atta* are among the more serious pests of New World tropical agriculture as a result of their leaf collecting; in English, they are commonly referred to as leaf-cutting ants. Media workers carry the plant material piece by piece deep into their nests, where it is licked, cut into small piece, wet with an anal secretion, and formed into beds of moist pulp. The newly formed beds are then "planted" with fungal mycelia collected from established bed$. The fungus grows rapidly and soon produces on the tips of the hypae small spheres, which are fed to the ant larvae. The fungus beds are tended and harvested by smaller workers (minora) that never engage in the collection of leaf fragments. However, in the case of *Atta cephalotes*, these smaller workers may accompany the larger leaf-gatherers; they do not assist with the leaf cutting but ride back to the nest on the leaf portion, apparently warding off parasitic phorid flies with their mandibles and hind legs.

One of the more fascinating aspects of the fungus culture is how the ants are able to maintain a monoculture of their specific fungus;

abandoned beds are rapidly overgrown by alien fungi of various species. Weber discovered that the worker ants tending the fungus beds "weed" them of alien hypae with their mandibles. It has also been postulated, but not proved, that the ants employ fungicidal and bacteriocidal substances secreted by their salivary or anal glands. The cultured fungus is carried to each new nest by the foundress. Before departing on her nuptial flight, the virgin queen packs a small wad of mycelia into a cavity near the base of her labium. The wad is then deposited in the nest she excavates. The queen tends the initial fungus garden but does not consume any of the culture herself. However, the first workers to emerge feed upon it and fertilize it with their fecal material. Not all ants are as industrious as the aphid herders and fungus gardeners. The Amazon slaver ant, *Polyergus lucidus*, and the red ant, *Formica sanguinea*, are common species that make raids on other ant species to capture slaves. The raiders carry larvae and pupae back to their nests, where they are raised to assume the duties of nest maintenance. *Polyergus lucidus* is an obligatory slave keeper, as the workers cannot feed themselver. *F. sanguinea*, on the other hand, is a facultative slaver.

Termites

The order Isoptera is clearly more ancient and more primitive than the order Hymenopters, yet all living termites are eusocial and have achieved a level of caste differentiation and social organization equal to the wasps, ants and bees. The most primitive living termite, *Mastotermes darwiniensis*, seems closely related to the primitive, social, wood-eating roach *Crytocercus*, with which they share a number of intestinal protozoans that are essential to the digestion of the cellulose they consume in a wide variety of forms. Some termites also lay eggs in small groups that are protected by a covering similar to the oothecae of roaches.

Termites lack cellulose-digesting enzymes and rely on their intestinal fauna of mutualistic flagellate protozoans, which must be passed continuously between individuals, to aid in the breakdown of their food. The passage of gut symbionts from old to young individuals. seems to have set the stage for the development of more advanced social behaviour. It has been postulated, therefore, that the social order of termites began with trophallaxis and led to the addition of brood care, rather than the reverse, as in the social hymenoptera.

The natural history of termites is quite varible. The so-called dry wood termites that frequently cause damage to the dry, seasoned wood

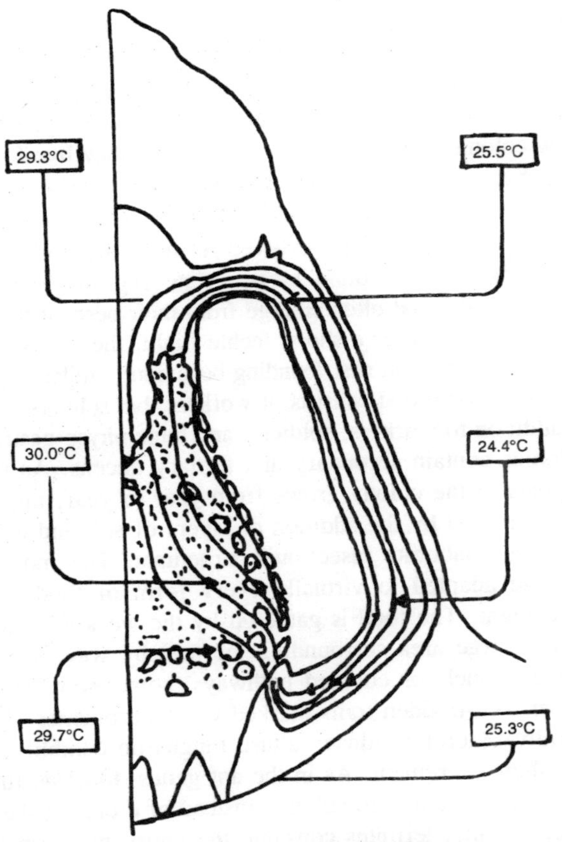

Fig. 2.14. Section through the mound nest of the African termite Macrotermes natalensis showing the basic design of the air conditioning runnels.

of buildings usually live in colonies consisting of only a few hundred individuals. Winged reproductives are produced seasonally, usually during the warmer months, leave the nest soon after emergence, and spread rather ubiquitously. When a female completes her flight, she divests herself of her wings and runs about in an excited fashion until joined by a male. After pairing, the king and queen engage in a behaviour called tandem running, in which the king closely follows his queen in search of a nest site. Whenn the pair locate a suitable place to begin a nest, they take turns excavating an initial tunnel with a small chamber at the boottom. The king and queen seal themselves in by plugging the tunnel with chewed wood, and the queen proceeds to produce a small batch of eggs. When the young nymphs emerge, they are fed regurgitated food until they are able to feed upon the surrounding wood and thereby enlarge the nest. As the colony grows in the ensuing years, soldiers are produced that provide the colony with defense. When the colony is several years old, more alates are produced, which found new colonies. The foundress of a colony may live for more than 10 years, but if her egg production declines or she is killed, her duties may be assumed by secondary queens.

Most of the more highly evolved termites are soil dwellers that construct conspicuous mound nests. Alate reproductives are agian produced seasonally and often emerge from their nests at the beginn-ing of the rainy season. After a short, feeble flight, the reproductives shed their wings and engage in nest founding behaviour similar to that already described. The first brood consists of workers, but soldiers are produced later. In addition to workers, soldiers, and huge physo-gastric females, some colonies contain secondary and tertiary queens. As the number of individuals in the colony grows from year to year, the size of the mound is increased by the addition of layers of soil and excrement to form the most elaborate insect-built structures. The mound-building termites have adapted to virtually every form of food with a high cellulose content. The food is gathered by the workers, which forage throughout a large area surrounding the nest by way of a system of subterranean tunnels or covered pathways; some even forage on the surface over well-trodden trails marked with pheromones. Termites of the subfamily Macrotermitinae culture fungus on comblike structures built from their excrement. As in the ant genus *Atta,* the fungus beds of occupied nests are a pure culture. Instead of cropping the fungus in the manner of ants, termites consume the entire mass, including the excrement substratum. The colonies of fungus growing termites may contain as many as 2 million individuals. The primary queen is capable of laying up to 30,000 eggs a day; in a life span of up to 10 years, a single queen may produce tens of millions of offspring.

The gigantic mounds of tropical termites often have distinctive forms that are characteristic of particular species. In addition to providing protection, termite nests are often constructed in such a way as to provide a well-regulated interior microclimate. *Amitermes* orient their nests in such a way as to maximize sloar warming in the morning and evening. Microclimatic control is most highly developed in the fungus-growing genus *Macrotennes*. Its members construct a series of passageways, through which air flows according to its own density, so that temperature and carbon dioxide conditions favourable to the growth of fungus are maintained in the culture chamber.

The evolution of the termites is far removed phylogenetically from that of the Hymenoptera; yet, two grops have, by quite different routes, arrived at a complex level of social organization. There are some striking similarities in the social biology these distinct ordres display, but there are some fundamental differences as well. The most important differences markedly affect the make-up of the colonies. In the case of the termites, the queen cannot control the sex of her offspring, so the workers are both males and females (genetically). Furthermore, after the first few instars, the termite immatures engage in work for the good of the colony. In the social Hymenoptera, the workers are all female, and the queen can regulate the production of males; the larvae are, in a sense, parasitic and contribute nothing to the well-being of the colony.

3

Hormones in Social Activities

The insect hormones are those which are effective and also well known by entomologists influence pre and postembryonic development. Three of these so-called metamorphosis hormones are recognized: (A) the activation hormone manufactured by specific neuroscecretory cells in the insect brain, which conditions the reactivation of the body after each moult and the production of the other two metamorphosis hormones; (b) the moulting hormone produced by the prothoracic glands or corresponding tissues, which conditons the moulting process and thus, indirectly, growth and morphogenesis; (C) the juvenile hormone produced by the corpora allata, which conditions the growth of the larval parts in all metabolous insects (pterygota), the function of the ovarial follicles of adult females in most insects, and many other structures and functions of the insect body, which are unable to evolve in its absence. All three hormones are essential for the normal course of postembryonic development. The activation hormone is a typical neurohormone, the other two are true glandular hormones order, and sometimes even class unspecific. They are dependent on one another both in their production and function.

Terminology

Metamorphosis

Metamorphosis implies that the single part of hte postembryonic development which takes place in the absence of a morpho-genetically active concentration of *juvenile hormone*, i.e. the last larval instar of

Hemimetabola and both the last larval and the pupal instars of Holometabola. This is corroborated with most other physiological works but in disagreement with some morphologists. But it must not be denied that a part of the adult differentiation is recognized at the beginning of each larval instar, there is no more reason to call this metamorphosis than there is in the case of the similar successive postembryonic differentiation in ametabolous insects (Apterygota).

Postembryonic Development

The postembryonic development of insects can thus be divided into two parts or stages: the larval stages (or larval development) beginning with the hatching of the larva and ending with the penultimate larval instar (the ecdysis from the penultimate to the last larval instar); and the later metamorphosis stage (or metamorphosis), ending with the development of the adult insect, i.e. the last larval instar of Exopterygota and the same including the pupal instar in Endopterygota.

Instar

Both the larval and the metamorphosis stages are sub-divided into instars. An *instar* is that part of the postembryonic development which takes place between two successive ecdyses. The term is used in both the morphogenetical and the chronological sense of the word. Yet another definition was suggested by Hinton (1958), and consequently accepted by many other entomologists, according to which the instar begins at the moment of deposition of the new cuticle and ends with the moult, which is the next separation of the cuticle from the epidermis. based on the morphological description this has the advantage that the term 'instar' refers to one and the same cuticle whereas in the earlier definition, each instar starts inside the old cuticle which is replaced by a new one during the instar. It has also the advantage of drawing hte attention of morphologists to those stages in the moulting process which take place beneath the opaque cuticle. Notwithstanding the fact that postembryonic development is further divided into a succession of overlpapping cyclic processes, so that the determination of the limits of individual instars is to some extent arbitrary, the first-mentioned definition of the instar seems most appropriate for the purposes of this book for the following reasons:

1. The strcture of the body, even though determined by the newly deposited cuticle, particularly that of heavily sclerotized parts such as the head capsule, is also changed at each ecdysis but not at the moults.

2. The detachment of hte cuticle at apolysis can only be observed in histological sections (rather difficult to make because of the hard sclerotized cuticle). The moment of ecdysis is therefore as much preferable as the dividing point from a practical point of view.
3. Ecdysis is linked with a series of other processes, such as the resorption of the moulting fluid, the occupation of the new shape of the body at the distention of the newly formed cuticle (even though this was laid down some time previously), the deposition of the wax and cement layers of the epicuticle, the tanning and hardening of the new cuticle, etc. It is therefore the most significant and visible section of each moulting cycle.
4. Inspite of the fact that successive moults are connected with an uninterruped succession of various physiological processes, most of these are, nevetheless, clearly arranged in cycles within each interecdysis, but not respiration and hormone production, resulting in swelling of the during moults. By selecting the detachment of the cuticles as the start of an instar, all of these clearly logical processes are divided between two different instars.

Most of these reasons remain acceptable even in such cases as the last larval instar of higher Diptera. There, although the chief characteristic of ecdysis, the shedding of the exuvium, is eliminated, as is the tanning of the newly formed pupal cuticle, other distinguishing features, such as, for example, the removal of the moulting fluid and the formation of the epicuticular layers, remain; and these are enough to enable the instars to be distinguished.

Pharate instar

If we do not accept Hinton's definition of the term 'instar', the term 'pharate instar' has a restricted meaning. It seems to be justified only in cases in which a new instar has been formed completely and the moulting fluid has been absorbed, as in the puparium of higher Diptera, for example. The determining factor is thus the extistence of air between the old and the new cuticle.

Development Stage

Generally the term stage of development is applied for a section of ontogenetical development characterized by significant biochemical or morphological processes or by both, such as the protoped, polypod,. oligopod, and postoligopod stages in embryogenesis, the larval stage (comprising the first to penultimate larval instar), the metamorphosis stage (comprising the last larval and the pupal instars), the maturation stage, the stage of sexual maturity, etc.

Moulting Process

There are several biochemical, physiological and morphological changes preceding each ecdysis constitutes the moulting process or moulting. This begins with the swelling of the epidermis followed by detachment of the old cuticle, usually referred to as the moult or apolysis. Then the moulting fluid is produced, in which the inner layers of the cuticle are gradually digested, and a new cuticle is laid down by the epidermal cell. The process ends in ecdysis which is preceded by the resorption of the moulting fluid and followed by the tanning and hardening of the exocuticle and the secretion of the wax and cement layers of the epicuticle. the laying down of the inner layers of the endocuticle by the epidermal cells persists after the ecdysis and usually does not stop until immediately before the start of the next moulting process. Thus the moulting process is terminated.

Few years back researchers began to use the terms 'larval-larval moult' in place of 'larval moulting', 'larval- pupal moult', instead of 'pupal moulting' and 'pupal-imaginal moult' instead of 'imaginal moulting. Apart from the linguistic aspect, this seems completely unnecessary, in so far as we taken into account that the above terms express the result of the relevant moulting process. In any case the meaning is usually amply evident from the cortex. However, the term 'emergence' for imaginal ecdysis is exitrely adequate.

The term 'moult' does not mean 'moulting process' since it has been accepted as the English term for 'apolysis', i.e. only one of the initial phases of the moulting process.

Larval moulting-any moulting which results in larval form.

Pupal moulting-moulting to pupal form. When a second pupa is produced as a result of experimental JH treatment, we tern it as 'second pupal moulting'.

Imaginal moulting or *emergence-moulting* which culminates in the adult form.

Growth

By growth is generally accepted as the increase in the amount of living matter of the body (protein molecules) unlike, e.g., the increase in weight by accepting food or ingesting water. The following lines piont the differnet types of growth may be distinguished in insects:

Isometric growth (harmonic or proportionate growth)

The type of growth where there is an equal rate of increase in all parts of the body, i.e. there are no morphological changes. The shape of the body at the end of an isometric growth period is a homothetic figure of that at the beginning.

Anisometric growth (disharmonic or disproportionate growth)

The type of growth where there is a different rate of increas in different parts of the body, with a resulting changes of shape. two kinds of anisometric growth are distinguished in this book. These are:

(i) *Allometric growth* (or anisometric growth in the restricted sense). In this kind of growth parts of the body grow although at different rates, as for instance in post-embryonic development in mammals.

(ii) *Gradient growth.* In this type of growth, only few parts of the body, the gradients, grow. Growth of the remaining parts is stopped and eventually they are wholly or partly removed either by histolysis or phagocytosis. This occurs, for example, in the embryogenesis of most animals an in the meta-morphosis of insects. Gradient growth is the significant feature in morphogenesis.

Fig. 3.1. Stink bug (Hemiptera). Stage of life history exhibiting heterometabolic metamorphosis or to say gradual metamorphosis.

Differentiation

The important aspect of development, the sequence of changes which occurs in the body in the given period of ontogenesis, is irrespective of growth. The three following types are discerned (1) chemical (biochemical) differentiation, the sequence of changes in the chemical composition of the body irrespective of changes of weight, shape, function, etc., (2) morphological differentiation or morphogenesis, the sequence of changes in the shape of the body, (3) functional differentiation, the sequence of development of various functions of the body. These three types of classifications are all closely related in normal development. In ontogenesis, chemical differentiation usually precedes morphological differentiation, which in turn precedes functional differentiation. On the other hand in phylogenesis, functional differentiation usually precedes chemical and morphological differentiation of the corresponding part of the body. (This follows from the action of natural selection and its role in evolution: it means that structures of the body can be favoured by selection and thus evolved and specialized for a given function only when this function already exists and is to some extent useful for survival to some extent.)

Larval structures

They are the parts of the body which can grow only the larval stage (see above), i.e., in the presence of a certain minimum concentration of juvenile hormone. They lose the ability for further growth during metamorphosis, and usually undergo partial or complete histolysis at that time. They include the greater part of the larval body is most pterygote insects.

Imaginal structures

Those parts of the body the growth of which is independent of the presence of juvenile hormone in the beginning of metamorphosis. Their development is accelerated durign metamorphosis when the growth of the larval structures ceases. They include the greater part of the imaginal body.

Larvo-imaginal structures

Are fully differentiated and functional in the larval period, but they remain and continue to function in adults.

Larva

The term larva is used here in its broadest sense, as in Wigglesworth (1939) and Hinton (1958), but not Jeschikov (1941) and Snodgrass (1954), to include the entire postembryonic stage of development which precedes metamorphosis in all insects. The term 'nymph', as a special type of larva, is retained for the larval stage of Hemimetabola and 'larvula' is kept for the first larval instars of Odonata differing from the later ones. 'Eonymph', mesonymph' and 'pronymph' in the sense of Berlese (1913), and Slama (1960) refer to the pre-pupal stages of sawflies. Pre-pupa is the term for the latter portion of the last larval instar of Holometabola, which is distinguished by the advanced pupal moulting process with plenty of moulting fluid and retraction of the body.

Pupa

The term pupa is used for the last postembryonic instar (second metamorphosis instar) in Holometabola, which is the period between the pupal and imaginal ecdyses. Internal morphological changes, which vary considerably in extent according to the insect, occur during this instar, whilst most of the external changes are occur in the first metamorphosis instar (last larval instar).

Hemimetabola and Holometabola

These terms are used as in the taxonomic sense of Imms (1946) and Obenberger (1952), but not Weber (1952), but not Weber (1954), Snodgrass, etc. *Hemimetabola* (= Heterometabola, = Exopterygota) have only one metamorphosis instar, *Holometabola* (= Endopterygota) have two. The prefix 'hemi' does not refer to the degree of morphogenesis which, on average, is notmuch different in the two subclasses, but to the number of metamorphosis instars.

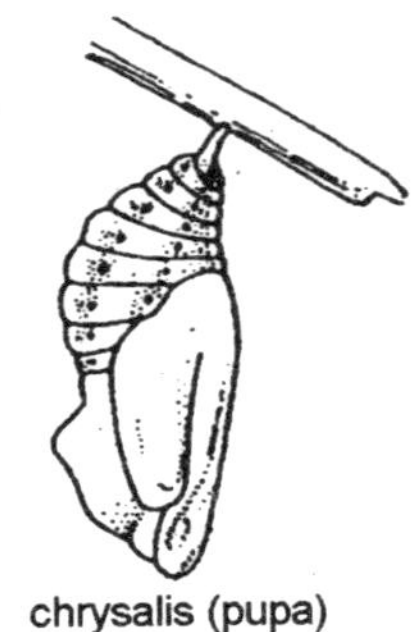

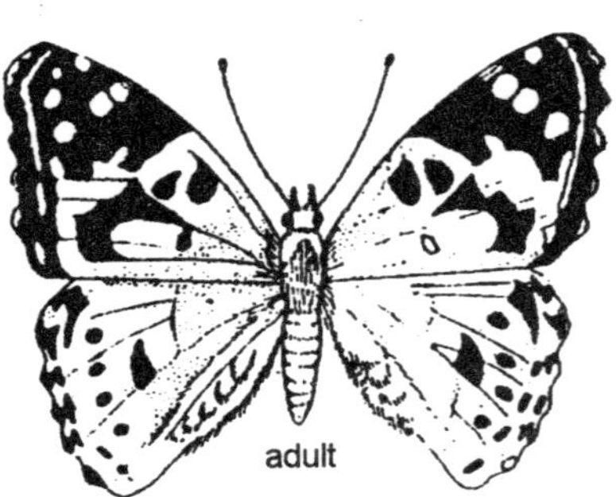

Fig. 3.2. Butterfly (Lepidoptera). Stages of life history exhibiting holometabolic (complete) metamorphosis.

ABBREVIATIONS

Abbreviations are generally used for frequently recurring terms, for the purpose of saving space and also for greater clarity. The advantage of this was soon relaized by chemists, who introduced a generally recognized and employed letter-symbol for every element (chemical compound). The same need is increasingly felt by specialists in insect endocrinology, but no rational abbreviation system which would be generally accepted has so far been created.

The system applied throughout this book, has the following rules Capital letters are employed for the hormones forming the main subject-matter and small letters for the often no less useful abbreviations of various endocrine organs. Differences in the origin of the same principle are expressed by adding a small letter or letters after the hormone symbol. The same primary symbol is always used for one principle (substance, effect). Consequently, instead of CAH, used by some authors for the corpus allatum hormone, and JHa for JH analogues, the common symbol JH is used for both, JHca denoting corpus allatum JH and JHa juvenile hormone analogues. JH is used alone where effects of the principle are discussed in general, irrespective of its origin. The abbreviations used by other authors are given in the list of synonyms at the beginning of each section dealing with the relevant hormones.

ca—corpora allata (corpus allatum).

cc—corpora cardiaca (corpus cardiacum).

nsc-neurosecretory cells.

ncc I, II, II—nervi corporum cardiacarum, first, second, and third pari.

agl—apolytic glands.

pgl—prothoracic glands.

rgl—ring gland.

vgl—ventral gland.

nho—neurohaemal organs.

mnho—metameric neurohaemal organs.

cns—central nervous system.

No abbreviations are used in this book for organs not associated with endocrine activity.

AH—activation hormone.

JH—juvenile hormone.

JHca—corpus allatum hormone.

JHa—juvenoids (juvenile hormone analogues).

MH—moulting hormone (ecdysone)

MHpg—prothoracic gland MH.

MHd—ecdysoids (ecdysone derivatives).

GH—gradient factor ((the distinction formerly made between the GF of insect metamorphosis and the general gradient factor gf has been abolished, because it is nwo assumed that they are identical).

BF-blastomo-factor

To facilitate easy reading the terms are also given together with the abbreviations by which theya re represented, wherever they appear for the first time in each chapter.

The Activation Hormone (AH)

The initial statement regarding a hormonal function of the insect brain was made by Kopec (1917, 1922) whose experiments with Lymantria dispar caterpillars proved conclusively that the brain is necessary for pupation. This was the first proof for an internal secretion of any kind in insects. Tauber (= Taabor) (1925), working independently, concluded, from several years' work on blood transfusion and ligaturing in Deilephila euphorbiae caterpillars, that insect metamorphosis is controlled by endocrines. More indication for the presence of the humoral control of moulting in other species of Lepidoptera is contained in the papers by Koller (1938) and Buddenbrock (1950).

But the confirmation of the existence of a special hormone, produced by the brain which controlled moulting, was achieved by Wigglesworth (1936) from his classic experiments with decapitation, parabiosis and transplantation in the blood-sucking bug Rhodnius prolixus. This was the genuine beginning of insect endocrinology which has now evolved into a special, very extensvely studied, biological discipline. It was the same author who, in collaboration with Hanstrom (1943), first discovered the source of the activation hormone in the neurosecretory cells of the brain. Another comtemporary entomologist Fraenkel (1935) demonstruted a similar hormonal control of pupation in Calliphora larvae by his well-known ligaturing experiments. Similar onclusions were reached by Bounhiol (1938) from his experiments. Similar conclusions were reached by Bounhiol (1938) from his experiments with Bombyx mori; by Bodenstein (1938), on the basis of several years' experimental work on the transplantation of appendages from young to fully grown caterpillars; by Kuhn and Piepho (1940) for Galleria mellonella, and by Hadorn (1939), Hardon and Neel and by Hadorn and B. Scharrer (1938) for Drosophila. Since then the number of papers on endocrine control of moulting has increased in number.

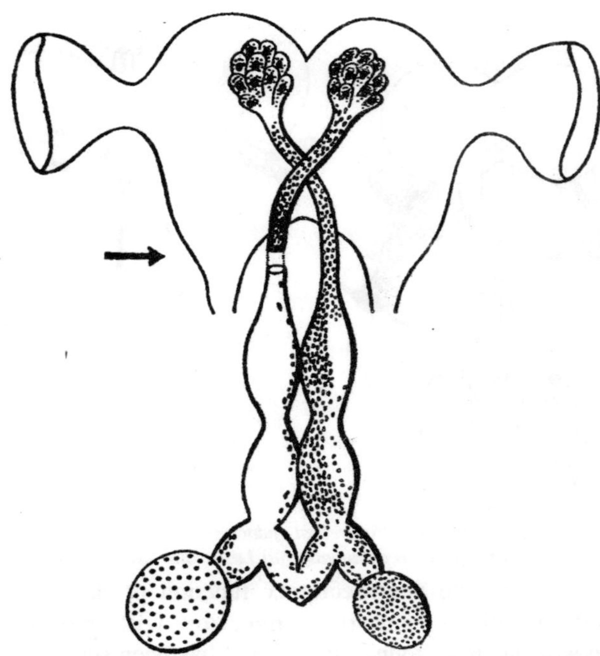

Fig. 3.3. The effect of breaking one of the two nervi corporum cardiacarum. Note the proximal accumulation of neurosecretory granules in the nerve slump, its disappearance from the corpus cardiacum and the enlargement of the corresponding corpus allatum.

Fukuda (1944) was the first to prove that the AH does not affect the moulting process directly but does so by controlling the secretion of the prothoracic glands. These findings were confirmed and closely examined by Williams (1952) and his colleagues.The suggestion of the possible function of the corpora cardiaca in neurosecretion was initially made by De Lerma (1933, 1934) and its connection with the neurosecretory cells of the brain was elucidated by Hanstrom (1942), by Pflugfelder (1937, 1938c, d) and, particularly, by B. Scharrer (1937, 1941) and B. and E. Scharrer (1937, 1944) who confirmed it by a detailed histological and experimental study, that the neurosecretory cells of the brain with their axons, together with the corpora cardiaca, form a single neurosecretory system corresponding with that of the neurosecretory cells of the brain with their axons, together with the corpora cardiaca, form a single neurosecretory system corresponding with that of the neurosecretory cells of the hyopthalamus with their axons and the neurohypophysis, in vertebrates.

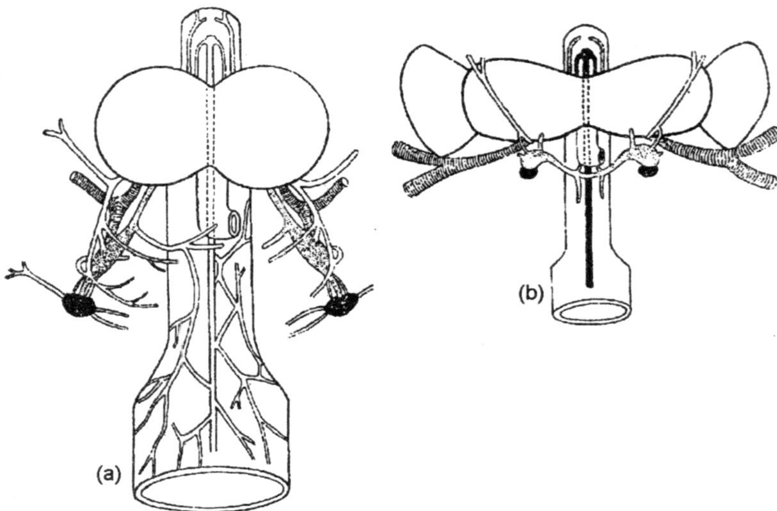

Fig. 3.4. Scheme of the brain and retrocerebral endocrine system in developmental stages of lemon butterfly (Paillio demoleus). (a) Last instar larva; (b) Early pupa.

Since that time the neurosecretory function of the insect brain and corpora cardiaca, and its connection with moulting and metamorphosis, has been found whenever it has been seriously looked for, that is, in practically all the chief orders of metabolous insects (pterygota). There is also histological evidence for its existence in Apterygota. Number of resarch papers by Gersch and his school on the chromatophorotropic and myotropic activity of insect brain extracts has raised the question as to whether some of the newly discovered (e.g. neurohormone C and neurohormone D of Gersch and Unger, 1957a; cf. 1960) could be identified with AH. This problem has been discussed compreensively by Raabe (1959). The possibility of the neurohormone D and AH being identical was suggested by Gersch (1962) on the basis of experimental evidence.

Recent investigations by the same author and his colleagues showed that the activation hormone is different from neurohormone D, and that it is composed of two components, with-different effects. These are the activation factor I (AH_1), which stimulates the synthesis of every kind of RNA, and the activation factor II (AH_{11}), which raises membrane potentials and thus has a positive effect on permeability. Both effects seem to be quite general in character, so that they might account for most of the specific AH effects, e.g. stimulation of the secretory activity of various glands metabolism, water balance, diapause development. Nevetheless, it is yet probable that AH_1 and AH_{11} could be two separate neurohormones produced by different nsc.

A number of other functions have been attributed to many known nsc, both of the brain and singel ganglia of the ventral nerve cord; these will be discussed later. B. and E. Scharer (1944) came to the conclusion that a close analogy exists between the neurosecretory system of three large groups of animals, Crustacea, Insects and Ver tebrates, are in full agreement with such a possibility. They confirm the earlier views of the above mentioned researchers, Hanstrom (1939), and others. The relationship between AH production and secretory activity of hte corpora allata, which regulates the activity of the ovarial follicle cells (studied in detail by Johansson, 1958), is in agreement both with the general nature of the AH effects and with the features of neurosecretory system common to the three groups of animmals mentioned.

The Neurosecretory Cells of the Brain (NSC)

Morphology

The neurosecretory cells of the brain, the source of the AH, are usually arranged in two groups, placed symmetrically on the upper surface of each hemisphere near the median furrow in the parts intercerebralis protocerebri. There are 4 to 15 or more main nsc in each group (4 in Bombyx mori, 6 in Calliphora larva, 7 in Pyrrhocoris apterus, 7 to 8 in various Hymenoptera, 15 in cockroaches and up to several hundreds in Orthoptera). They are often quite visible in a living brain dissected in Ringer because of thier milk-white or slightly bluish colour. The transparent nucleus often appears as a dark spot in the middle of each cell. The cells are more obvious when examined under dark-ground illumination, appearing shining white against the dark background. The axons of these two groups of cells form a pair of nerves-the nervi corporum cardiacarum 1 (interni). These two nerves cross inside the brain in the mid-line and each enters the corpus cardiacum on the opposite side the body. External to each of the two above-mentioned groups of NSC is, in most insects, a smaller group, usually of 2 to 4 cells, the axons of which form the nervi corporum cardicarum II (externi), each nerve entering the corpus cardiacum on the siimilar portion of the body.

The neurosecretory material has been detected in histological preparations and under dark-ground illuminationg. Severing the nervi corporum cardiacarum led to an accumulation of nurrosecretory material in the distal end of the portion connected to the brain and the

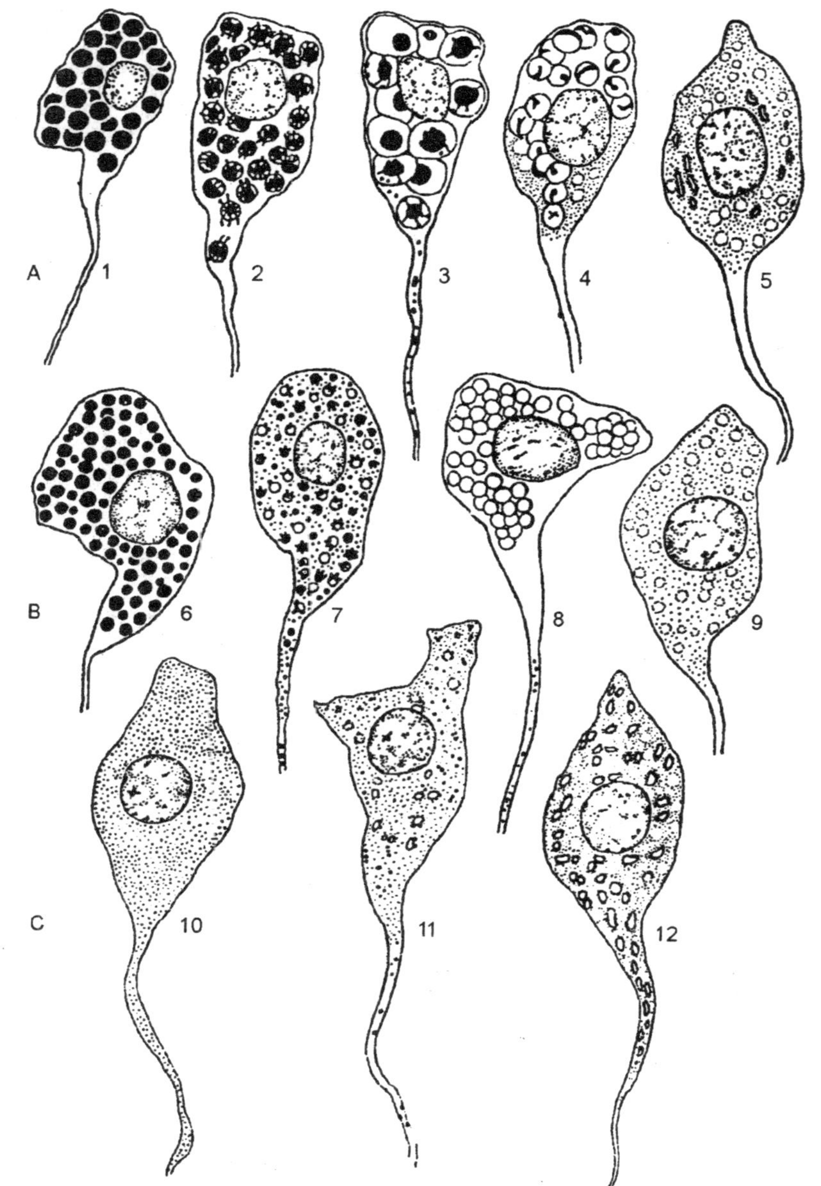

Fig. 3.5. Scheme of developmental changes in three types of neurosecretory cells of the pars intercerebralis in Lampyris noctiluca. A–nsc with large granules; B–nsc with medium sized granules; C–nsc with fine granules.

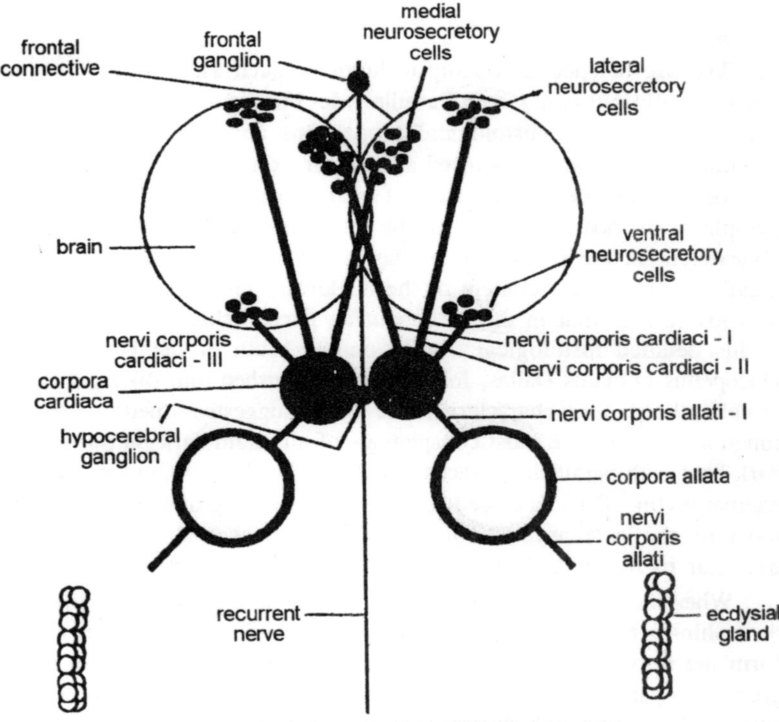

Fig. 3.6. Cephalic endocrine system (diagrammatic).

disappearance of granules in the end on the other side of the cut. Resembling discovery was made by E. Thomesen (1954), using dark-ground illumination, when the cardiac-recurrent nerve of adult Calliphora females was ligatured with a silk thread. The passing of part of the neurosecretory material from the corpus cardicaum via the nervus allatus into the corpus allatum, where it is expected to stimulate the gland and increase its size, has recently been studied by Nayar (1958). E. Thomsen (1954) also observed that the neurosecretory granules pass via the nervus recurrence into the walls of the dorsal aorta, which is innervated by this nerve. From here the neurosecretory granules appear to more directly into the blood stream. In many circumstances distinct cyclicity of the amount of neurosecretion was detected. For example Baehr (1969) found accumulation of neurosecretory granules in one of the three different types of A- nsc in about 12 min after a blood meal and their release at the of ecdysis, immediately after sucking blood and during egg laying.

Histology

The use of special staining techniques, such as Gomori (chrome haematoxylin-phloxin;) and Paradlehyde Fuschsin, renders the nsc exremely visible in histological prepartions and the neurosecretory materail may readily be followed along their axons. Using histochemical methods, a considerable amount of acid phosphase was found in the cytoplasm of most of the nsc in the bug Iphita limbata, whilst it was nonexistent in the cytoplasm of the other nerve cells. On the other hand, the phosphates content of the nuclei of the neurosecretory cells was lower than that in the neighbouring nerve cells. Based the basis of his detailed histological studies on the brain of the Iygaeid bug Oncopeltus fasciatus Dallas, Johansson distinguished four different types of cells, the staining characteristics of which suggested a neurosecretory function. A-cells, the most conspicuous of all, stain dark purple-red or dark-blue with paraldehyde fuchsin and blue-black with Gomori chrome haematoxylin. They enclosed a large number of granules which full most of the perikaryon. The cells are pear-shaped or sometimes irregular in shape, and occur immediately beneath the neurilemma.

When the live cells were examined under dark-ground illumination, their shining bluish-white colour makes them quite visible. Their axons form the nervi corporum cardiacarum I. B-cells stain green or blue-green with paraldehyde fuchsin and red with Gomori. They are somewhat smaller than the A cell and more irregular in shape. They are without granules and occur deeper in the mass of ganglion cells. C-cells stain purple-red with paraldehyde fuschsin and reddish with chrome haematoxylinphloxin. They are found in the same layer as the B-cells, are irregular in shape and contain a flaky cytoplasm. D-cells have a fine granular cytoplasm which stains plae purple-red with paraldehyde fuchsin and pale blue-black with chrome haematoxylinphloxin. They are to some exten larger than the A-cells. Removal of the A-cells resulted in reduced fertility, and delayed, but did not prevent, egg laying. After extirpation the corpus allatum bwcame smaller. Baehr, distinguished seven different groups, used different letter for the individual types, and so did Girardie and Girardie (1966), and Joly (1970) who followed Girardie. Changes in the histological appearance of the neurosecretory system of the Cholorado Beetle Leptinotarasa decemlineata were similary detected by Schooneveld (1966-1972).

Ultrastructure

Because the ultrastructure fo the pars intercerebralis was first described by B. Scharrer (1962) in the cockroach Leucophaea moderae, it has been well-considered by various other researcher e.g. Willey and Chapman (1962) in Blaberus craniifer, Normann (1965) and Bloch et al. (1966) in Calliphora erythrocephala, and Girardie (1967) in Locusta migratoria, etc. Diverse types of cells and neurosecretion were described.

Embryogenesis

The lone information on the origin and development of the neurosecretory cells dring embryogenesis are available in the papers by Jones (1953, 1956a, b). He discovered that in the embryos of Locustana pardalina, after diapause when the mitotic activity has occured again, several large cells can be viewed in those places in the brain where typical neurosecretory cells will later occur. These cells have a particular connection with the function of the ventral glands of the head and with moulting. They are probably the promary active source of the hormone in the insect embryo.

Phylogenesis

There is plenty of research material on the source and evolution of the neurosecretory cells during phylogenesis. It has been the neurosecretory cells that are found in all insects including Apterygota and also in other arthropods such as Crustacea Xiphosura Chilopoda Symphyla in Annelida and, phylogenetically 'lowest', in Turbellaria Polyclada. Hanstrom (1940, 1953) presuppsed that neurosecretory cells originated with the so-called lateral frontal organs in the lower Crustacea, which lie in the epidermis quite separate from the brain. In the higher Crustacea they are very close to the brain and form the well-known pair of frontal organs. The neurosecretory cells of these are lined by their axons to the corresponding sinus gland at the side of the brain, where the secreted material is accumulated.

In Apterygota, they form two group of cells on the surface of the protocerebrum which are enclosed by a connective tissue membrane. From all these groups a nerve arises which passes through the brain in a typical chiasma and reaches the corpus cardiacum of the opposite side. The neurosecretory cells nad the granules in their axons give the same Gomori-positive reaction as the neurosecretory cells of Pterygota. The typical neurosecretory cells of Pterygota develop from these two groups by a process of inclusion into the brain mass during which the connective tissue membrane vanishes.

It is clear from the published research work that the activation hormone is phylogentically the oldest of the three metamorphosis hormones. As regards the origin of the neurosecretory cells of the pars inter- cerebralis, Clark (1955) derives their function from the original epidermis cells from which they have evolved as nervous tissues and accepts their secretory function to be primary and thier nervous function secondary. He indicates oints out that they are neurosecretory cells found in the most primative, and phylogenetically oldest parts of the brain.

The Corpora Cardiaca (cc)

Morphology

It was Lyonet (1762) who was the first mention of cc in the literature is without doubt that Lyonet (1762) in this work on the anatomy of Cossus cossus. They include of a pair of bodies situated immediately behind the brain, between the anterior end of the dorsal vein and the oesophagus, in front of the corpora allata with which they are linked by the nervi allati. They are sometimes fused medially and in some cases they are also fused with the corpora allata. In the ring-gland of Diptera Cyclorhapha, they form the lower median part of the ring, which is characterized by the small size of the cells, and fuse in the middle with the hypocerebral ganglion and at the sides with the R-cells. They are as mentioned earlier, innervated by two pairs of nerves from the brain, the nervi corporum cardiacarum I and II (interni and esterni) which are constituted respectively of the axons of the median and lateral groups of neurosecretory cells of the pars intercerebralis protocerebri.

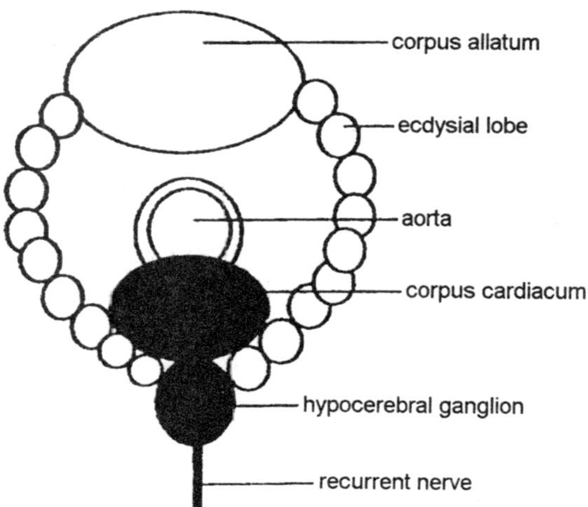

Fig. 3.7. Weismann's ring or ring gland in cyclorrhaphan larva (Diptera).

The cc connected medially with the hypocerebral ganglion by a nerve bridge and with the suboesophageal ganglion by the nervi allati. They send one or paired visceral nerves along the digestive tube to the visceral ganglia. The structure of the nervus allatus in Gryllodes sigillatus was described by Awasthy (1968) and in Gryllus domesticus by Belyaeva (1966), and by Theodorescu and Novak (1973). The cc is composed of two parts, a nervous part (originally, cc were supposed to be normal ganglia, the pharyngeal ganglia'), and a glandular part. They differ, at first glance, from ca and from all nervous ganglia by their bright, milky-white, opalescent colouration remembering that of the A-nsc.

Histology

Including the neurones (with scarce intrinsic neurosecretion) two other types of cell can be distinguished: (1) The chromophobic cells whose cytoplasm is to stained by the usual staining methods. They consist the main connective tissue in cc in which there are, usually, scattered ceiks if the other type. (2) The chromophilic cells which are deeply stained by various stains and furnished with characteristic pseudopodia-like processes. Cazal (1948) accepted them as modified neurones which had lost their nervous function and intensified their secretory function. In Plceoptera, Arvy and Gabe (1954) found small granules in the cytoplasm of the cc cells and in the nerve endings bringing the neurosecretion, which stained deeply with phloxin. Numerous ramifying fibers of the nervi corporum cardiacarum are present in the cc in addition to the alone-mentioned cells.

Some of the granules produced by the neurosecretory cells of the brain are stored here, but the remainder reach the surface where they are washed off by the haemolymph. A small amount of neurosecretory matrial has been shown to pass through the cc into the ca, via the nervi allati in several species. Besides to the neurosecretory granules, Nayar (1956, 1957b) found a secretion, which he did not accept to beleive to belong to neurosecetory origin, in the chromophilic cells of Iphita limbata. De Lerma (1956) also advocates the existence of a special secretion of the cc in Hydrous piceus. It occurs in the form of homogeneous acidophillic inclusions which are somewhat abundant in the chromophilic cells of the cc. No intracellular Gomori-positive granules' were found within these two types of corpora cardiaca. Arvy ad Gabe (1953a, b, 1954) also insisted on the occurrence of a true intracellular neurosecretion in the neurosecretion in the cc on the basis of their great knowledge of many group of insects. Johansson

(1958) was able to differentiate three different types of celll in the cc of Oncopeltus fasciatus. Couple of these are large chromophilic cells which occurs in different parts of the cc and differ in their staining characteristics with respect of paraldehyde fuchsin. Those of the frontal and lateral portions of the gland stain green where as those of the hind and median portions take on a pale red colour. The third type are interstitial chromophobic cells. These are dispersed through-out cc but are most abundant medially. It is remarkable that Johansson did not find any accumulation of neurosecretory material in the cc of Oncopeltus, where the granules are stored only in the walls of the aorta dorsalis as in Iphita limbata changing from the condition in Rhodnius prolixus where they are stored in the cc.

In Hydrocyrius columbiae granules are also stored in the walls of the aorta as they are in Pyrrhocoris apterus. Phylogenetically, it is apparently a secondary utilization of the tissue which capacites the neurosecretory material to pass into the haemolymph at a quicker rate. Mayer and Pflugfelder (1958) discovered only two kinds of cells in cc of Carausius morosus: the osmiophilic cells containing granules of neurosecretory material and generally forming well-delimited groups in the corresponding part of the gland; and the osmiophobic cells containing a system of vacuoles, special lamellated granules and peculiar ablong mitochondria. In the nerve endings inside the gland, granules 500 to 2000Å in size were found, which conforms to those known from the neurohypophysis of vertebrates and are of neurosecretory origin. M. Cazal (1971) studied the structure of the cc in Locusta migratoria in detail, while Divakar and Novak made an exact study of thier morphology and histology in Schistocerca gregaria. Their histology during the penultimate and last larval instars, and the pupal instar in Galleria mellonella were indicated by Karachali and Novak (1973).

Ultrastructure

A much of the research to papers has been published on the ultrastructure of the cc in various insects species, the first being those by Pflugfelder (1958) in Dixippus morosus and by B. Scharrer (1963, 1968) in Leucophaea maderae. Other authors who examined the ultrastructure of the cc were Johnson (1966) in Calliphora stygia, King, Surrinder, et al. (1966) in larval Drosophila melanogaster, King, Aggarwal et al. (1966) in adult females of the wild type and a mutant of the smae species, Normann (1965) in Calliphora erythrocephala, M. Cazal (1971), in Locusta migratoria and Cassier and Fain-Maurel (1971), in the same species, in is probably the most comprehensive and recent study on this subject. There are two methods in which neurohormones can be released into the circulating fluid, Cassier and Fain-Maurel regard the one which attributes the passage of intact granules by exopinocytosis as the more likely in Locusta. They observed no fragmentation of the granules nad found four types of neurosecretory granules.

Function

Johansson (1958) assumed accepted the cc to has two functions: the of the products of diverse groups and types of neurosecretory cells, and in addition, the production of a separate hormone of their own. A similar suggestion was made by Cameron (1953) previously he demonstrated that the cc produce a special active substance (orthodiphenol) which is quite different from neurosecretion and affects the heart-beat and peristaltic movements. Such a result was obtained by Altmann (1956), who examined the myotropic effects of corpora cardiaca extracts of the bug Iphita limbata on the gut peristalsis of another bug Aspongopus janus Fabr, and their chromatophorotropic effects on the red pigement cells of the decapod Coridina laevis Heller. The neurosecretory cells of the brain of the same species had no such effect. ohansson's (1958) experiments on the influence of the cc on reproduction failed to give postive results. Recent discovery clarifies as to the intrinsic endocrine activity of the cc both in their glandular and their neurosecretory (intrinsic nsc) components, also as their neurohaemal function for the brain neurosecretory cells. These three functions are perrhaps all mediated by more than one active substance.

In addition to AH_1 and AH_{11}, cc were discovered; by many authors, to contain the following substance; polypeptides with myotropic and chromatophorotropic activity (e.g. the neurohormones C and D of Gersch and his school), hyperglycaemic and hypertrehalosaemic action produced by extracts of both neurohaemal and glandular components (so that at least two different substances are involved), substance affecting nervous function, substances having both a diuretic and an antidiuretic effect on water metabolism (most of them undoubtedly of peptide nature) and a number of pharmacodynamic substances such as serotonin, orthodiphenol, etc More information is required to identify all these substances and to consider their activity spectra.

Embryogenesis

The functioning of cc during the embryonic period was studied by Weismann (1926) and Plugfelder (1937) in Carausius morosus; by Mellanby (1936) in Rhodnium prolixu, by Roonwal (1937) in Locustana paradalina; by Poulson (1937) in Drosophila, and by several other research. Their findings, cc, together with the hypocerebral ganglion, originate as unpaired dorsal evaginations of the oesophagus. Only later are they are separated and their cells differentiated into nervous and secretory ones. Initially, they are espected to have been normal visceral ganglia which secondarily attained a neurosecretory function and became innervated by the nervi corporum cardiacarum.

Phylogenesis

The cc are well formed in the Apterygota with the possible probable exception of Collembola (cf. Cazal, 1948). They are more developed in this group than the ca and have been described in Hapygidae and Lepismatidae. They are rather reduced in Campodeidae and are found in other Diplura and in Thysanura. It is probale that cc have developed during the phylogenesis of Apterygota, but earlier than ca. If Cazal's (1948) obsevation that the corpora cardiaca are absent in Collembola is confirmed, this will agree with the conclusion of B. Scharrer (1952) who comfirmed it from experimental evidence, that the cc merely store temporarily the secretion from the neurosecretory cells of the brain and are not necesary for the action of the *activation hormone*. Thus, for instance, transplantation of a brain with active neurosecretory cells may break diapause in cecropia pupae without simultaneous transplantation of the cc. There is as yet no specific evidence for the presence of a special cc hormone as was stated by L'Helias (1956b).

Direct Effects of Activation Hormone (AH)

AH is more complex in contrast with other hormones. It is difficult to determine whether particular action of AH is related to that of hte other brain neurohormones. This is often unasweable problem, so that the present classification is arbitrary and is based on the author's opinion, but supported by the date on AH_1 and AH_{11} characteristics.

The activation of the prothoracic glands

The first rescarch papers by Kopec (1917) and Wigglesworth (1934, 1940) pointed out the necessity of the brain for moulting, pupation and metamorphosis. The discovery by Fukuda (1940) of the function of pgl in *Bombyx mori* demonstrated that the effect on moulting is an indirect one and depends upon the activation of these glands. This was confirmed by Williams (1952), who found that active pgl alone are able to induce moulting in isolated pupal abdomens of cecropia while an active brain can only produce the same effect in the presence of (inactive) pgl. Wigglesworth (1952b) came to the same conclusion with *Rhodnius prolixus* and by several other authors with various insects. Apparently some of the recent finding seem to contradict this. For example, Johansson (1958), discovered that complete extirpation of all ten of the type A neurosecretory cells of the pars intercerebralis in *Oncopeltus fasciatus*, carreid out not more than four hours after the last larval moult, did not prevent imaginal moulting, although this process started two or three days later than in the controls.

Undoubtedly, this is because of the secretion of the A-cells, which is stored in the form of granules in the walls of the aorta, as mentioned above, and, also may be, in the A-nsc of the ventral nerve cord. According to Raabe (1959b), the way in which pgl are stimulated by AH has not yet been explained. A possible explanation was suggested by Church (1955), who considers it might be directly concerned in the syntesis of ecdysone. The findings of E. Thomsen and I. Moller on the influence of AH on intestinal protease activity in *Calliphora erythrocephala*, as well as other effects, would however suggest an action or more ordinary nature, as would also its relationship with the *antidiuretic hormone* of the vertebrate hypothalamus.

The activation of tne corpora allata

AH's effect on the function of ca is now adequately demonstrated. AH is necessary for the reactivation of ca in a manner similar to that of pgl. This was suggested in some of the earlier accounts and is quite evident from the work of Johansson (1958a, b) de Wilde (1958a-c) and his collaborators, Nayar (1957). Johansson (1958a) observed the effect of extirpation of the A-cells of the brain on the volume of ca and found that complete exitrpation of these cells results in extraodinarily small ca. The same process was observed by E. Thomsen (1952) in Calliphora only a few A-cells are left, the ca volumes reamin normal. The proper understanding of how the ca are activated by AH has proved difficult for two reasons: (1) The action of the ca seems to depend upon nervous stimuli; (2) the activation hormone reaches the ca not only through the haemolymph, but also directly from the brain, in the form of granules, via the nervi corporum cardiacarum cc and nervi allati. Niether of these factors appears to play a primary role in the regulation of ca activity, except in the case of diapause, but such has been attributed to both by different authors. The activation of ca produce the primary effect of the activation hormoneon the ovaries.

The effect on ovarian development

Including its indirect effect on ovarian development through the activation of the ca, the existence of a direct effect of AH on the development of eggs has been hinted. A detailed examination of the endocrine mechanisms controlling ovarian development in adult *Calliphora erythrocephala* showed that removal of the neurosecretory cells of the brain can be extirely compensated for by the implantation of active nsc from another specimen but not by active ca. The effects of AH deficiency are different from those of allatectomy. Implantation of cc from an active specimen has a similar, though lesser, effect than the reimplantation of neurosecetory cells. The cc alone in this species. Thomsen concludes that the function of AH in the insect organism is quite complicated and that the nsc of the brain form an 'all-over controlling centre of the endocrine system'. The removal of median nsc from the ovaries Indian Housefly *Musca nebulosa*, the ovaries failed to develop.

The effects were not similar with those of allatectomy; instead of the hypertrophy of the fat body which follows allatectomy, the fat body was reduced in size after nsc removal compared with the accessory glands, and the accessory glands were devoid of secretion. This is probably absence f the general stimulant effect of Ah. No yolk was deposited after median nsc removal in the flesh-fly Sarcophaga bullata, although yolk deposition started after their reimplantation, but not after the implantation of ca or of brain tissue minus the nsc. Removal of brain in female Bombyx mori pupae often caused the laying of diapause, instead of non-diapause eggs, which were distinguished by low carbohdrate and lipid concentrations. The impact of the nsc of the pars intercerebralis on the development of the ovaries has been.

Highnam (1962) has demonstrated that in *Schistocerca gregaria* they exert a positive control over occyte development and that copulation as well as electrical stimulation, drasrtic wounding, or enforced activity may all bring about a release of material from the nsc of the pars intercerebralis and accelerate the development of the terminal occytes in 14-day-old females reared without males. Hill (1962) found that in the same species a there is positive control of haemolymph protein concentration by neurosecretion during ovarian development. The presence of mature males to accelerates the release of material from the female neurosecertory system, resulting in rapid development of terminal occytes and an increase in the number of eggs. A positive effect of the pars intercerebralis on the development of the ovaries, as well as on postembryonic development, on the function of the ventral glands and ca, and on metabolism, water balance, and chromatic adaptation was described by Girardie (1964) in *Locusta migratoria*.

Mordue (1965a-c) too, obtained similar result with adult females of *Tenebrio molitor*.

The effect on the accessory glands

The impact of cerebral neurosecretion on the male accessory glands, initially discovered by Wigglesworth (1936), and since then confirmed by various researcher working on other species of insects is, as in the case of the ovaries, primarily an indrect one through activation of the ca. No direct effects have obeen till now observed, though they may be assumed.

The control of diapause

As indicated by E. Scharrer (1952), the author of the concept of neurosecretion, the specific role of neurohormones in the animal body depends on a long-term co-ordination between the nervous system and the ininternal secretion. A classic example of this interdependence is the direct influence of AH on both larval (pupal) and imaginal diapause. In each case it is a nervous stimulus induced by external conditions which induces a temporary interruption of AH production whcih in turn results in the inhibition of the functions concerned. As demonstrated by Williams (1948) and several other entomogists a stage corresponding to natural diapause can be caused by the removal of the brain, the source of AH.

Similarly, a natural diapause can be broken at any time by implanting an active brain or other source of AH (corpora cardiaca or corpora allata; see below). It seems that this dependence is the result of a phylogenetic utilization and improvement by diapausing insects of a tendency present in all insects. This results in the control of AH production in response to a variety of unfavourable environmental conditions and it is highly adjustable. Thus the process of diapause generally occurs in the winter period in temperate regions, whereas it is common in the hot dry season in the tropics and subtropics. The same is true of the dependence of AH production on the nervous impluse produced by the distension of the abdomen in blood-sucking insects as described by Wigglesworth (1936, 1948) and by Detinova (1954). The condition in blowfly larvae first observed by Cousin (1933) may be assumed as an transitional stage in the developing process diapause.

The effect on morphogenesis

In the experiments with the cricket Pteronemobius haideni, Sellier (1956) found that the implantation of an active brain into the larva (VIIIth instar) at the start of the diapause not only in suppression of diapause, but also in a lengthening of the wings when they become adults. But there were no changes when the implantation was effected at a more advanced stage in diapause. The implantation of brains from other species of the genus *Gryllus* gave similar results. However, this seems to be much more an indirect effect depending on the time of activatoin of the ca. The entire problem of both direct and indirect effects has been discussed by Gilbert (1964) and Buckmann (1969), but many questions are yet unanswered.

The effect on the fat body

An observation made by E.Thomsen (1952) proves that the fat bodies of *Calliphora* females, whose AH source had been removed, had a higher glycogen content though the quantity of fat was reduced. The glycogen content corresponded with that following allatectomy but the reduction in the quantity of fat was grater.

The effect on intestinal proteinase activity

The two entomologist known as, E. Thomsen and I. Moller (1959a, b) studied the effect of extirpation of the neurosecretory cells of the pars intercerebralis on intestinal proteinase activity in Calliphora females. The colorimetric method of Day and Powning was used to determine this activity in gut homogenates. The proteinase active in the operated specimens was 5 to 8 times lower than in the control. The effect of the removal of AH was identical with that caused by lack of proteins in the food. As proteinases are themselves proteins, their production may be considered as protein synthesis in the gut epithelium cells, and the researchers conclude that AH affects protein synthesis in gneral. This, But this, is not the only possible explanation— a more general activation effect corresponding to that observed in other tissues, such as, for example, some kind of membrane activation, would be equally satisfactory. Corresponding results were reached independently by Strangways-Dixon (1959, 1961) using the same species. He showed that by feeding flies selectively on sugar and protein that when deprived of their neurosecretory brain cells the flies digested no proteins even when forcibly fed on them for six days; their ca and their eggs showed no increase in size.

The effect on the hind midugt esterases

E. Thomsen (1966) studied the esterase distibution pattern in the hind midgut cells of *Calliphora erythrocephala* females by applying 5-bromoindoxylacetate (as substrate) to fresh tissue. The enzyme appeared in the form of granules, filaments and 'caps' similar to those described by Wigglesworth (1958) in Rhodnius prolixus. Generally in meatfed flies, the pattern of enzymes in contact with fat droplets followed a cycle correlated with ovarian development. In flies fed on a protein-free diet, the esterase level was low. The same fining was made in flies deprived fo thelr medial nsc, even when they were fed on meat. It was comfirmed that the medical nsc control, or influence, esterase production in the cells.

The effect on on water balance

Altmann (1956a, b) experimented on studied the effect of extracts from different endocrine glands on the intake, retention, and excretion of water by the honeybee. Injections of ca extracts increased water consumption and decreased the viscosity of the haemolymph; cc extracts increased haemolymph viscosity and decreased water consumption. Excretion was increased by ca extracts and decreased by extracts of cc. Injections of adrenaline produced similar effects to corpora cardiaca extracts. The effects of the corpus allatum extracts could be due to juvenile hormone or to neurosecretory material reaching the gland via the nervus allatus; the latter is more probable.

In *Iphita limbata*, when distilled water was provided for drinking for some time an increased amount of neurosecretory material was found in the walls of the aorta, and nsc were practically free of granules. While in the insects which were forcibly fed on salt water for the same length of time, the release of neurosecretory material through the walls of the aorta almost stopped and the neurosecretory cells were filled with granules. There were similar results were obtained with specimens where the wax layer of the epicuticle was removed by washing with benzene or chloroform. In insects kept in an atmosphere dried with calcium chloride, neurosecretory material accumulated in the nsc; in the those kept in a more humid atmosphere, the material was released and enlarged quantity of granules was found in the walls of the aorta. Same result were acquired by Gutmann and Novak in Pyrrhocoris apterus. It may be concluded that nsc enable the animal to maintain the water balance within certain limits. The comparison with the antidiuretic hormone of the vertebrate hypthalamus is striking.

Raabe (1959a) reached to the same conclusion with *Carausius morosus* where the cc had the same effect as the neurosecretory cells of the brain. This agrees with the findings of Stutinski (1952a, b). He injected rats with extracts of the pars intercerebralis protocerebri and cc of the cockroach *Balbera fusca* and found that the urine was reduced in quantity in the same way as after the administration of pitressin. The opposite result was obained by Nunez (1956) using the beetle *Anisotarsus cupripennis*. There, the extracts of brain nsc and cc appeared to produce diuretic effects. These experiments, are yet to be analysed under comparable conditions.

Control of phase differentiation

Conforming with Girardie (1966) and Girardie and Girardie (1966), Joly (1970) assumed that phase differentiation in Locusta migratoria was controlled by special nsc in the pars intercerebralis. Three types of cells were recognized in the pars intercerebralis: A-cells, which are small and contain a large amount of Gomori-positive neurosecretion, B-cells, which are similar, but have highly phloxinophilic cytoplasm and no neurosecretory granules (A- and B-cells are regarded as being possibly two different stages of the same type of cell) and C-cells, which are much larger and contain only a few neurosecretory granules. But, no comparison with various types of cells in other insects, similarly experimented by other authors, is available. Destruction of the C-cells by electrocagulation had the same effect as ca extirpation. If performed at the beginning of the IVth (penultimate) larval instar, the result would be a well-developed adult. Destruction of the A- and B- cells had the reverse effect, i.e. it acted like ca implantation. If carried out at the beginning of the last larval instar, a green adultoid developed. The implantation of extra A- and B- cells acted like allatectomy if carried out at the beginning of the penultimate instar. The researchers accepted that the C-cells activate the ca under normal conditions, while the A- and B-cells inhibit them. There is, however, a simpler explanation, consistent with findings in other insects, which has not been sufficiently ruled out, i.e. that it is injury to the C-cells which inhibits the ca in the first case. In the second case, removal of the A- and B-cells would prolong the interecdysial period of the last larval instar, so that the individual's own JH would, unlike in normal development, remain active; while the implantation of extra cells in the IVth instar would shorten the intermoult period, with the result that the ca would not attain an active concentration in the penultimate instar. Joly's inability to find any difference between ca ultrastructure in normal specimens and those with destroyed A- and B-cells supports the second theory.

Other effects

Many Confirmed effects of AH, as well as the possibility that they may be due to a common and very general action, leaves no doubt that further research will reveal a number of other effects. It also remains to be determined whether some of the observed effects of brain and cc extracts, such as the various chromatophorotropic and myotropic actions, are due to AH or to other neurosecretory products. There is little experimental evidence for any definite information on this connection.

Indirect Effects of AH

Probably a hormone may act both directly and indirectly on the same tissue, e.g. it may stimulate RNA production by the tissue directly and at the same time induce the production of another hormone stimulating tissue growth or function. Another aspect to be considered is that the direct or indirect effects may be of different degrees. For instance, a hormone may act on the target tissue directly or may induce another hormone to do so, or the other hormone may cause a metabolic change in which only the tissue is affected. The control of the moulting process must be considered as an indirect effect of the activation hormone.

Moulting is restricted by removal of the AH source and re-evoked by its reimplantation. For this reason, the brain hormone was originally described as the moulting hormone by Wigglesworth (1934). It was called a 'growth and differentiation hormone' (partim) by B. Scharrer (1948) and Williams (1947, 1952a). The ca hormone was also initially interpreted by researchers as a moulting hormone. Confimating of an indirect as opposed to a direct effect is provided by the fact that the effect can be produced by another hormone but not by the given hormone alone. While a mere finding that removing another organ inhibits the effect is in no way proof of an indirect effect. In such a case there is always the possibility that the effect of the hormone was prevented by breaking another link in the chain between the hormone and the part of the body responding. With the metamorphosis hormones it is particularly important to differentiate between the direct and indirect actions of a given hormone. The interaction of thier effects is very complicated, as in metamorphosis and moulting. View, a revision of many of the false on the metamorphosis hormones and their mode of action is required.

Control of AH Production

The functional cycles of neurosecretory cells. The existence of particular 'critical' periods in each instar as far as the hormonal activity is concerned was demonstrated experimentally shortly after the discovery of the first metamorphosis hormones. During these periods, the presence of the source of the given hormone is necessary for normal development whereas after wards it is no longer necessary. This shows that there is a certain cyclicity in the function of the AH, which has been confirmed by subsequent histological research on the neurosecretory system. The course of neurosecretion and the function of cc during postembryonic development has been studied in detail by Herlant-Meewis and Paquet (1959) in *Dixippus morosus*. They discovered a decrease in the amount of neurosecretory granules in the pars intercerebralis and cc at the time of each moult. Neurosecretory activity reached its maximum after about one-third of the next intermoult period had elapsed. Even at this time also, the largest amount of neurosecretory material was seen to pass into the corpora cardiaca. This period of maximum secretory activity agrees well with what has been observed regarding the critical period for AH release in other species.

The quantity of the neurosecretion in cc decreases makedly about half way through the intermoult period, i.e. when the critical period for the juvenile hormone has been reached. Another increase in neurosecretory material was observed in the last quarter of the intermoult period. Cyclical changes in the amount and character of the neurosecretory material were observed in Phasmids and other insects by a number of other authors while a report from Fuller (1959) discribes recurrent cyclical changes in the amount of secretion in the individual neurosecretory cells of the pars intercerebralis of Periplaneta americana without any noticeable cyclicity in the neurosecretory system as a whole. There have been suggestions about the factors which govern the changes observed in neurosecretory activity.

One of the first answers was provided by Wigglesworth (1936a, 1940a) in his classic experiments with the blood-sucking butg Rhodnius prolixus. He demostrated that production of the brain hormone and thus also the start of the moulting process, and each the interrelated developmental processes, is induced by a nervous stimulus produced by distension of the abdomen due to ingested blood. This stimulus depends on the quantity, not the quality, of the fluid intake, since the same effect is caused by a corresponding volume of pure water. Several small blood meals do not, however, produce any effect, even though thier sum exceeds the necessary minimum quantity. Severing the ventral nerve cord anywhere between the brain and the abdomen prevents the passage of this stimulus. Therefore it can be accepted as a typical-case of the transformation of a nervous stimulus into an endocrine impulse effected by the nsc of the pars intercerebralis protocerebri, as claimed by E.Scharrer (1952). A very similar neurosecretory effect was found by Detinova (1954) in Anopheles.

On the other hand, no such connection was observed byNovak (1951b) in the plant-feeding bug *Oncopeltus fasciatus* and many other insects where poorly fed specimens can undergo a normal moulting process. It is also recognized that meal-worms undergo several extra moults when left without food, and starving clothes-moth larvae can moult as many as 40 times. The above-mentioned dependence of AH production on a nervous stimulus from the distended abdomen is probably a secondary phylogenetic adaptation in blood-sucking insects. It enables them to survive for long periods in the absence of a suitable host in a state of reduced metabolism corresponding to that of a true diapause. In most other insects, AH producting is automatically renewed at the beginning of each instar with the passing of the first digested food into the haemolymph.

The phylogenetic character of this dependence on the stimulus from the abdomen was shown by Larsen and Bodenstein (1959) in the mosquitoes *Culex pipiens* and *Aedes aegpti*. Whereas *A. aegypti* and normal *C. pipiens* show the usual dependence of AH production (and through this juvenile hormone production which regulates egg development and oviposition) on the distension of the abdomen by blood. AH prodution by the autogenic form, *Culex molestus*, is free of this stimulus. An exhaustive study of the function of the neurosecretory cells in insects which feed continuously throughout their growth has been made by Clarke and Langley (1962) using Locusta migratoria L. No changes were discovered in the amount of neurosecretory material in the median nsc or in their axons (nervi corporum cardiacarum), or in cc during postembryonic development or at different times during the intermoult period. The neurosecretory material is apparently produced continuously throughout growth in this species.

As the AH is presumably not used during the ecdyses, the authors assume it accumulates in the haemolymph at these times. The increased concentration of the hormone in the haemolymph at these periods would then reactive agl. A very remarkable association between the frontal ganglion and the production of neurosecretion was also found. Its extirpation resulted in the complete inhibition of further growth and moulting. The same effect was attained by cutting the frontal connectives, whereas cutting the recurrent nerve or the ventral nerve cord in front of the first abdominal ganglion had no effect. Histological examination of neuroendocrine system five days after the operation revealed a marked accumlation of neurosecretory granules in cc, the nervi corporum cardiacarum I but scarcely any in the nsc. No neurosecretory cells were discovered in the frontal ganglion. The authors suppose that the forntal ganglion plays a part in transmitting the nervous impulses from stretch receptors in the oesophagus to the nsc of the pars intercerebralis which would thus connect the release of the hormone with the intake of food.

A momentary inhibition of AH production caused by various external impulses, the mechanism s of which are not yet entirely apprehencled, is the direct internal cause of all types of insect diapause except the early embryonic one. In addition to photoperiodism, temperature and the above-mentioned distension of the abdomen in blood-sucking insects, several other factors controlling AH production have recently been described. Therefore the inter-relationship between juvenile hormone production and the carrying of oothecae by female cockroaches, incorrrectly interpreted as a direct effect on corpus allatum secretion, is undoubtedly governed by neurosecretion. Similarly, neurosecetion, or AH production, appears to be the chief mechanism influenced by the quality the food as shown by E. Thomsen (1959) and Strangways-Dixon (1959); and the effect of feeding on protein synthesis is also controled in this manner.

Thomsen and Lea (1968) examined cyclic changes in the medial nsc of *Calliphora erythrocephala* under various conditions. The nuclei and nucleodi displayed cyclic changes in volume and in the amount of neurosecretion. Neurosecretory activity rose on the first day after emergence, with acceptance of a sugar diet; it then fell again but rose once more after the fourth day in connection with the beginning of meat eating and fell again after eff laying. Allatectomy reduced nuclear and nucleolar volume, but the implantion of active ca renewed neurosecretory activity in such specimens. The authors concluded that the activity of the neurosecretory cells was regulated by ca, but the effect is undoubtedly an indirect one.

Chemical Characteristics of the AH

The researcher also who was first to identify the chemical nature of the brain neurohormone was L'Helias (1955a, 1956a). Her pterinederivative theory suggested the existence of a close relationship between AH and JH on a pterine basis. A few years later, relationship between AH and JH on a pterine basis. A few years later, however, Gersch and Unger (1962) showed that pterines had nothing in common with either of these hormones and suggested, on the basis of earlier paper that neurohormones were possibly of a peptidenature. This accords with phylogenetic considerations on their relationship to vertebrate neurosecretory material, the polypeptide character of which has now been fully accepted.

The discoveries by B. Scharrer (1952) and others appear to speak equally against the hypothesis of L; Helias (1956) on the interaction of the brain secretion with those of cc or ca (1) ext\irpation of the cc has no qualitative effect either on moulting or metamorphosis; (2) the effects of brain extirpation may be at least partly compensated by the implantation of active cc (3) the careful histological studies on the neurosecretory activity of the brain and cc in some Hymenopytera and Diptera carried out by M. Thomsen (1953a, b) give support to the conclusion that the neurosecetory material passes through cc directly into the haemolymph. Few rearchers ignored the findings of a Gersch's school or phylogenetic relationships into account, however, and expressed entirely different views on the chemical character of AH and other neurohormones. For example, isolation of the active ingredient of the AH has been reported by Kobayashi and Kirimura (1958). They used 8500 silkworm (*Bombyx mori*) pupae preserved in methanol 24 hours after pupation and centrifuged three times after homogenization 200 ml of the methanol solution obtained were concentrrated to 30ml and extracted with 145 ml ethyl either. On evaporation, the either solution yielded about 2 mg of an oily yellowish brown material; the evaportaion temperature did not exceed 38C. A injection of this substance into decerebrated permanent pupae, in which no ecdysone has previously been found, caused these to moult to adults 16 to 20 days later.

It has been claimed by Kirimura et al. (1962), that the active principle of these extracts is cholesterol and its identity with AH was taken for grantend. This view has, not much assistance from other facts known about AH. Nevertheless, when correlated with the recent finding of Karlson and Hoffmeister (1963) that cholesterol is the precursor of ecdysone, it becomes of remarkably. Carlisle and Ellis (1963) injected IVth instar nymphs of *Locusta migratoria* migratoriodies with cholesterol dissolved in either. The purified perparation and one of two samples of commercially available crystalline cholesterol were ineffective whereas the other commercial sample, hastened moulting by about 18 hours (PL V. 01). The authors found that pure cholesterol as such has no prothoracotropic effect, which they assume to be derived from impurities present in the sample concerned. They conjectured that the active ingredient is a steroid related to cholesterol, and that the natural brain hormone is similarly a steriod of this group, or possibly a triterpenoid of related configuration.

Even Gillbert (1964) expressed his opinion in favour of the steriod character of the brain hormone in his review of the question. He also agreed with the earlier authors in assuming 'a number of neurosecretory substances produced by the brain, the prothoracotropic effect being produced by one of the them and the other effects of the brain neurosecretion by others. Another researchertook a different stance Novak agreed with Gersch (1962, cf. 1964) in his assumption of the polypeptide nature of AH and claimed that most of the known effects of the brain secretion could will be produced by one and the same substance influencing membrane activity and thus the water metabolism and secretory activity of the cells. The only exceptions are neurosecretion from the lateral nsc, shown to be engaged in inducing circadian rhythms of activity and the effects of neurohormone C. the effect of cholesterol would be that of a vitamin supplying the steroid skeleton necessary for the production of ecdysone. This is necessarily lacking in decapitated or decerebrated insects unable to accept food. In such cases its supply by injection can reinduce the production of MH when at least a small amount of AH is present. but this, is not the case in late diapause when AH is absent.

Williams (1963, 1968), first thought that AH was hyaluronic acid or a related substance and later expressed the view that it might be a mycopolysaccharde. Neither of these hypotheses was confirmed by further investigations.

Contemporary research work by Gersch and his colleagues submitted definitive evidence in support of their original opinion. First of all they showed that the prothoracic gland stimulating agent is different from neurohormone D, previously described, and that it actually affects the prothoracic gland. Using electrophoretic separation on polyacrylamide gels, they further succeeded in demonstrating that the actual AH, i.e. the prothoracic gland-stimulating hormone consist of two components- one with high molecular weight, stimulating RNA synthesis, and the other, AH_{11} with low molecular weight, which raises the membrane potential and hence membrane permeability. Both components are of a peptide nature. However, It was discovered that Neurohormone D, is a peptide with a molecular weight of about 2000, to be thermo- and acid labile and to be decomposed relatively quickly by trypsin.

Mode of Action of AH

The movement of the neurosecretory granules through the axons of the nervi corporum cardiacarum from the nsc of the pars intercerebralis protocerebri into cc is well known, but their role in cc and their passage from there into the haemolymph is yet unknown. Significant experimental evidence for the transport of AH from the cc to the pgl in Rhodnius by the haemocytes was produced by Wigglesworth (1956b). He showed that blocking the haemocytes by injecting Chinese ink, trypan blue or iron saccharate, the particles of which are phagocytosed, results in a significant delay of the next moult if the injection is carried out before the end of a specific critical period which corresponds approximately with that of AH. However, when at the same time he implanted active pgl or injected a sufficient amount of a solution of crystalline ecdysone, no such delay occurred. Wigglesworth concluded that under normal conditions the haemocytes transport AH from the cc to the pgl, or, it is possible, that they secrete a further substance for the activation of the pgl under the influence of AH.

As suggested by Wiggleswortha, other explanation are so possible, for example, that AH is adsorbed on to the injected material and removed with the material from the haemolymph by phagocytosis; or the haemocytes, damaged by phagocytosis, discharge some AH inactivating substance into the haemolymph. An alternative some AH inactivating substance into the haemolymph. An alternative explanation could be that, in normal insects, the role of the haemocytes is to phagocytose neurosecrtory granules, thus freeing the AH whilst digesting the carier substance. This would be in agreement with the observed occurrence of neurosecretory granules in the aorta dorsalis of various insects and with the observation of Hodgson and Geldiay (1959) who found that hyperactivity in *Blaberus cranifer* and, to a lesse extent, electrical shock treatments, resulted in bloods cells overflooding of all parts of the brain.

Hardly any information is one available regarding the mechanism of activation of pgl and other organs influenced by AH. But, it has been found by Williams (1952) that the same activating effect may by obtained by implanting another, active pgl. This is to be interpreted, together with Wigglesworth's (1957) conclusions, as meaning that continuous activity by AH is must for the pgl to provide an effective amount of moulting hormone. This is based on the discovery that the first change in the epidermis definitely attributed to pgl hormone commences about two days before the end of the critical period for AH, during which time the removal of the AH-source by decapitation stops the process of the moulting process.

Moulting Hormone (Ecdysone) (MH)

The initial indication of the existence of a humoral factor controlling moulting can be found in the experiments of Kopec (1917, 1922), and later Wigglesworth (1934). During time of Wigglesworth, Fraenkel (1934, 1935) showed that similar hormonal factor, necessary for puparium formation, was present in the brain region of *Calliphora erythrocephala* larvae. Prior to this, however, Hachlow (1931), on the basis of his experiments with butterfly pupae (*Vanessa* and *Aporia crataegi*), suggested the existence of a 'thoracic centre' which was necessary for development. Bodenstein (1933a, b, 1934) in his transplantation experiments with the legs of caterpillars got the same result.

The honour of being the first person to clearly distinguish between the hormone of the brain cells and that of a new hormonal source, the prothoracic glands, belongs to Fukuda (1944) who showed the importance of pgl for moulting by transplantation experiments in silkworms (*Bombyx mori*). After this, papers on the moulting hormone appeared at an increasing rate. The role of pgl, or the analogous peritracheal glands in lower Diptera, or the pericardial glands or ventral head glands in Hemimetabola, in the moulting processes has been demonstrated in all the chief groups of insects.

Earlier the research recognized and were mainly concerned with the importance of MH for inducing the moulting process, but most of the later investigators stressted on its indispensability for growth and morphogenesis and used for it the less suitable terms 'growth and differentiation hormone', or 'metamorphosis hormone' (see above). A certain amount of confusion appears to have been caused by the discovery of the effects of the ring gland in flies, which is a composite structure that contains the sources of all the three metamorphosis hormones also, the role of MH in diapause and imaginal differentiation in Cecropia pupae as resolved by Williams seemed at first to support the concept of the growth and differentiation concept.

The most significant stage in MH invitations, after its separation from AH by Fukuda, was its isolation in crystalline form by Butenandt and Karlson (1954) from extracts of silkworm pupae. A significant requirement for this was the discovery by Becker and Plagge (1939) of a suitable test organism and a quantitative measure for judging concentration, the so-called Calliphora-unit. This method was improved by Butenandt and Karlson in their isolation experiments. While the specificity claimed by Williams (1951a, b) for the so-called spermatocyte-test appears to be questionable in the light of the findings of Laufer (1960). Becker and Plagge (1939) were also the first to show the broad inter -Order non-specificity of MH, which was confirmed by Wigglesworth for such widely separated orders as Hemiptera and Diptera. Another significant step was the successful extirpation of the ventral glands in migratory locusts by P. Joly, L. Joly and Halbwachs (1956).

The theory of the pgl hormone as a MH, which corroborates with the original findings of Kopec and Wigglesworth (1934) as well as with the gradient-factor theory of the author, has been approved by two independent pieces of work: Halbwachs and Joly (1957) showed, using Locusta migratoria, that transplantation of the pgl accelerates the moulting process without utilizing any positive effect on growth differentiation; Luscher and Karlso (1958) observed moulting but no and growth or imaginal differentiation following the injection of a large quantity of ecdysone into the nymph of Kalotermes flavicollis.

These discoveries have been confirmed by many other researchers. For instance, Zdarek and Slama (1972) proved that the injection of a large dose of ecdysterone at the outset of the last larval instar in *Calliphora* and *Sarcophaga*, inhibits morphogenesis and results in a supernumerary larva, whereas small amounts at later stages simply depress growth and result in small puparia. Administration at interval times produce forms between the larva and pupa.

The impact of ecdysone on colour change in *Cerura vinula caterpillars* (Buckmann, 110) also conforms with this theory; the greater the dose of hormone injected the quicker becomes the process of moulting whilst the effect on colour change decreases with an increase in the amount of hormone.

The function of MH (together with that of JH) at the cellular level was examined in detail by Wigglesworth (1963c) in the epidermal cells of Rhodnius prolixus. He came to the conclusion that MH is not a necessary ingredient for growth in insects in general, but it is necessary for the activation of the epidermal cells to produce their secretion (the moulting fluid and the chitinous cuticle) and to grow and divide. But precisely the same response is obtained by 'wound hormones' from injured tissue (cf. Wigglesworth, 1937). And the cells of the fat body and haemocytes do not need MH at all—their activation and mitotic activity is brought about by nutrition alone. The first discernible effects of MH on the necleolus of the epidermal cells is discussed and constrasted with the effects on the puffing patterns in the salivary glands in Chironomus.

Significant evidence supporting the view that MH only indirectly stimulates growth and morphogenesis is provided by the distribution of DNA synthesis during insect development. These authors conclude that ecdysone should be viewed primarily as a moulting hormone that initiates biosynthetic activities which caused moulting.

In some tissues, such as all chitinogenous epithelia, the Malpighian tubules and the nervous system, DNA synthesis occurred soon after MH secretion started. In others, e.g. pgl, midgut and haemocytes, it continued throughout the whole of the larval moulting cycle, but acquired a maximum at the peak of ecdysone production. In further tissues, e.g. the imaginal wing discs and muscles, no correlation between DNA and ecdysone production was discovered.

Schaller and Andries (1970a, b,) got the same results while studying regeneration and metamorphosis of the midgut cells in Aeschna cyanea and by Mouze and Schaller (1971) and Mouze (1971) in a study of development of the eyes in the same species, based on observations of mitotic activity. Agui et al. (1972) showed that an explanted pgl of *Mamestra brassicae* clearly stimulated the induction of moulting in the integument of the diapausing Stemborer (*Chilo suppressalis*) in vitro. On the other hand, MH failed to stimulate spermatogenesis in naked cysts of Cecropia pupae, but was active in a culture of intact testes.

Morohoshi and Iijima (1969) and Morohoshi et al. (1972) also reached the same conclusions with larvae and pupae of *Bombyx mori*. The only contradictory evidence is the report by Sondhi (1968), who claims to have found that in *Drosophila melanogaster* the injection of active ring glands into inbred larvae of the same sex and age did not affect either the time of puparium formation or of adult emergence, but raised the recipients' wet and dry weight by up to 10 per cent. The rate of the increase and the number of individuals are too few to allow definitive conclusions to be obtained from the results.

MH has a primary, direct effect on tissues of ectodermal origin only, i.e., the epidermis, stomodeum and proctodeum, the epithelia of the tracheal system, the ectodermal parts of the imaginal discs and the nervous system, etc. Its presence in the minimum active concentration, in the absence of JH, determines growth of the imaginal parts and degeneration of the larval parts according to the species specific morphogenetic pattern. This, together with induction of the moulting process, may per se be enough indirect stimulus to initiate metabolic and growth activity in other, MH-independnet, tissues also.

Apolytic Glands (agl)

Many glands in the insect body have been considered to produce MH. they have some commonn characteristics. They are paired, laterally localized glands, mostly ribbon-like in form, and sometimes more or less branched. Their main component is large parenchymal cells with 'intricately interwoven cytoplasmic processes extending towards the glandular surface' (Scharrer, B., 1964), where their ultrastructure has been examined, different quantities of smooth tubuli of the endoplasmic reticulum has been found. These are supposed to be connected with steriod production . The glands also seem to have the same embryonic and phylogenetic origin.

There are six classifications of these glands have described in different parts of the body in various groups of insects.. Perhaps they can be extended to include the larval oenocytes. It was felt that a common term, applicable to all these glands, was needed and some authors suggsted the collective designation 'ecdysial glands'. Wigglesworth (1962) however, pointed out that this term has already been applied to a certain type of epidermal gland and might thus cause confusion. Another contrary reason is that ecdysis is precisely that part of the moulting process which is uncontrolled by the hormone from these glands. I would therefore suggest calling them 'apolytic glands' and have employed this common term in the present book. It is derived from 'apolysis', the first stage in the moulting process under the direct control of the glands in question.

Many researches has expressed their doubts (Locke, 1969; Romer, 1971, as to whether the agl actually produce MH. But these doubts seem to be vitiated by the findings of Agui et al. (1972) and Kambysellis and Williams (1972), who have demonstrated the activity of the glands in vitro. This discovery is substantiated by numerous authors in vast quantities of early and recent experimental data, whereas the contradictory observation can be explained by assuming that a primary MH source does exist.

1. Prothoracic Glands (pgl)

Morphology

The pgl of the Cossus cossus caterpillar were almost accurately described by Lyonett as early as 1762. A more complete description

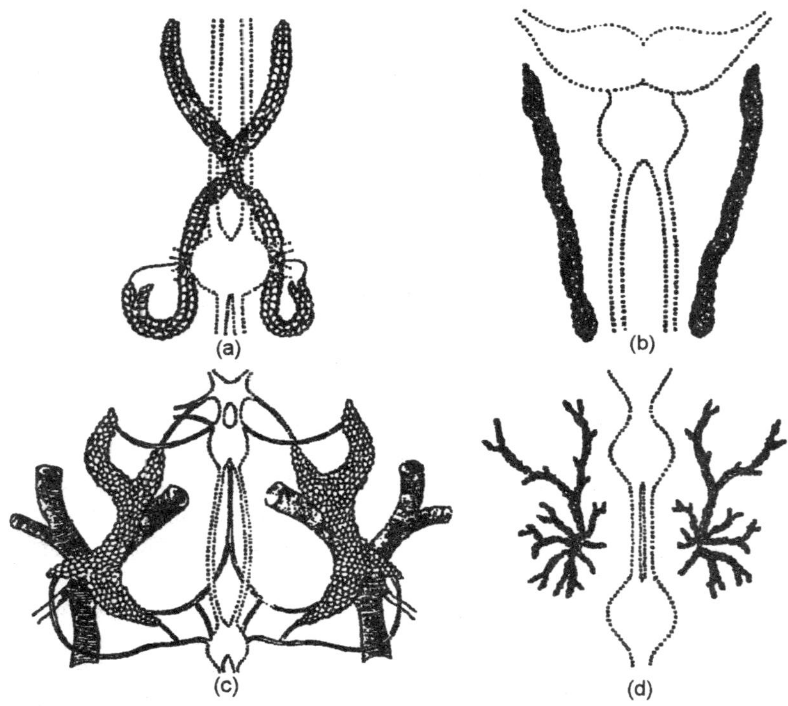

Fig. 3.8. Prothoracic glands in different insect orders: (a) Blattoptera, (b) Hemiptera, (c) Lepidoptera, (d) Hymenoptera.

along with data on embryonic development is given by Toyama (1902) who called them 'hypostigmatic glands'. Ke (1930) used the term pgl for the first time. Ever since because of the interest exicted by the discovery of their function, they have been described in detail for all the important groups in insects.

Many comparative morphological studies of the pgl and analogous organs in various groups of insects have also appeared. For example, they were described by Lee (1948) in many lepidopterous larvae, by Pflugfelder (1947d) in various orders of Hemimetabola, by M. Thomsen (1951) in Diptera, by Wells (1954) in Hemiptera, by Srivastava (1959) in Coleoptera, etc.

Although they change a great deal in shape, the pgl have certain features in common in all groups where they occur: they are paired glandular structures located in the ventrolateral areas of the prothorax (extending into the mesothorax in some species), and are sometimes partially, fused in the mid-line. They are generally closely linked with the main lateral tracheal branches near the prothoracic spiracles. In some bugs, like Rhodnius, they do not form independent compact organs, but chains of cells within the inner pair of fat body lobes. In cockroaches they are fixed in the body cavity by one or more muscles fibres and by a nerve ring from the prothoracic or, less often, the suboesophageal ganglion. Four types of pgl (apart from the ring gland) were differentiated on the basis by Joly (1968): the massive type (Apterygota, Ephemeroptera, Odonata), (2) the ribbon-like type (Orthoptera), (3) the blattoid type (Blattoidea), and (4) the diffuse type (some Heteroptera, Lepidoptera, Hymenoptera). They are supplied in great number with trachea in many groups they were till now unobserved by the researchers.

In each larval instar there is a distinct cyclicity in the secretory and mitotic activity of the pgl which agrees with the experimentally dertermined critical periods for MH activity. The secretion cycle in Pieris brassicae was described comprehensively by Kaiser (1949). The greatest size and adult insect they degenerate two to ten days after the imaginal moult. The changes in the pgl of Tenebrio molitor during development were critically studied by Srivastava (1960). In each larval instar they reach their maximum size at the time when feeding is interrupted, i.e., about two days before ecdysis (cf. the critical period for MH); they produce most of their secretion at this time and afterwards become reduced. A new cycle begins with the feeding of the next instar. The critical period occurs less than 24 hours after the interruption of feeding in the last larval instar. The glands are completely reduced at the commencement of eye pigmentation. However, they persist throughout the life of insect in Apterygota.

Herman and Gilbert (1966), accomplished a detailed anatomical and histological study of the pgl in *Hyalophora cecropia*, and found numerous, very diffusely arranged chain of cells, reaching to the posterior portion of the thorax, which arose from the four branches (anterior, dorsal, ventral and posterior) of the relatively compact tissue of the gland round the first spiracle. Each gland contained some 200 to 290 cells, which were larger and usually more in number in females and degenerated immediately after adult emergence

Histology

The pgl are usually formed of two strips of glandular tissue. The essential components of these are glandular cells equal to those of the corpora allata. Their cytoplasm is basophilic staining deeply with methylene blue, neutral red and other stains. Numerous deeply staining granules are found in the cytoplasm. Abundant black granules, probably lipoid in character, are found after fixation with osmium tetroxide. The glandular cells are joined by intercellular bridges which stain blue with azan and are connected by their anastomoses with the membrane envelopine the axial muscle fibres. a rich supply of thin tracheae was observed by Wigglesworth (1952a) in *Rhodnius prolixus* in contrast to the feeble tracheation of the surrounding fat body. In contrast to the pgl in cockroaches and butterflies, no nerve fibres were observed in the pgl of *Rhodnius*, either in dissected glands or histologically.

In *Hyalophora* each pgl cell consists of peripheral with striated border through which secretory substance is released. Cycles of activity are evident, characterized by nuclear changes followed by cytoplasmic vacuolization and correlated with the moulting cycle. A low level of seceroty activity is discovered in young pupae and a higher one in older pupae. Discernible variability of secretory activity could be seen among the cells of the same gland. a detailed histological study of the

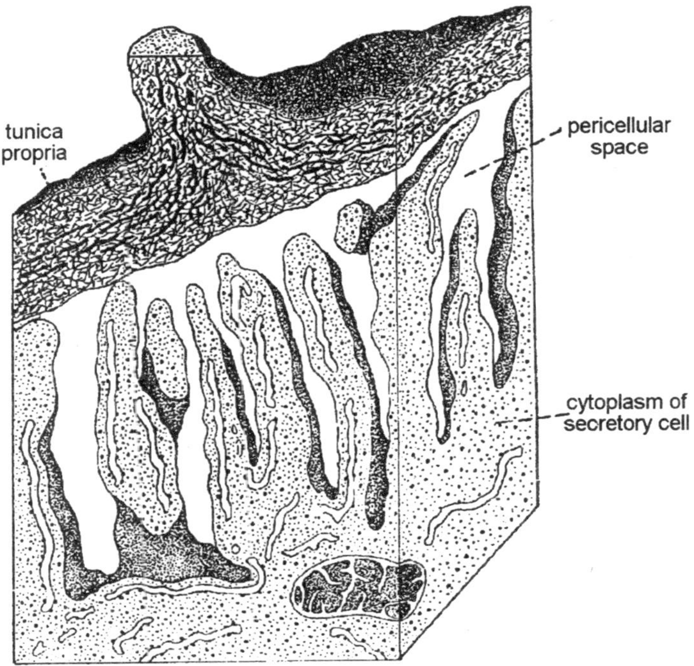

Fig. 3.9. Block diagram of the surface layer of a prothoracic gland cell of Antherea pernyi as seen under the electron microscope.

pgl of Galleria mellonella was carried out by Mala et al. (1972) with the aim of investigating JH effects at both the histological and the ultrastructural level. The more compact gland is composed of about 55 polyploid cells with giant nuclei and different layers of cytoplasm. The secretory cycles connected with the moulting process were examined comprehensively in the penultimate VIth and last VIIth larval instar and the pupal instar and differences between them were ascertained

The amount and distribution of DNA and RNA in the pgl of Samia cynthia and several related species was considered by Oberlander et al. (1965) in the penultimate and the last larval instars and in pupae. DNA synthesis was found to be quite high in the IVth larval instar and lower in the last larval instars and pre-pupa. No DNA synthesis was observed in the pupa during diapause or during adult development. In these two stages, even JH injection failed to induce synthesis. However, RNA synthesis was several times higher during diapause than during adult decelopment, and was easily stimulated by JH injection. Gersch and Sturzebecher (1970) observed a similar increase in RNA synthesis in the pgl of Periplaneta americana nymphs following induction by AH.Eight hours after injecting the hormone, a noteworthy increase in the mixture of stearic, oleic, linoleic, linolenic, palmitic and myristic acids, but not MH (ecdysone) was detected in the glands.

The phases of pgl secretory activity were examined histologically and ultrastructurally by Hintze (1968) in *Cerura vinula*, in relation to the corresponding cycles of the nsc of the brain, ca and the dorsal thoracic gland. The characteristics of the pgl in the active state at both the light and electron microscope levels are described and the penetration of haemocytes across the tunica propria into the pericellular space is reported. The innervation of the pgl and the possibility that brain neurosecretion has been transported to the glands along these axons is discussed.

Ultrastructure

The inital electron microscopy study of insect pgl was accomplished out by B.Scharrer (1966) with cockroach (*Leucophaea maderae* and *Blaberus craniifer*) nymphs at various stages of the moulting cycle. She considered the following to be their most distinctive cytological features: (1) long and intricately interwoven parenchymal cell processes, the cells sometimes being separated from each other by extracellular channels with an average width of 0.5μ ; (2) a specific surface membrane (external lamina) of extracellular origin; (3) numerous micropycnotic caveolae and vesicles (thought to facilitate the transport of material from the glandular cells to the haemolymph or vice versa) covering a large area of the cell surface; (4) Clear but never visible Golgi units; (5) a small amount of smooth endoplasmic reticulum characteristic of steroid - producing cells, sometimes with a striking array of microtubules; (6) a varying incidence (according to the phase of the moulting process) of other structures, such as nucleoli, mitochondria and lysosomes. Degeneration of the glands started within a few days after adult ecdysis. Early signs of their destruction were observed in the nuclei, in the form of patches of contrasting electron density characteristic for pycnosis. later, large heterogeneous inclusion bodies, thought to be autophagic vacuoles, appeared in the cytoplasm. In the advanced stages of degeneration, the plasma membrane and nuclear membrane were slowly autolysed and the protoplasm was invaded by phagocytic haemocytes.

An exhaustive electron microscopial study of pgl in Lepidoptera (*Bombyx mori*, *Antherea pernyi* and certain other Saturniidae) was carried out by Beaulaton. The structute and origin of the tunica propria, the pgl surface membrane, was investigated. It is perhaps matches with the basement membrane of the epidermis, since it has a similar microfibrillar structure and several types of haemocytes take part in its formation. A pericellular space between the tunica propria and the surface of the glandular cells is explained. In agreement with Herman and Gilbert (1966), a neurohaemal control, different from AH control, was found to participate in the cyclical secretory function of the glands. Many kinds of cells are described in relation to these glands.

A remarkable increase in the external surface area of the cell membrane, caused by the projection of finger-like processesinto the pericellular space, a large quantity of free ribosomes and the presence of specific alveolar bodies within the nucleus of female glands are regarded as characteristic for the glandular cells. Rough endoplasmic reticulum and ergastoplasmic saccules are abundant, but smooth endoplasmic reticulum is somewhat rare and is regarded by the author as unimportant. The giant nuclei are differentiated by specific alveolar inclusions of a chromatin character (in females only) and by numerous polymorphous nucleoli consisting almost entirely of RNA. The distribution of phosphatase activities in the pgl of Antherea was examined by Beaulaton (1966). Their accumulation in the lysosomes (dense bodies) and vacuolated bodies are confirmed.

The histology and ultrastructure of the pgl in Tenebrio molitor and their changes within the moulting cycle were investigated comprehensively by Romer (1971). The use of ^{3}H-cholesterol showed that the glands absorb cholesterol from the haemolymph. Neurosecretory granules were discovered in the nerves innervating the pgl. Absorption of lipid granules by both rough and smooth tubules of the endoplasmic reticulum was observed. The extensive Golgi system was expected to be concerned with the production of ecdysone from cholesterol. Extracts of isolated pgl were shown to invite pupation in Calliphora.

Mala et al. (1974), Novak et. al. (1974) and Blazsek et al. (1975) made a detailed study of the structure and ultra-structure of the pgl in the last two larval instars and pupal instars of *Galleria mellonella*, for the purpose of determining JH effects. According to them all structural differences between the penultimate and the last larval instar as being due to JH deficiency in the latter. In consequence, the cytoplasm is vacuolated and less compact, and at first an increase in the amount of secretion is effected in the gland.

The impact of JH and MH on nuclear RNA synthesis in the pgl and corpora allata of saturniid pupae (*Philosamia cynthia* and *Hyalophora cecropia*) were examined in an autoradiographic study by Siew and Gilbert (1971). The pgl were stimulated by MH within three hours, ca three hours later. JH also stimulated both glands, but in this case the ca were activated sooner than the pgl. The administration of MH to pupae in which the adult moulting process had just started, resulted in drastic re duction of nuclear RNA synthesis. This is in agreement with the findings of Socha on hormonal interactions in various phases of the instar. It has already been demonstrated by Toyama (1902) that pgl appear at a very early stage in embryonic development as an epidermal invagination of the lateral portion of maxillary segment and from here later extend into the prothorax and, where this is reduced, even into the mesothorax such result was also obtained by Wells (1954) for the bug Dysdercus cingulatus. Here, two pair of invaginations arise in the second maxillary segment; the anterior, larger, pair develops into labial glands whilst the posterior pair gives rise to the prothoracic glands. If, however, their supposed origin form the nephridian tubules and their innervation (in most insects by nerves from the prothoracic ganglion) is taken into account, it appers more probable that the invagination mentioned does not start in the second maxillary segment but in the most anterior portion of the prothoracic segment.

Phylogenesis

Pgl, or their equivalents are today known in practically all groups of metabolous insects (Pterygota). The conclusion, however, that their occurrence is not restricted to the Pterygota is well founded. Pfluggelder (1958) suggested the possibility of a homology between the ventral glands of Hemimetabola and the so-called head nephridia of Apterygota which are also developed from the second maxillary segment. The papers by Gabe (1953b, 1956) and Echalier (1955, 1956) suggest the possibility of homology between the pgl and the moulting gland ('laglande de mue') or organ Y in the Cruatacea Malacostraca. The view appears to be strangthened by the recent findings of Karlson (1957), who was able to induce puparion in ligated Calliphora larvae with a concentrated extract obtained from Cranton vulgaris. The earlier conclusions of Pfygfekder (1947, 1952) and B. Scgarrer (1948), that both the pgl and vgi, as well as the ca, originate from the nephridia of the ancestral Annelids during the evolution of the Class Insecta are also in agreement with this. It may be expected that, if suitable techniques are used, the equivalents of the pgl will be found in all groups of arthropods such as the "Crustacea. Arachnida. Myriapoda, etc. If this is so, production of the moulting hormone may be looked upon as phylogentic adaption, which has developed in the close relationship with the chitinous cuticle as a mechanism to ensure the simultaneous moulting of the whole body surface, This is necessary to allow the organism to escape from the inextensible chitinous covering.

2. Ventral Glands (vgl)

Morphology

Vgl, often called ventral head glands or tentorial glands, were first found by Pflugfelder (1947a), in the hind ventra region of head in Phasmidae. Boisson (1947) described corresponding structures in Bacillus rossii as corpora incerta and discovered a cuyclicity in their secretory activity, which was connected with moulting, Pflugfeelder (1947d) assumed that they were absent in Hemiptera and in all Holometanola.

An important feature of the vgl is that they degenetate at the end of metamorphosis or in the first few days after the imaginal moult in the same way as the prothoracic glands (= tentorial glands) of the worker and soldier castes of termites., which are less developed than in the sexuals but are active during the whole life period (Springhettin and Bernardini, (1955; Kaiser, 1956). This appears to be connected with the neotenic character of these castes insome lower termite groups, evidence by the underdeveloped state of their ovaries, in which they are maintained byinhibitory substances produced by the reproductives. The influence of the vgl on the moulting process was shown by Strich-Halbwwachs (1954, 1958) and by Halbwachs, Joly and Joly (1957) in

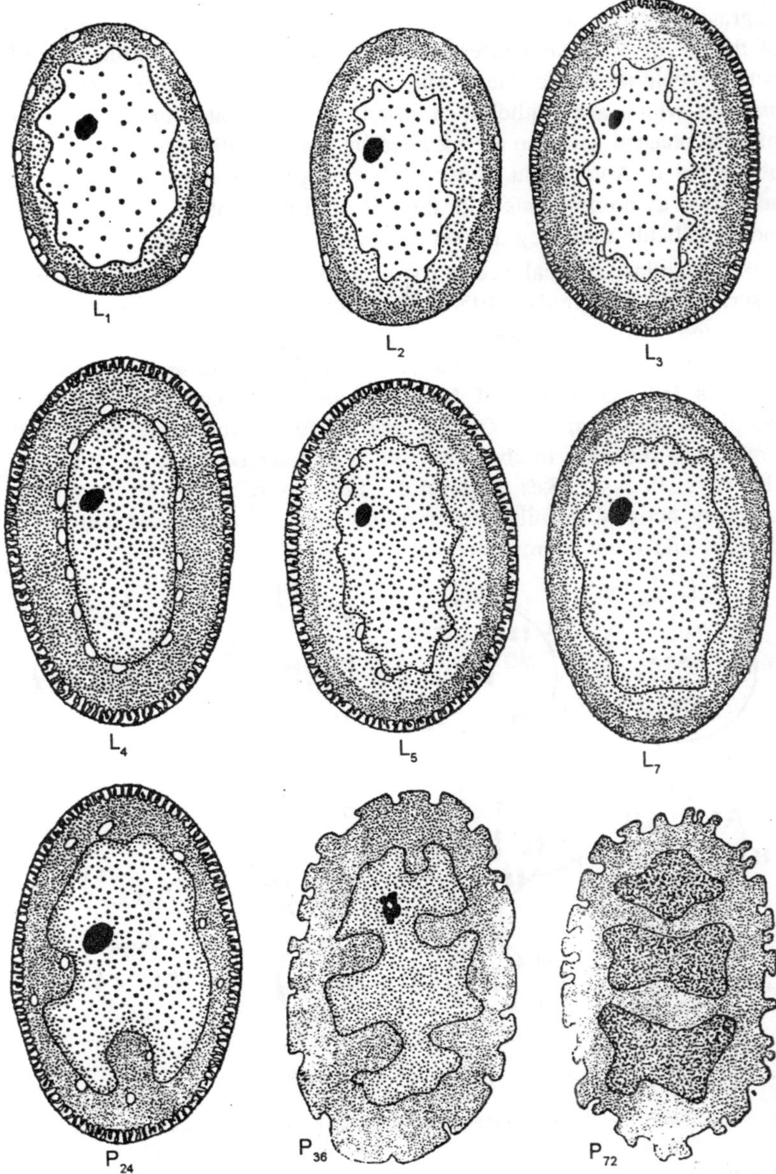

Fig. 3.10. Scheme of developmental changes in a prothoracic gland cell during the last larval and pupal instar of Galleria mellonella. L_1-L_7–Ist to 7th day of VIIth instar lava; P_{24}-P_{72}–24th to 72 hour of pupal development.

migratory locusts. Tyhe implantation of an active gland at the geginning of the IV of V instar causes an acceleration of the moulting process with a feeble prothetlic effect. The so-called corpora adnexa in *Dinjapyx marcusi* which are epithelial in structure and situated caudally to the suboeso phageal ganglion are viewded by Marcus (1951) as a homologue of the vgl in Apterygota. Some authors suggested that the vgl can be uderstood as homologous with the pgl of Holometabola. But, it seems more probable that they are serial homologues of the pgl.

A regular cyclical secretory activity of vgl in *Carausius* was described by Pflugfelder (1958). The drops of secretion apper, closely connected with the nuclei, during and for a few days after ecdysis. This period of excessive secreetion is followed by a phase of exhaustion. Between 3rd and 7th day of the intermoult period both the nuclei and cytoplasm grow rapidly. On 8th and 9th days, there are many cell dimensions resulting in abundant cell aggregates between the 10th and 11th days. Ecdysis takes place on the 12th day relatted with the strong secretory activity. Similar observations have been made in Odonata, termites and other Hemimetabola.

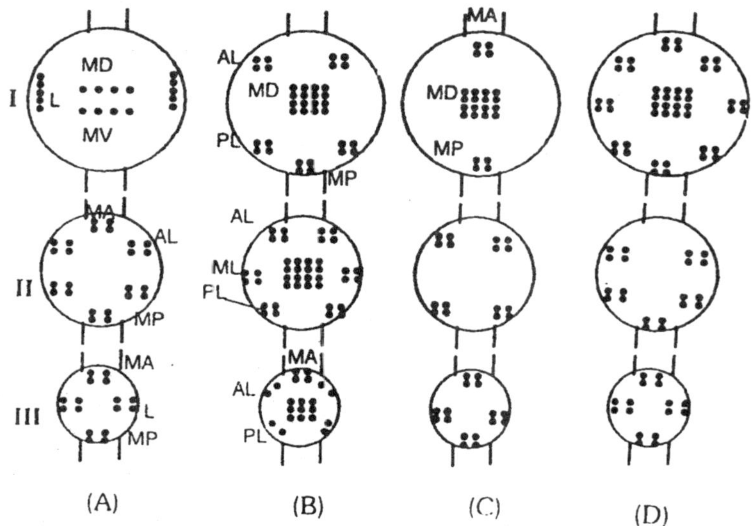

Fig. 3.11. Neurosecretory cells in the ventral ganglia. A–dragonfly, Orthetrun chrysis; B–cockroach, Periplaneta; C–locust, Schistocerca and D–moth, Othreis materna.

Histology and embryology

The histological structure of the vgl is quite similar to that of the pgl. Their embryogenesis was examined by Pflugfelder (1958) in

Carausius morosus. They develop from epidermal proliferations in the ventrocaudal part of the head, but are soon separated from the epidermis and become bladder-like. Jones (1953, 1956) found vgl in *Locusta pardalina* and *Locusta migratoria* as invaginations in the head region of embryos at the end of the kataterpsis stage. He considered them to be homologous with the pgl of the other insects. He examinied their function in relation to the beginning of the secretory activity of neurosecretory brain cells and the first embroynig moulting process.

Degeneration of the vgl of the earwig Anisolabis maritima during metamorphosis was studied histologically and analysed experimentally by Ozeki (1968, 1970), who transplanted them to various other stages of the same species. He concluded that both the small amount of JH in the last larval instar and the latter's internal state were determining factors in the regression of the vgl. The implantation of active vgl from larvae to adults led to a supernumerary moult in the adult stage

Ultrastructure

Joly et al.(1969) studied the vgl of Locusta migratoria using the electron microscope. The major structures found in the cytoplasm were the Golgi apparatus, ergastoplasm and structures of lysosomal or autophagic character; no smooth endoplasmic reticulum was observed. The authors concluded that a substance which was probably protein nature, and not MH, was produced. In this they agree with Locke (1969,) the vgl produces a protein substance, which merely controls ecdysone production elsewhere in the body. Joly et al. (1969) described special structures consisting of degenerating rough endoplasmic reticulum. In another research paper from the same laboratory, published soon afterwards the authors formed a different opinion, as in the oenocytes they found an abundance of smooth endoplasmic reticulum, which may be related with the occurrence of a steroid component in the cuticular lipids. Unlike Locke (1969), they did not consider cytological structure alone as correct evidence for steroid production.

In an ultrastructural study of the secretory cycle in Locusta migratoria vgl, Cassier and Fain-Maurel (1968) based their conclusions on the belief that the glands participate in ecdysone production. They demonstrated that, during the vgl activity, cycles the tubules of the smooth endoplasmic reticulum were extruded from finger-like cytoplasmic evaginations into the intercellular spaces of the gland. They stressed that this extrusion coincides with maximum MH activity and concluded that it was the most effective way of discharging the hormone from the glandular cell. They suggested the use of isotopically-labelled cholesterol in order to solve this question. Differences in the ultrastructure of vgl in the active state and during diapause were examined by Guellin (1971) in the grasshopper *Pyrgomorpha conica*. The ultrastructure of diapausing glands is indicative of remarkable depression of metabolism : the pericellular system is reduced. lysosomes are apparently absent and free ribosomes and eragastoplasm at the end of diapause, the author concludes that active vgl produce a protein type of substance, as did Joly et al. (1969) in Locusta.

Fain-Maurel and Cassier (1968, 1969) examined the involution of the vgl in the last larval instar of the gregarious phase of *Locusta migratoria migratorioides* (in adult, the glands persist in the solitary phase) and compared it with the secretory cycle in earlier instars. They claim ' an early and primordial role of Golgi apparatus; in the elaboration of the 'vacuolar bodies' of 'cytosomes' and participation of rough endoplasmic reticulum in the process of degeneration, through the formation of 'cytosegresomes' and 'particular microtubular areas'. The process begins with the autolysis of the named structures followed by fragmentation of the cells and nuclear lysis. Specific 'Golgian vacuoles' are formed in the Golgi apparatus, which afterwards undergo autolysis or are enclosed in autophagic vacuoles. degeneration can be avoided by the daily application of carbon dioxide from the IIIrd instar onwards.

The structure and development (degeneration) of the apolytic glands have been studied in several other insect groups and species, e.g., Mantis religiosa Coenagrion angulatum (Odonata), and Nauphoeta cinerea (Blattoidae) Hoffmann and Joly (1972) paid special attention to the physiology of the vgl in Locusta migrtatoria, and found complete inhibition of moulting following extirpation of the glands, but this was fully compensated by the implantation of an active ecdysone, however large the dose, was ineffective. The authors reached the conclusion that vgl produce a hormone which is different from ecdysone.

Phylogenesis

Rae and O'Farrell (1959) analyzed the formation of vgl and the retrocerebral complex in Grylloblatta campodeiformis, a represent- active of the primitive Orthopteroid Order Notoptera (= Grylloblattidae). They explained the structures obviously endocrine in function in the ventrocaudal region of the head and in the cervical area which resembled very closely the so-called 'head lobes' of Blatella germanica and the 'cervical glands' of Periplaneta americana. These researchers accepted that the coxal muscles they describe in Grylloblatta are the remnants of the degenerated pgl of cockroaches, and homologous with the axial muscle cord in Blattidae depicted as Scharrers' organ by Chadwick (1955). The presence of both pgl and homologous of vgl in the cervical structures of cockroaches seems to be well accounted for by the morphological and experimental evidence of Rae (1955) and Chadwick (1955, 1956). Thus complete etirpation of the pgl of *Periplaneta americana* does not prevent moulting. Cyclical changes in the volume of the 'ventral lobes' in *Blatella* are related to the moulting process and regeneration

3. Pericardial Glands

The initial indication of the existence of ductless glands associated with the dorsal vessel was made by Verson(1911a,b). The pericardial glands in various species of Phasmidae (Phyllium, Carausius etc. Were fully explained by Pflulgfelder (1938a-d, 1946b). Besides the vgl they occur in the form of paired glands, the histological structure and function of which, unlike origin, is equal to that of the pgl and vgl. They also decline in the adult.

4. Peritracheal Glands

Matching structures were described by Possompes (1949 a-c, 1953a,b) and other researches in the Chironomidae, Simullidae, Tabanidae, and other groups of Diptera under the name of peritracheal glands. They have not been discovered in Tipulidae. Both reseatchers agree their homology with the pericardial glands of Phasmidae. The relation of the moulting process was demonstrated experimentally by Pflugfelder (1949b). No moulting was observed in specimens in which the glands were extripated.

Histology, embryology and phylogenesis

In their histological structure the pgl of both Phasmidae and Diptera correspond closely with the pgl and vgl of other groups of insects. The pericardial glands were initially accepted to be of mesodermal origin by Pflugfelder (1938). In a later work however, Pflugfelder (1958) mentions their ectodermal-like structure and emphasized the difficulty in distinguishing between mesenchymal and ectodermal cells in the early embryonic period of Dixippus morosus. This would remove the only possible objection to their identification with the peritracheal cells, which are of ectodermal origin according to Possompes. The problem of their homology has, however, been rather complex because of by their confusion with pericardial cells of obvious mesodermal origin, which is not at all simmilar to the glands.

5. The Ring Gland (rgl)

In higher diptera special endocrine organ occurs, which combines the source of all three metamorphosis hormones, the first signs of which can be seen in the Tipulidae. It was initially explained by Weismann (1846) in *Calliphora vomitoria* and is therefore referred to by some authors as 'Weismann's ring'. The first to suggest its endocrine character was Hadorn (1937 a-c) although the experiments of Fraenkel (1934, 1935). In these papers the rgl was expected to exert a positive influence on metamorphosis. But Burtt (1937, 1938) postulated that the rgl produced an opposite, i.e., inhibitory, effect on metamorphosis, and he identified the rgl with the ca. Hanstrom, on the other hand, attempted to equate the rgl with the cc on the basis that nerve cells are found therein, which do not exist in other endocrine glands.

The first suggestion made by B. Scharrer and Hadorn (1938) who first suggested the ring gland as a composite structure, consisting of both ca and cc. The structure of the rgl during metamorphosis and the development of the adult cc and ca in *Calliphora* were thoroughly studied by E. Thomsen (1941, 1947). She recognized the lateral portions of the rgl, formed of large glandular cells, with the pericardial glands of Phasmidae by showing that they disintegrate during metamorphosis. She also found that the hypocerebral ganglion is associated with this composite structure. Her findings were confirmed and elaborated by Vogt (1941c, d) for Drosophila, M. Thomsen (1951) for many species of Diptera, and by Possompes (1953) in a monograph chiefly concerned with the structure and funcstion of the rgl in Calliphora erythrocephala based on the individual implantation of parts of the ring.

Four major components of the rgl may now be differentiated : (1) a corpus allatum of the small cell type situated in the dorso-median part of the ring, often covered by large cells of the lateral portions; (2) pericardial (peritracheal) glands formed by the large glandular cells of the lateral portions of the ring; (3) a corpus cardiacum in the ventromedian part of the ring, with only one a couple of nervi corporum cardiacarum formed by the coalescene of the nervi corporum cardiacarum interrni and externi before they enter the ring gland fusing with the nerve recurrens in some species; (4) the hypocerebral ganglion consists of transparent ganglionic cells and closely joined with corpus cardiacum.

Ultrastructure

An exhaustive study of the ultrastructure of the rgl of a normal and a 1/2/gl mutant strain of Drosophila melanogaster was carried out

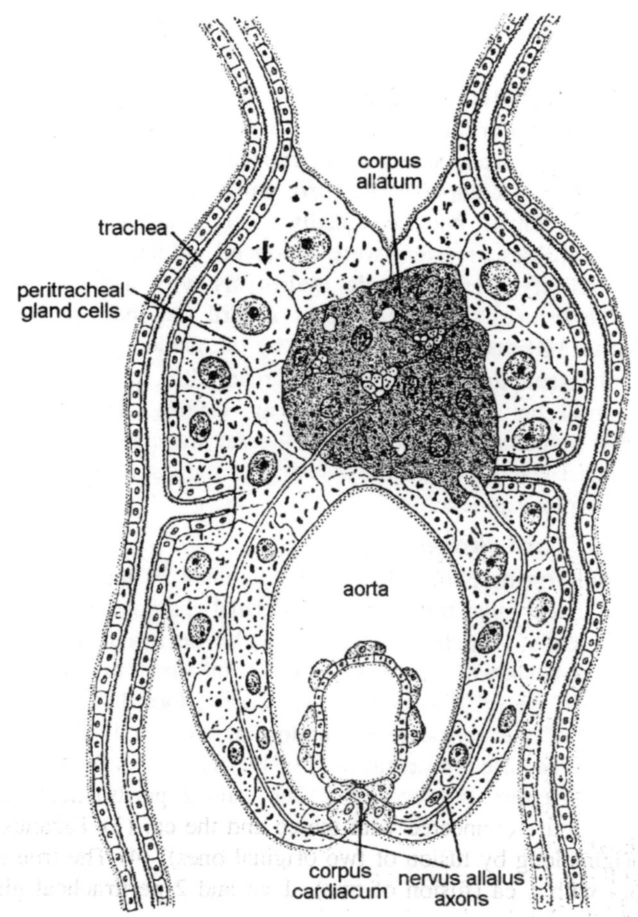

Fig. 3.12. Cross-section of ring gland of Drosophila melanogaster IIIrd instar larva.

by King et al. (1966) and Aggarwall and King (1969). The quantity of smooth surface endoplasmic reticulum in the peritracheal gland cells increased about tenfold in the pre-pupal period in normal larvae, but did not exceed 1 percent in mutant larvae. Metamorphosis of homozygous larvae of the lethal mutant was impossible unless an active gland from a normal larva is impainted. The researchers reached the conclusion that the relevant gene (the + allele of the 1/2/gl mutant) was indirectly concerned with the synthesis of smooth endoplasmic reticulum, where cholesterol from the food is transformed to ecdysone. In the intracellular spaces between the gland cells they encountered numerous cytoplasmic vesicles, which were pinched off from the cell projections.

Willig et al. (1971) examined ecdysone and ecdysterone biosynthesis in 7-day Calliphora erythrocephala larvae by injecting the larvae with 14- C-Cholesterol. The same kind of experiment was performed with explanted brain-rgl complexes in vitro. In the latter case, no acative hormones were found, but occasionally a very small amount of active ecdysone glycosides or esters were produced, which exhibited lot of hormonal activity after enzymic hydrolysis with a-glucosidase of esterase. Therefore, the researcheres regard the inactive hormone glycosides as MH precursors produced by the gland.

The ultrastructure of the rgl of Drosophilia melanogaster was studied comprehensively by King et al. (1966). Typical smooth endoplasmic reticulum was found in the lateral branches, corresponding to peri-tracheal glands, which the researchers suggest as being involved in steroid hormone production.

Embryogenesis and phylogenesis

The embryogenesis of the rgl was examined in detail by Poulson (1954a). Amongst other things, the ectodermal character of the lateral, large gland cells, which correspond to the peritracheal glands of Possompes, was shown. There is not much information regarding the phylogenesis of the rgl. Possompes distinguishes four different types of development of the retrocerebral endocrine system in Diptera: (1) Tipula type: 2 ca, 2cc, (no equivalent of the pgl has so far been found here), (2) Chironomius type : 2ca, 2cc, and 2 peritracheal glands, each of which are connected with the cc and the ca. (3) Tabanus type : I ca, originationg by fusion of two original ones). (4) The true rgl of Calliphora with 1 ca (fusion of two), 1 cc and 2 peritracheal glands.

The Direct Effects of MH

The initiation of the moulting process

The implantation of active pgl on their equivalent results into moulting, not only where secretory activity is naturally suppressed, as in diapause, or where the glands have been experimentally removed, but also in cases where neither the pgl nor the moulting process normally occur, as in adault insects. The moulting process and consequent growth can be rosuced by active pgl along, without a simultaneous impl antation of an active brain or cc along is not able to induce multing in the absence of an effective concentration of MH. The resufe was flly confirmed by experiments with isolated MH. the resuf was fully confirmed experiments with isolated MH prepared in crystal form by Butenandt and Karlson (1954).

The circulation of labelled ecdysone injected into the insect body was analyzed by Karlson and Sekeris (1963). They discovered that soon after injection the hormone accumulated in the epiderims and later in the fat body. The nuclei of the epidermal cells exhibited the greatest ratio activity. The authors realized that the sites of action of MH are the nuclei of the epidermal cells and that microsomes of the fat body eliminate the excess of the hormoes from the haemolymph.

The regulation of growth

The most important requirement for postembryonic growth in insects, as in any other arthorpod, is the loosening and subsequent freeing of the body from the inexpansible cuticle. MH is therefore indispensable, though indirectly so, for growth, as, for instance, the influence of Mh on the mitotic activity which precedes the detachment of the old cuticle. While, there is an increasing amount of evidence to the contrary: (a) Growth is possible without any moulting process and thus without MH (e.g., the growth of ovaries and other internal organs as well as of regenerating tissues). (b) The implantation of agl as well as the injection of the ecdysone induces moulting, often without any effect on growth or even with a negative one). (c) As distinct from JH, MH alone is not able to induce mitotic activity in larval parts of the body. Without further evidence the presence of MH is thus to be accepted as one of the conditions of growth but there is no cause for accepting it as a true growth hormone of the JH type.

The discovery, that the deposition of endocuticle and the secretion of wax also took place during the intermoulting period in the absence of any MH, led to the conclusion that these processes are not controlled solely by MH. MH action after the attainment of an active concentration of the hormone in the haemolymph is perhaps responsible only for a change in chitinase activity (or concentration), resulting in the chage from chitin synthesis to its decomposition, causing to moulting and subsequent gradual dissolution of the old endocutile.

Investigations by Morohoshi and Iijima (1969) were concerned with the effects of injection of phytoecdysoids (inokosterone, ecdysterone, cyasterone) into starved, ligated last instar Bombyx mori larvae and into pupae. In some cases supernumerary larval instars (a VIth instar) resulted. The researches confirmed that the MHd were concerned ony with the control of moulting and that the structure of the moulted insects depended on JH action. In another study, Morohoshi et al.(1972)

injected Vth instar larvae of the same species with varying amounts of ecdysterone at different intervals after the preceding moult. They investigated the effect on the length of the intermoult period

Scrutiny of adaptive changes in larval behaviour prior to pupariation demonstarated interesting correlations with MH action. During 30 hours under dry conditions after the larvae have left the nutrient medium, they become increasingly sensitive to injected ecdysone. If the dry period is interrupted by a further exposure to moisture, however, their sensitivity diminishes. The researchers stated that this observation might be due to a 'convert' effect of ecdysone. It, seems practical enough however, to assume that leaving the moist enviornment induces MH production (undoubtedly by activating AH release) and as more endogenous MH is produced, less exogenous MH is needed to induce pupariation. Replacement in the wet medium inhibits AH production, and with it the amount of MH produced up to a given moment, sothat more has to be injected to induce puparium formation. The researchers concluded that this was an significant form of phylogenetic adaptation to the given conditions.

In another research paper, Zdarek and Fraenkel (1972) explained that injection of larvae with a large dose of ecdysone under wet conditions shortened the pupparіation time to 12 hours, as against 30 hours in the controls. Tanning occured at various stages of puparium formation (longer, less contracted body, and less smooth surface), again demonstrating the independence from MH and the moulting process of the individual process of puparium formation. Earlier experiments with the facefly (Musca autumnalis), in which the puparium undergoes calcification instead of tanning, proved that this process was equally independent of MH. MH is, of course, the essential releasing factor in this series of processes.

The evolution of typical puparia following the injection of ecdysone into Calliphora erythrocephala larvae deprived of their rgl confirms that MH controls contraction of the puparium as well as its tanning and other associated processes. During other set of experiments, Fourche (1969) starved Drosophila melanogaster larvae from the 48th hour of the last instar and then supplied them with ecdysone in pure gelation. If the concentration and time of action were adequate, the larvae formed puparia at the proper time. The relationship between endogenous MH and exogenous ecdysone and their correlation with pupariation were discussed. The above results prove that, as in morphogenesis, these processes are only indirectly dependent on MH.

The action on larval and pupal diapause

But it is quite clear that MH acts directly to break both larval and puppal diapause. Here, however, it must be noted that it is the absence of MH consequent upon the absence of AH which is the immediate internal cause of diapause. a state identical with diapause can be induced experimentally by removal of the source of AH.

The effects on the puffing patterns

The influence of the puffing patterns of the giant polytene chromosomes in the salivary gland cells and other tissues of Chironomous and Drosophila will be discussed later.

The action of MH

The impact of MH on the amounts of RNA and DNA during metamorphosis in Calliphora erythrocephala was studied by Berreur (1965a) in whole animals and in their separate organs and tissues. The resultant changes, which were tisuse specific for different parts of the body, were not discussed with reference to whether the action of the hormone is direct or indirect. In another investigation, using autoradiography of imaginal discs of the same species, Berreur (1965b) demonstrated that no DNA was synthesized in the absence of MH caused by removal of the ring gland, and that RNA synthesis was limited. Reintroduction of the hormone resulted in nuclear RNA synthesis and, somewhat later, in DNA replication.

Another group of researches, Sekeris et al. (1965) demonstrated that the injection of ecdysone raises the incorporation of radioactive orthophosphate (H^{33} $PO^2{}_4$) into nuclear ribosomal and messenger RNA in the epidermal cells of Calliphora erythrocephala. They accepted that MH stimulated DNA -dependent RNA synthesis. Gillot and Daillie (1968) found a break in DNA production in the silk glands of the Bombyx mori at the beginning of each moulting period. They observed asynchrony of the DNA cycle and differences between the two parts of the silk glands, but the maximum amount of DNA per nucleus in the two parts at the beginning of the Vth instar was identical. Neufeld et al. (1968) analyzed the effects of injected ecdysterone on Calliphora stygia larvae and the conditions for its action on protein and RNA synthesis. Gorell and Gilbert (1969, 1971) demonstrated a similar positive effect by ecdysterone on RNA and protein synthesis in the hepatopancreas of the crayfish Orconectes viridis. The particularity of all these effects is yet to be throughly examined.

In Philosamia cynthia and Hyalophora cecropia a nuclear RNA synthesis in the pgl was activated 3 hours after injecting ecdysterone, whereas similar activity did not start in the ca until 6 hours after administration of the hormone. The administration of MH to pupae which had just began adult development culminated in a drastic decrease in nuclear RNA synthesis.

The action of enzyme activity

Various researchers have expressed that MH interferes with the reaction chain of some of the principal enzymes of the insect organism that regulate normal metabolism and growth, as for instance the cytochrome oxidase system. However interesting and important these findings may be in the study of the corresponding enzyme activity showing amongst other things the functional importance of insect hormones in research on any question of insect physiology and physiology in general—they do not bring any definite proof in favour of the assumption of a direct participation of MH in the corresponding reactions.

Marmaras et al. (1966) discovered an inverse correlation between 5-hydroxytryptophanase activity and the ecdysone titre during the pupal instar of Calliphora erythrocephala. Kobayashi and Kimura (1967) illustrated that labelled glucose was converted to haemolymph trehalose is decerebrated Bombyx mori pupae injected with ecdysone, but that in the controls (injected with water) it was converted to glycogen. Adenyl cyclase in the pupal wing epidermis of Hyalophora gloveri is stimulated after the injection of ecdysterone.

Injected ecdysone has a positive effect on different types of metabolism in ligated last instar larvae of Musca domestica. Time and tissue specific incorporatiion of tritiated thymidine was observed after the injection of ecdysone, but it was limited; however, tritiated uriding was incorporated intensively under the same conditions. Tritiated leucine had a different incorporation pattern. The incorporation pattern. The incorporation of labelled glucose was characterized, in various tissues, by a large increase 1 hour after the administration of ecdysone; it then fell rapidly, at 6-hour intervals until, 18 hours after injection, it reached zero.

The effects on metabolism

Bernardini and Laudini (1966) examined oxygen consumption by the prothoracic glands of young and last instar larvae of two species of cockroach (Leucophaea maderae and Periplaneta americana) and found that it has cyclical variations. Ultimate oxygen consumption was reached on day before, and minimum consumption one day after each ecdysis. The researchers discussed the possible significance of these changes. Fourche (1967) observed a typical U-shaped curve in Drosophila melanogaster pupae, with increases in respiratory metabolism accompanying pupariation, apolysis and ecdysis. In Thermobia domestica, Rhdendorf and Slama (1966) found specific oxygen consumption cycles associated with moulting and off-laying, which corresponded with those in pterygote insects. Oxygen consumption increased during actual ecdysis and fell about half way through the intermoult period.

Many researchers have examined the uptake, transport and utilization of cholesterol and other steroid derivatives by insects. Bergmann and Levinson (1966) tested 15 cholesterol derivatives and isomers in Musca vicina larvae and some of them in Dermestes maculatus larvae. Both the requirements and utilization of steroids in these insects was discussed. Chino and Gilbert (1971) fed silkworm (Philosamia cynthia) larvae on an artificial diet containing 14 C-cholesterol. Practically all the substance found in the haemolymph in vitro. It was finally accepted that the midgut is necessary for the incorporation of 14 C-cholesterol. A qualitative and quantitative study of haemolymph proteins in *Dixippus morosus* demonstrated a remarkble relationship between them and the moulting process (or morphogenesis), and to yolk formation in adult females. In the final instar larvae, the haemolymph protein concentration reached the maximum on the 13th day after the preceding ecdysis, i.e., at about the time of the new adult apolysis. After that the protein concentration declined abruptly, largely owing to the decline of two glycoproteins, both of which practically disappeared during yolk formation in the adult females.

The effects on rhythmical activity of muscles

Roussel (1970,1971) explained specific acceleration of the cardiac rhythm in locusts following the implantation of vgl. In last instar nymphs, the implantation of active glands accelerated the cardiac rhythm only in the presence of the ca. In allatectomized adults the implantation of four vgl also acclerated the heart rate. While extripation of the vgl at the beginning of the last larval instar stabilized the cardiac= rhythm at a relatively high level compared with normal specimens. But even here the direct nature of this efect still requires verifacation.

The indirect effects of MH

The effect on differentiation

Including the effects of MH on growth, many researchers consider that it affects differentiatiion. While the question of the direct nature of an MH effect on growth may still be considered as a matter for debate, there is no longer any cause for assuming a direct influence of MH on differentiation. However, the dependence of any surface morphological change on the moulting process is unquestionable. The slightest morphological changes on the surface of the body can only be realized during the relatively short interval between the detachment of the old cuticle and the deposition of the new.

Any break in the process of moulting therefore necessarily means the inhibition growth and differentiation. But the initiation of the moulting process does not in any way mean a simultaneous initiation of differentiation. On the contrary, a premature introduction of MH, as in the ecdysone experiments, results not in more differentiation but in less. This was shown in Fukuda's experiments (1944) with prothoracic gland implantation, and in the experiments of Wigglesworth (1940a, b, 1948) with the parabiosis of moulting specimens of Rhodnium combines to those which has passed the critical period for the moulting hormone.

The same effect implies that the result necessarily results from the effects of juvenile hormone on growth and differentiation. By implanting active ca the gradient growth, resulting in differentiation, may be changed into an isometric one without change in form. Such an inhibition of differentiation does not mean any inhibition of the moulting process which is distinctly accelerated. There is no basis for assuming any antagonism between MH and JH. Differentiation can be completely separated from both moulting and growth and there is no connection between the effects of Mh on one or the other. It is confirmed that the influence of MH on differentiation is therefore purely indirect.

Joly (1958) when analyzed the effects of the ventral gland hormone on the mitotic activity of the epidermal cells and an allometric growth in Locusta during metamorphosis, and found that MH only affects the moulting process.

The postulation that MH has only an indirect effect on growth and protein synthesis is supported by the following experiments. Chilled Antherea polyphemus about to start adult development and ligated abdomens of Hyalophora cecropia and Samia synthia adudlts were injected with crystalline ecdysone together with mitomycin C (which inhibits DNA replication). Ecdysone was found to cause apolysis in the pupae of these insects, in the presence of mitomycin, but growth, morphogenesis and the rest of the moulting process were inhibited. If ecdysone were administered after DNA synthesis, the epidermal cells also responded by depositing new cuticle. Similarly process took place in adult insects, in which no more DNA synthesis takes place, but the entire moulting process can be activated experimentallywith the same ecdysone.

Courgeon (1969, 1970) analyzed the effect of MH on the imaginal discs of eyes and antennae in tissue cultures. Generally mitotic activity stopped a few hours after explanation, but if an active ring gland was implanted at the same time it continued for up to five days. When the nervous system (the cerebral hemispheres and the ventral ganglion) was also added, there was a stronger effect.

The impact of MH on morphogenesis were examined in detail in ligated Galleria Mellonella larvae. Repeated dipping of isolated abdomens in methanol solutions of ecdysterone in different concentrations resulted in morphogenetic changes of varying degrees after ecdysis. Different types of cuticle were obtained in different changeing from a larval type, via a more or less normal pupal type, to an imaginal type with more or less developed scales. Taking into consideration the amount of hormone applied (the concentration and number of dippings) and the interval between treatment and subsequent moulting, the following explanation is submitted, in agreement with the gradient-factor theory. The degree of morphogenesis depends upon the length of the interecdysial period, or to be precise, upon the time between apolysis and deposition of the new cuticle. If this is shorter than normal (after a large dose of the hormone), the degree of differentiation is smaller and the cuticle retains a more or less larval of character. If the MH concentration is low, the interecdysial interval is longer and morphogenesis can progress further. This is the condition with omission of the pupal type of cuticle. Moulting is not caused by small does, but certain developmental changes in the internal organs do occur.

The action of ovarian development

The indirect character of the morphogenetical impact of MH is quite clear from its effect on the development of the ovaries. It was demonstrated by Strich - Halbawachs that the removal of the vgl at the beginning of the last larval instar in Locusta migratoria, which prevents moulting, also suppresses the development of ovaries but to a limited extent. The oocytes, which average 0.7 mm in length at the time of the operation, reached 1.8 mm within one month in the permanent larvae thus obtained, while they reached 7.2 mm in the metamorphosed controls at the moment of egg-laying. few ooctyes in the experimental series were abnormal or had even degenerated. It is evident that the same factor is afecting here, i.e., the inhibitory influence of the inextensible cuticle, the only difference being that its effect on the internal organs is less direct than on the epidermis. The relationship between the moulting process and the development of the ovaries also permits a rational explanation of the findings of Bodenstein (1953a-d) and Wigglesworth (1957) in which the induced nature of this type of MH effect is also evident.

The effects on ovarian development

The injection of ponasterone A affects the ovarian development (the weight of the ovaries and the number of eggs) in isolated pupal abdomens of *Bombyx mori*. It also had other morphogenetic effects. Ovarian weight and the number of eggs increased proportionally to the dosage up to an amount of 2.5 μg per specimen. No change was observed after larger doses, which resulted in the formation of a scaleless adult cuticle. The implantation of vgl into adult Locusta migratoria females, stimulated growth of the oocytes, but caused the more developed oocytes to degenerate.. The activity of the accessory glands and the genital apparatus was stimulated in allatectomized locustus of both sexes.

One may even include that the independence of the development of the ovaries from the moulting process and MH is quite evident from the fact that it normally takes place in adults where no prothoracic glands exist and no moulting occurs.

The action on the corpora allata and reproduction

The entomologist Englemann (1959) examined the effect on the function of the ca and the development of ovaries of implanting active larval pgl into adult females of the cockroach Leucophaea maderae. He obtained moulting of the adult in some cases and nearly always a partial inhibition of ca secretion. He describes this as a direct effect on the ca and an indirect one on the neurosecretory brain cells. Both these inhibitory effects were avoided by the simultaneous implantation of a brain and suboesophageal ganglion. An injection of ecdysone had an effect similar to that of the implanted prothoracic glands. In both cases cell division was found in the ca with a subsequent increase in tthe number of cells. As suggested by, Englemann, the results for a related species, Periplaneta americana to some extent contradict those obtained by Bodenstein (1953a, e). There is also a remarkable discrepancy in the conclusions if both these researches since the prothoracic gland secretion is produced during larval instars almost simultaneously with that of the ca without producing the slightest inhibitory effect. Some results in recent years have even shown that the ca may cause a direct positive effect on the induction of moulting which has no connection with JH Nevertheless, there is the possibility of a single simple explanation of all these facts provided in the 'law of correlation in consumption' by the different tissues inside the body. When comparing the findings of Englemann with those of Bodenstein, two major points emerge. Firstly that the moulting process, even when not accomplished, necessarily consumes part of the available body reserves. This deficiency becomes clear in the reduced activity of the corpora allata and in the restricted growth of the ovaries. Secondly, such a serious interference with the physiological processes of the body as that of allatectomy and the resulting regeneration processes may in itself produce an impulse strong enough to prolong the activity of the pg.

In Locusta migratoria, the disconnection of ncc and the resulting increase in the volume of the ca has no inhibitory effect on the moulting process. In this context it is interesting to note the effect of folic acid on diapausing pupae of Pieris brassicae, where it increases the amounts of ribonucleic and desoxyribonucleic acids causing tumour growth in the absence of MH. The researcher attempts to explain these findings on the principle of artificial induction of endogenous viruses.

The action on regeneration

Other aspect of differentiation is also true for regeneration : that is, it is also independent of moulting. However, there is thought to be an indirect interdependence, at least as far as more extensive regenration is concerned. It can either be acclerated by the action of MH, or suppressed or greatly delayed by the removal of MH. Less extensive regeneration, including mitotic activity, may occur independently of MH.

A study of contrasting features of the activity of the pericardial glands in normal and regenerating (with the mesothoracic legs removed) Ist instar nymphs of Carausius morosus was made by Vietinghoff, Penzlin and Spannhof (1964). A statically significant difference in the volumes of nuclei was found between the glands of regenerating and normal specimens. The decrease in volume of the nuclei in normal specimens connected with the onset of the moulting process, was delayed by two days in the regenerating individuals.

Many researchers whve examined the effect of injury and wound healing on local cuticle formation on the general moulting process (Krishnakumaran, 1972). In both cases the effect was positive. The effect of injury on the general moulting process seems to be mediated by the brain, since brains from injured specimens, unlike those from the intact controls, displayed AH activity on both the pgl and ca. In the case of local cuticle formation, which was also observed in adults, the presence of MH does not seem to be absolutely necessary for the cuticulogenic activity of the epidermal cells. This is also pointedout by the findings of Gersch (1972), who observed chitin hypertrophy in the salivary glands of *Periplaneta americana* following injury in the thoracic region (pgl extirpation).

Schaller and Andries (1970a, b) and Schaller (1972a, b) analyzed the regeneration cells in Aeschna cyanea (Odonata) larvae deprived experimentally of their vgl resulted in an abrupt and significant increase in the activity of these cells. In another research paper, Schaller (1970) concluded that MH was responsible both for the activation of remaining regeneration cells and for regeneration of the midgut epithelium in the intermoult period, although these cells, like the cells of the eye imaginal discs, also displayed lower mitotic activity in the absence of the hormone. On the other hand, no mitosis was observed in the absence of MH in permanent larvae. A detailed study of the eyes and optic lobes in permanent larvae of the same species fully confirmed the above conclusion. Keeping in consideration the gradient-factor theory, it emerges as an interesting observation.

The influence on the morpholigical colour change

Remarkably visible morphlogical colour changes occur in various insect species during metamorphosis. Few exhibit this in the adult stage, but in the majority it is a transient phenomenon. These changes may be quite affected by the suppression of premature introduction of MH and can either be suppressed or prematurely induced. Because of this, many researchers have assumed that the moulting hormone exerts a direct effect on the chemical synthesis of the corresponding pigments. Buckmann (1956-1959), for example, in his extensive studies of morphological colour changes in *Cerura vinula* larvae during metamorphosis reached the conclusion that these changes are caused by low concentration of the MH, whilst large doses of the same hormone induce pupation. Generally the green caterpillars turn dark at the time the cocoon is made due to the production of a red ommochrome, which first appears in the epidermis and one day later in the fat body. The morphological process of the pupal moult does not start until five days later. The colour changes are completely halted by ligaturing the part of the body lacking the MH. When 66 *Calliphora* units were injected into such a permanently green abdomen, they cause reddening of the epidermis. Three hundred and thirty Calliphora-units caused reddening of the fat body only, whilst very large doses, 330 to 6600 units caused pupal moulting without any colour change. Buckmann (1947a) concluded that there the effect of ecdysone could be due to a simple reduction of the ommochrome pigment and he suggested that 'das ware ein idealer Modelfall einer chemisch einfach definierten Hormonwirkung'.

Even though an alternate, direct dependence of a pigment reaction on the hormone cannot be excluded, the following explanation seems to be the most probable : the effect of MH consists, as in other-metamorphosic processes, in the conditioning of species specific chemical and morphological changes, in the normal course of development. The discovery that removal of the ca (by decapitation) in the earlier, larval instars result in similar changes in the larval periodfullly accords with this.

Doubtlessly a similar dependence would be found in all other cases of morphological colour change described for various other insect species, with an experimental approach similr to that used by Buckmann with *Cerura*. It is confirmed from the experiments of Wigglesworth (1940) with *Rhodnius* that there is no direct influence of MH on pigment vanishes from in some places where it was earlier present.

The correlation between the prothoracic (= ventral) glands and phase development

This was examined in *Schistocera gregaria* and *Locusta migratoria*, The prothoracic glands which gemerally disappear in adults were found to sruvive for a longer period in the solitary than in the gregarious phase. Besides this the ca are larger in solitary than in gregarious nymphs. Extirpation of the greater part of one of the pair of the prothoracic glands in solitary nymphs resulted in the assumption of the gregarious habit. It is accepted that a complex interaction exists among all three metamorphosis hormones in fluenceing phase differences.

The action on the development of internal parasites

The group A virus *Sindlois arbovirus* is known to reproduce during the metamorphosis of *Calliphora erythrocephala*. In larvae deprived of their MH source by ring gland extirpation, no multiplication of the virus takes place.

The Control of MH Production

The secretion of the MH by the pgl and their equivalents is influenced by the presence of a certain effective concentration of AH in the hemolymph. It is therefore the control-system of AH which, initially, also controls the production of MH. It was shown by the experiments of Williams (1952) that pgl may be activated to produce MH even in the absence of AH by the introduction of MH. Thus, diapause in *cecropia* pupae may be broken by the implantation of active pgl or by the injection of ecdysone into pupae with inactive pgl, even in an amount which itself would not be sufficient to initiate the moulting process (in specimens with their pgl removed). One the basis of these experiments Williams suggests that co-ordination of secretory activity between the left and right pgl is definite. Furthermore, extensive or repeated injuries and the regeneration resulting from these injuries are brought about by the reactivation of pgl. They may even in some conditions cause supernumerary moulting in adult insects. But, such effect is probably restricted to the phylogenetically 'lowest' groups of Hemimetabola such as cockroaches, not yet every far remote from the Apterygota where moulting persists in adult insects.

New information on this subject is available in recent years. Zdarek and Fraenkel (1971) studied neurosecretory control of MH release during pupariation of *Sarcophaga argyrostoma*. In an earlier paper they demonstrated that there is strict environmental control of the initiation of the pupariation process in Cyclorrhapha. The larvae do not pupariate unless they leave their wet feeding medium and spend one or two days in a dry environment. However, implantation of the cerebral and ventral ganglion, together with the rgl, induced pupariation even in wet larvae. If the rgl alone was implanted, only larvae which has spent some time (insufficient by itself) under dry conditions pupariated. Neural connection with the cns proved to be essential for an independent pupariation effect. The activity of the rgl consistently increased in the pre-pupariation period in active flies.

Gersch and Sturzebecher (1968, 1970,) carried out a detailed examination of AH and its effect on the pgl in *Periplaneta americana*. The pgl were experimentally activated by one larval brain of *Antherea pernyi* of three brains of *Periplaneta americana*. After 8 hours' exposure to AH, a substance was produced in the pgl. Thin-layer chromatography and mass spectrometry have confirmed this substance to be a non-separated mixture of stearic, oleic, linoleic, linolenic, palmitic and myristic acid. According to further discoveries on this subject, some juvenoids were found at activate diapausing or decerebrated *Antherea polyphemus* and Hyalophora cecropia pupae. Their effects on the pgl were proportional to their juvenilizing activity.

Sehnal and Edwards (1969) examined the inhibition of pgl activity body constraint in *Galleria mellonella* larvae. Constraint inhibited AH secretion and the authors interpreted this a nervous inhibition (possibly owing to the elimination of an environmental nervous impluse). MH production was also inhibited, although the epidermis remained capable of its moulting function (demonstrated by the administration of ecdysterone). This indicates that an appropriate external or internal stimulus is needed for activation of the neuroendocrine system.

Ohtaki *et al.*, (1968) examined the rapid decrease in MH activity in the insect organism after hormone administration. They concluded that a special inactivating mechanism was present in the tissues, but not in the haemolymph. When injected into mature *Sarcophaga peregrina* larvae, 1 Mg ecdysone lost 50 per cent of its activity in 1 hour and by 8 hours only 2 per cent remained. In the critical phase for pupariation, the larva contained only 7 per cent of one *Sarcophaga* unit. According to the researchers, this could have been because of accumulation in the tissue of convert effects determining the overt response. No accumulation was observed in the haemolymph. A more practicle explanation, however, is that early change in the target tissue (the ectodermal cells) are reversible up to a given point of time, until then the presence of MH is indispensable; but not later than that.

Chemical Characteristics of MH

Isolation of MH

The first prothoracic gland hormone of the insect metamorphosis hormones to be prepared was in crystalline form. Butenandt and Karlson (1954) used 500kg of silkworm pupae to prepare an extract using the following technique.

The pupae preserved in methanol were subjected to a pressure of 50atm and the resulting extract reduced by evaporation in *vacuo*. The 30liters remaining from about 650 liters of the original extracts were shaken several times with butanol. The separated butanol extract was then washed successively with 1per cent H_2SO_4, 10 per cent Na_2CO_3 and 5 percent acetic acid. It was evaporated and the remaining red-brown oily substance (79g) was redissolved in butanol, and after filtration through about 800g alluminium oxide, following Brockmann (activity V), re-evaporated and dried in *vacuo*.. It was then dissolved in water, shaken successively with ether and ethyl acetate and the water phase again evaporated. The remaining 1.15g of red-brown oil showed and activity of 1 *Calliphora* –unit in 0.5g. It was then purified by chromatography in aluminum oxide activity IV and, using ethyl acetatebutanol (3:1) and ethyl acetate-methanol (3:1), 167 mg of a highly active fraction (1 Calliphora-unit in 0.1μ 1) was obtained. After recrystallizing from ethyl acetate and methanol, 25mg of crystalline substance was obtained which depicted an activity of 1 C.u. in 0.0075 Ml. It is called endysone and has been identified with MH.

Chemical characteristics

This crystalline preparation formed silky fibrous needles. These change their appearance at 161-162°C probably because of the loss of water of crystalization, and melt at 235-237°C. Within these limits the substance is heat stable; it is soluble in water and very soluble in methyl alcohol; it shows distinct absorption bands in the ultraviolet and infrared parts of the spectrum. The ultraviolet spectrum reaches its maximum at 244nm (in ethyl alcohol) and at 249 nm (in water); and the infrared at 612 nm. It is optically active:

$$(d)\frac{20^\circ}{D} = 58.5^\circ \pm 2^\circ \qquad (d)\frac{21}{546} = 82.3^\circ \pm 2^\circ$$

(at a concentration of 7 mg in 2ml ethyl alcohol).

Depending on these findings the presence in the molecule of α and β unsaturated keto groups was assumed. The infrared part of the spectrum also suggests the presence of at least two hydroxyl groups. It was further inferred that a $_2CH_5$ group and perhaps also a Ch_3 group are present. Both acids and bases easily inactivate the substance. There is another band in the spectrum at 265 nm which may be assumed to have originated by the splitting off of water which led to the formation of the double unsaturated system.

The hormone decomposes quickly in alkaline solutions, the decomposition being accompanied by a successive change of spectrum. The first analysis of the dried substance showed elements in the following ratio: C_4, $_4H_7$,$_3O_1$. The molecular weight was originally shown to be about 300; and the emperical formula $C_{18}H_{31}O_4$. Concentrates from Crangon vulgaris which were found to be more active in crustaceans than in insects point to the existence of several different chemical and physical characteristics (e.g. increased polarity) in the crustacean hormone. Further research has confirmed that at least two differently substance are present with identical physiological activity, which are separable into α- ecdysone and β-ecdysone (now ecdysone and ecdysterone, by extracting with a cyclohexane-butanol-water mixture. The two substances may be separated by their melting points and solubilities though their spectra are very similar.

Karlson and Hoffmeister (1963a) worked incessantly on the chemical properties of ecdysone. 1000 kg of dried pupae of *Bombyx mori* removed from cocoons obtained from the silk factory were used. From these 250 mg of crystalline ecdysone with a specific activity of 100 000 *Calliphora*- units per mg was obtained.

With the help of X-ray analysis and mass-spectroscopy a molecular weight of 464 was confirmed. This corresponds to the empirical formula $C_{27}H_{44}0_{6}$. Evidence was obtained of the steriod nature of ecdysone, one oxy-group and five hydroxyl-groups were found. On the basis of the empirical formula a relationship with cholesterol was suspected. Injection of tritium-labelled cholesterol into *Calliphora* larvae yielded crude radioactive ecdysone after extraction and purification. It was then confirmed that cholesterol is a precursor and ecdysone. A similar observation was made by Kanazawa and Teshima (1971) with C^{14}-labelled cholesterol in the spiny lobster *Panulivus japonicus*. The activation of the prothoracic glands in decerebrated *Bombyx* by ether of brains and by cholesterol, found by the Japanese authors, could probably be explained by this. It is no way contradicts the fact that cholesterol is inactive in diapausing sawfly larvae.

The substance proved to be active not only in various Lepidioptera but also in Hymenoptera, and Hemiptera Karlson (1957) later succeeded in isolating a similar substance from the shrimp *Crangon vulgaris*, using the same method. The *Calliphora*-test shows a distinct, but much lower activity in this substance (1 C.u. = 50 mg of the dried oil;. Similar preparations have been obtained from *Carcinus moenas* and *Portunus holstaus*.

Almost similar and yet more active MH preparation has recently been obtained by Stamm (1958) from the grasshopper *Dociostaurus maroccanus*. The researcher used 10 kg of adult specimens which yielded 11 mg ecdysone and 12 mg ecdysterone in crystalline form. The substances were shown to be chemically identical with the silkworm α-and β-ecdysone and were found to be more active in *Calliphora*–tests. The reduced activity of the silkworm preparations is assumed to be due to the presence of kynurenin and others substances harmful to *Calliphora*.

A number of research papers by the Karlson school deals with the question of the chemistry of the MH and the biochemistry of its production. On the basis of their work the definitive solution of the chemical structure of ecdysone was made by Karlson (1965). As stated earlier, it is a steroid molecule of the following structure:

It is a 2β, 3β, 22β F. 25=pentahydroxy-Δ7-5β-cholesteron-(6). The C_{20} has the normal steric configuration, i.e. 20β F-methyl 6 –keto-0 =0 –1; 14α - 0 = 0-2, 2β-0=0_3, 3b0=0-4,25-0=0-5,22-0 = 0-6.

Five various biologically active fractions of the ecdysone are assumed to have been obtained by chemical treatment of *Bombyx mori* extracts, three of them being undescribed and two known from the work of Karlson and his colleagues and earlier from that of Burdette and Bullock (1963). A positive effect on tissue growth in mammals is claimed together with stimulation of protein synthesis in the cytoplasm.

Immediately after Karlson's original formula for a-ecdysone was discovered, the compound was simultaneously synthesized in laboratories in West Germany and the U.S.A. (Furlenmayer *et al.*, 1966a, b, 1967; Siddall *et al.*, 1966a,b) and its structure fully confirmed. Furlenmayer *et al.* started with ergosterin and used 14 steps. The main steps were (20S)-2, 3β-diacetoxy-20-formyl-5α-pregn-7-ene-6-one and (22R)-2β, 3β-diacetoxy-14-22-dihydroxy-25- (tetrahydropyran-2-yloxy-)-5α-cholest-7-ene-6-one. Later ecdysterone was also synthesized.

As soon as the chemical structure of ecdysone and its artificial synthesis were resolved, the number of studies on its various chemical properties and those of its many natural and synthetic derivatives rapidly multiplied.. The most complete survey is the latest one by Slama *et al.* (1973), which is yet to be published.

The Mode of Action of MH

Inspite of the fact that the chemical composition of the moulting hormone has now been fully determined, the mode of action of this, the best known insect hormone, is still not properly understood. The way it affects the cytochrome oxidase system has already been mentioned, as has it action on the ommochrome changes. Its effect on development of the epidermal glands in *Ephestia kuehniella* has been examined experimentally by Hanser (1957), using the ecdysone preparation of Karlson and Butenandt. His paper shows : (1) The injection of MH at any time during the intermoult period induces the moulting process. This commences with the loosening of the old cuticle and the secretion of the moulting fluid, and continues with the formation of a new cuticle (pupal cuticle). (2) A single dose, even if rather strong (40 C.u. or more), is inadequate for the completion of the moulting process. If no further dose is supplied the moulting process stops in one of its later stages and can only persist when more ecdysone is injected.

It can be confirmed from the findings of Hanser (1957) and the results of earlier experiments with parabolses, and pgl transplantations that MH influences the activation of all the enzyme of the epidermal cells. The first phase of this activation consists of the swelling of the cells, which may or may not result in mitotic activity. It is followed by the second phase, the secretory activity of the cells, resulting in the production of moulting fluid, the initial effect of which is the loosening of the old cuticle by chitinase which subsequently dissolves the inner layers of the cuticle. In the third phase, the new cuticle is deposited and the secretory activity of the cells continues yo produce the first layers of the new cuticle, most probably under the action of the same enzyme (chitinase). The phases generally overlap one another. It seems that the presence of MH is imperative at least to begin all these phases.

The *consequence of thiourea injections* on moulting and pupation in silkworm (*Bombyx mori*) larvae were exmined by Chmurzynska and Wojtczak (1963). Injections up to 24 hours before larval moulting or cocoon spinning resulted in the retention of the old cuticle and accumulation of a haemolymph-like fluid between the old and the new cuticles. Partial inhibition and a delay in the sclerotization of pupal cuticle was observed in histological sections.

Advancement in the technique for isolating MH is small quantities have made it possible to compare the MH titre at various stages of an instar and in various periods of postembryonic development (Shaaya and Karlson, 1964; Shaaya and Sekeris, 1965; in *Calliphora erythrocephala*). There are two maxima-one just prior to pupation and the other on the 3rd to 5th day of the pupal instar. The first peak correlated with puparium formation, and the second with histogenetic processes. No ecdysone could be detected on the last two days of the pupal instar. The same method was employed to examine the activity of several enzymes was studied in a similar manner.

Many other research investigations were carried out to examine, the ecdysone titre in relation to body weight, in two lepidopteran species-*Bombyx mori* and *Cerura vinula*. In *Bombyx mori* a maximum was found in the IVth (penultimate) and Vth larval instar shortly before ecdysis; immediately before ecdysis occurred MH fell to zero. Two maxima were observed in the pupal instar. Some ecdysone was still present after adult ecdysis (about four times more in the female than in the male), but by the 8th day it had virtually disappeared. Two maxima were found in *Cerura vinula* - one at the beginning of the colour change and the other shortly before pupation. No ecdysone was detected in the pupal instar. But, the information may not corre;ate to the amount of active MH in the haemolymph. Adelung (1966) estimated MH titres similarly in the crab *Carcinus moenas*, and discussed their variation in the context of the crustacean endocrine system, i.e. the X organ-sinus gland complex and the rostral gland (Y organ).

The number of individual secretory and morphogenetic processes which take place in the epidermis during moulting cycle and their a hormonal dependencies were discussed by Locke (1965). From the results of experiments on the moth *Calpodes ethlius* (Hesperidáe), Locke asserted that whereas the moulting process is controlled only by ecdysone, intermoult events such as wax secretion and endocuticle deposition need the assistance of a neuroendocrine factor from the brain. A combination of recent views and findings on moulting control was published by Weir (1970), who on the basis of his experiments on *Calpodes ethlius* classified three phases of development in the last (Vth) larval instar. The first is the growth phase, from ecdysis up to 66 hours. After that the brain is no longer required for the induction of pupation, and the next phase, up to 156 hours, ends when the pgl are not longer necessary. Pupation takes place at 192 hours. From an analysis of the data on the pgl, the researcher concluded that certain abdominal tissues itneracted with the pgl in the induction of the moulting process and stated that the oenocytes might be invovled.

The Juvenile Hormone (Neotenin) (JH)

The metamorphosis-preventing activity of the corpus allatum was initially demonstrated by Wigglesworth in his classic parabiosis experiments with the blood-sucking bug *Rhodnius prolixus* in 1935. Wigglesworth (1936) demonstrated that joining a nymph in the penultimate or an earlier instar with a nymph at the beginning of the last larval instar results in the partial or total suppression of metamorphosis without preventing an increase in size. In this way VIth instar giant nymphs imagos. Conversely, when the ca of the younger nymph was removed by decapitation, a miniature precocious imago resulted.

Wigglesworth not only the source of this inhibitory effect but was also the first to provide definite evidence of its hormonal nature by demonstrated that the same results are produced when the specimens are connected by a glass tube so that only the haemolymph communicates. The original term 'inhibitory hormone' was soon changed to 'juvenile hormone' after the active role of the hormone in producing larval characters in adults was recognized. This discovery was confirmed by a number of other authors for various insect species,

Bounhiol (1936, 1938a) for *Bombyx mori*; Pflugfelder (1937a, b) for *Carausius morosus*; Piepho (1938b-d) for *Galleria mellonella*; Bodenstein (1938b) and Weed-Pfeiffer (1938) for *Melanoplus differentialis*.

Contemporary researcher, Wigglesworth (1935, 1936) also discovered the effect of the ca hormone on ovarian development. This was later confirmed for a number of other species whilst in ohters, such as *Bombyx mori*, no ovarian control bya corresponding humoral factor was observed. The other JH effect were ascertained subsequently. The apparent contraction betwee the inhibitory, metabolism increasing influence, suggested to a number of researchers that there were two or more ca hormones having different effects. However, this hypothesis theory was not subsequent experiments, and it was soon recognized that all the available data could be explained by the operation of a single substance, JH, possessing the character of a growth hormone. Since then experimental observational evidence has accumulated in favour of this explanation. Interordinal non-specificity of JH was shown by the transplantation of ca between Blattoptera and Hemiptera, and hasmida and Lepidoptera. The first attempts at elucidating the chemical nature of JH were made by Williams (1956), Schneiderman and Gilbert (1958, cf. Schneiderman, 1960), using ether extracts of male abdomens of the Saturnid *P. cecropia*. Its chemical structure was first determined by Roller and his co-workers (1965a, b etc.). Many other chemically different substance, i.e. the juvenoids behaved in the same manner.

The Corpora Allata (ca)

The first description of the insect ca is probably that of Muller (1828), in *Blatta orientalis*; however, he expected the glands to be nervous in character, believing visceral ganglia supply the dorsal aorta and the anterior portion of the gut. This analysis was accepted by a number of other researchers. The first doubts regarding the nervous character of the ca were expressed in 1899 by Janet, from his detailed anatomical study of the female of the ant *Lasius niger* and to whom we owe the term corpora allata. In the same year, Heymons (1899) also concluded that the ca in *Carausius morosus* were non-nervous. The first systematic, comparative morphological study of the insect ca was carried out by Nabert (1913) the number of research papers dealing with their structure and function has been published in great number.

Morphology

The ca are found in all groups of winged insects in which they have been searched for, and in some Apterygota, where they are very primitive. Several different types of ca can be distinguished, on the basis of structure, position, relationships with other endocrine glands and phylogenetic considerations. The simplest and most primitive type known are the bladder-shaped ca of Thysanura and stick-insects, which in some species still keep their original position on the lateral epidermis. They are followed by the paired lateral ca with cc, as in Ephemeroptera, Odonata etc. The direction of development at this stage is through the approximation and later fusion of the ecc, as in Blattoptera, Orthroptera, Mecoptera and some of the Diptera, e.g., *Culicidae*. The other tendency was to fuse the ca which can take place either below the dorsal aorta as in Hemiptera, or, above it, as in Tabanidae. A special type with both cc and ca fused and connected by the prothoracic glands fused with the nervi allati to form a single structure, is represented by the ring gland of Diptera Cyclorrhapha.

Cyclical changes occur in the ca in each larval instar as well as in the adult. The ca in the plant-feeding bug *Oncopeltus fasciatus* have their smallest volume at the beginning of each larval instar and increase gradually in size to reach a maximum at about the middle of the intermoult period. Then they reduce in size with an associated decrease in the secretory activity before the next moult, the minimum size at the beginning of each instar being distinctly less than the maximum in the earlier one. The increase in size of the ca in each larval instar thus has two components—an increase due to secretory activity (i.e. the difference between the maximum volume during an instar and that the next ecdysis), and growth of the gland itself (i.e., the difference between its volume at one ecdysis compared with the previous one). In the last larval instar a distinct increase in the volume of the gland also occurs, but is comparatively less than in the earlier instars. In adult stage a clear increase in the size of the gland byins 3 to 4 days after imaginal moulting and continues for the whole period of egg-laying (from the 8th and 30th day), when the volume of the ca increases by about 30 times compared with that immediately after moulting. Legay (1950) found a distinct increase in the volume of the gland in *Bombyx* at the time of each larval ecdysis and a very small increase during the intermoult periods. This type of growth is probably connected with the displacement of the critical period for AH and thus, necessarily, also of that for JH towards the end of the intermoult period. It is a secondary adaptation in the phylogenetically 'hightest' insects, connected with the shortening of the moulting process.

The development of the ca in *Carausius morosus* and in both larvae and adults of the honey bee (*Apis mellifica*) was systematically examined by Pfulgfelder (1958). He found a remarkable increase i the volume of the ca of workers bees in the first days of imaginal life, at the height of activity of the pharyngeal glands (between the 6th and 12th day), in connection with the first orientation flight (9th and 10th day), and at the time of maximum secretion of the wax glands (14th and 15th day), followed by a slow decrease in volume. The ca of the queen larvae are larger than those of workers larvae during their entire process of development but especially during the pupal instar, which perhaps is connected with the development of the ovaries, as confirmed by Lukoschus (1955). In the adult queen their volume remains practically constant, distinctly below the maximum in the workers.

The increase in the volume of the ca between the first and last larval instars in 18 different species of both Holometabola and Heterometabola was examined by Novak (1954). It is comparitively lesser than the increase in volume of the body, but greater than the increase in volume of the brain and nervous ganglia. Kaiser (1949) analyzed the relationship between the growth of the ca and the prothoracic glands (pgl) during larval development in Lepidoptera (*Pieris*). Whilst the ratio was 1:2 in favour of the pgl at the moment of hatching it increased to 1:29 at the time of cocoon spinning. During the last larval instar the volume of the ca increased very little whilst that of the pgl was doubled. The conclusions of this researcher was based on the hypothesis of antagonism between MH and JH have found little support. We must also keep in mind that even if the volume of the pgl does increase, compared with that of the ca, it definitely does not do so compared with the body as a whole. It was recently shown by Mala et.al (1975) that the initial indication of pgl degeneration appear soon after the last larval ecdysis because of the non-existence of JH, and can be reversed by applying JH. The moulting process is distinctly accelerated in the presence of active ca therefore it is not possible that should be any antagonism between the two hormones.

In *Hyalophora cecropia* adults, and extraordinary sexual dimorphism in the size of the ca was discovered by Schneiderman and Gilbert (1959) in contrast to that in bugs and most other insects. Whereas the pupal ca show the same weight of about 30 μg in each sex, those of adult males increase to about 400 μg whereas to those of the female adults only increase to about 65 μg. This is in agreement with the JH activity of ether extracts of male and female abdomens, that of the female beig only about 1/40th that of the male. A similar variation is observed in the ether extracted liquid (0.201 g in the male and 0.045 g in the female), which is correlated with the much greater flight activity of the male. Similarly, Rehm (1951) found a 20-fold increase in the ca adult male *Galleria mellonella*, whereas the increase in the female was only 3-fold. But the number of cells per ca is, almost equal in each sex.

Histology

The ca are sphericae, ovoid or pad-shaped bodies in most insects, connected with the cc by a nerve (nervus allatus) which may become reduced so that the two kinds of glands are closely approximated or, in some cases, partially fused. They are enclosed in a mesenchymatous connective tissue membrane formed by plate-like cells with an oblong nucleus and homogeneous surface layer. Mendes (1948) found the following three types of glandular cells in the ca of *Melanoplus*—undifferentiated cells, normal secretory cells and polyploid giant cells. A remarkable feature of the glandular cells of the ca is that they increase in size considerably during the period of their secretory activity and shrink during their inactive periods. The period with histological signs of secretory activity coincides with that of their hormonal activity as shown by parabiosis and transplantation experiments. There is no histological proof that more than one ca hormone is produced in any of the species studied.

The ca can be classified into four groups, according to their histological structure:

1. *The epithelial vesicular type*, consists, in a central cavity, secretion (as in *Aeschna*), connective tissue (as in *japyx*) of a system of concentric lamellae (as in *Phasmidae*). This is the original phylogenetically 'lowest' type (even though secondarily simplified in some cases), showing the ectodermal origin of the ca.
2. *The pseudolymphoid type,* structured by small cells with reduced deeply staining cytoplasm resembling that of lymphoid tissue in vertebrates. Ca of this type have been discovered in Ephekerida and Odonate.
3. *The small cell type,* which is the most common, is found in Blattoptera, Orthoptera, Hemiptera, Coleoptera, etc. It is structured by numerous small irregularly shaped cells, in the intercellular lacuane of which, secretion accumulates during periods of activity. The nuclei are egg-shaped to oblong their cytoplasm being relatively clear.
4. *The large cell type,* phylogenetically the most evolved, and it is found in Panorpata, Hymenoptera, Trichoptera, Lepidoptera and many Diptera. The gland consists of few large cells with large, often labular nuclei and numerous uniformly distributed chromatin granules.

The histophysiological changes in the ca in adult *Leucophaea maderae* were examined compreyhensively by B. Scharrer and Harnack (1958) and Harnack (1958a, b). No great changes in the low secretory activity of the male ca were observed during the whole adult life except for a short 'activation period' just after imaginal moult. While in females significant changes occur in connection with the ovarian cycles. A conspicuous increase in the amount of cytoplasm, both absolute and relative (in relation to the size and number of nuclei), occurs during the activation period. The researchers describe the changes minutely. Harnack (1958a, b) also observed the effect of starvation on imaginal moulting. The ca volume decreased markedly shortly after a limited increase at the begining of the instar. Feeding following a long period of starvation results in a marked increase in the volume of the gland which greatly exceeds that in normally fed specmens, not only in the first ovarian cycle but also in the later ones. The researcher indicated tthe possibility of resemblance with the adenhypophysis in vertebrates.

B. Scharrer (1956, 1958) examined the impact of ovariectomy on the ca in *L. maderae,* and observed an increase in the volume of the glands, the number of nuclei, and an increase nucleocytoplasmic ratio following the operation. Similar changes follow denervation of the ca (interruption of the nervi corporum cardiacarum). The researcher calls attention to the analogy with the neuroendocrine cycle in vertebrates. The succession of developmental changes in the ca in the adult bug *Oncopeltus fasciatus* was analyzed by Johansson (1958a) and that in the pupae of *Mimas tiliae* (Lepidoptera) by Hughnam (1958b).

The disparity between ca from adult females which inhibited metamorphosis (induced 'supernumerary' larval moults) and those from adult males, which failed to produce this effect, was found by Fukuda (1963) in the silkworm (*Bombyx mori*). This sexual difference in the activity of the ca persisted after gonadectomy, after both gonadectomy and reimplantation of the gonads of the opposite sex, and even in ca which were transplanted into another specimen after the biginning of the pupal period. The activity of the female ca was not affected by decerebration at the beginning of the pual period.

Akahira *et al.* (1967) examined the effect of the ca in 32 species of stingless bess (genera *Cephalotrigona*, *Trigona scatura*, *Plebeia*, *Partamona Nannotrigona* and *Melipona*) and two subspecies of the honeybee (*Apis mellifera adamsonii* and *A.. m. ligustica)* and observed their biometry (size, number of nuclei, volume, etc). and histological characteristics. They agreed in many characteristics, but differed in others. In general, diversity was less in species of the same genus than in related genera.

Secretory activity of ca was observed histologically during sexual maturation and imaginal diapause in *Leptinotarsa decemlineata* and *Grullus domesticus*. Changes in the size of the ca were observed in association with different stages of glandular secretion. In both species the secretory cycle was further classified into three periods- a pre-secretion stage, a stage of nuclear activity and a stage of production and release of the secretory substance. The ca of constrated females displayed hypertrophy and accumulation of hormone, which in later stages could result in degeneration.

A process of the volumetric measurement of the ca was expained and used to an^lyse the relation between the size of theca and their activity, determined by the *Locusta* colour test. A close correlation was found. Lea and Thomsen (1969), however, found no correlation between the size of the ca, their activity and egg maturation. Activity was judged by the neurosecretory cell nuclear test.

Ultrastructure

Electron microscopic examination of the ca and its fuctional changes began with the work of B. Scharrer (1961, 1962) on adult females of *Leucophaea maderae*. The secretory cells interlock by mean of long cytoplasmic processes. Their cell membrances are folded in a chracteristic way in inactive glands. In active glands the cytoplasm of the cells becomes more abundant, the number of mitochondria and ribosomes increases, an endoplasmic reticulum and very numerous granular inclusions appear. The gland is encased by a thick membrance from which processes of varying thickness branch off and penetrate the spaces between the secretory cells. These processes form a 'stroma', containing denser bodies, through which the neurosecuretory axons of the nervus allatus penetrate the ca. the 'stroma' seems to be continuous with the adjacent cc. The researcher suggests that such process serves as a pathway through which food and other chemical are exchange between the gland and the haemolymph, including diffusion of the hormone.

Subsequently, electron microscopic studies of various insect species of different orders have been published with reference to the secretory activity of the ca, on the original elaboration, storage and release of the 'C body', a specific cellular product in the ca of *Leucophaea maderae*. 'C. bodies are found or hexagonal structures of varying size, composed of highly electron dense material. They orginate in the Golgi zone and are slowly transported to various intracellular and extracellular compartments affording them access to the haemolymph. From their incidence, distribution and fat in the insect organism, it was assumed that C bodies might have a bearing on the secretory function of the ca. Special attention was paid to the acellular stromal matrix, which seems to act as a vehicle facilitating the transport of C bodies to the haemolymph. C bodies might correspond to the 'osmiophilic bodies' described in the ca of the adult *Bombyx mori* female.

Number of research of papers on the ca of *Locusta migratoria*, their activity and the impact upon the glands of the A, B and C ncs of the pars intercerebralis was published by Joly and his co-workers. Abundant ergastroplasm was observed in actively secreting cells, which disappeared from cells in the non-active phase while the smooth endoplasmic reticulum showed conspicuous accumulation.

With reference to the previous discovery, Panov and Bassurmanova (1970) found an entirely different structure in the ca of the bug *Eurygaster integriceps*. Here, the smooth-surfaced endoplasmic reticulum was entirely absent in inactive cells, which were chracterized by diminished size, reduced cytoplasm and numerous ribosomes. The most pronounced feature of active cells was a large nucleolus formed of hollow spheres like the one observed by Blazek *et al.* (1973) in pgl cells of *Galleria mellonella*, while in inactive cells it was irregularly shaped with small vacuoles. the researcher discussed the possible fuctional significance of these changes in connection with similar observations in vertabrates (in the egg cells of *Protopterus ethipoicus* and in certain blood cells and liver cells of the rat). In addition, ultrastructural changes in the ca cells of specimens infected by larvae of the parasitic fly *Clitimyia helluo* were described.

The ultrastructure of the ca was closely scrutinized by Deleurance and Charpin (1972) in cavernicolous beetles of the family Bathysciinae, by Odhiambo (1966a, b) in *Schistocerca gregaria*, by Waku and Gilbert (1964) in *Hyalophora cecropia* and by Fukuda *et al.* (1966) in *Bombyx mori*. The ca of *calliphora erythrocephala* female adults were examined ultrastructurally by E. and M. Thomsen (1969), with reference to hormone production. It was acceeeoted that the hormone or its precursoe originates in the smooth endoplasmic reticulum (which was found to be very abundant at the stage of incipient activity) and that it dissolves in lipids, finally forming fat droplete which are thought to be gradually extruded through the surface of the ca. The fat droplets seem to correspond to the osmiophilic bodies of Fukuda and the C bodies of B. Scharrer. An exhaustive comparative electron microscopic study of the ring gland in the larva and of the cc-ca complex in the adult female of *Drosophila mellanogaster* was carried out by King *et al.* (1966).

Embryogenesis

As opposed to earlier belief that the ca was of mesodermal or even of endodermal, origin (cf. Nussbaum, 1889), their ectodermal origin now seems to be confirmed. Mellanby (1936) proved that they originate as ectodermal invaginations in the area between the mandibular and maxillary pleurites in *Rhodnius prolixus*. The original paired rudiments fuse into a single median body in Hemiptera towards the end of he embronic period. Abnormalities due to incomplete fusion of the to embryonic invaginations, sometimes even showing two separate bodies, were often discovered by the resarcher in the plant-feeding lygaeid bug *Oncepelus fasciatus*. An increased number of such aberrations, amounting to 20 percent, was observed in animals kept at higher temperatures. It is suggested that they are the result of a selective influence on the process of fusion exerted by the increae in temperature. Similar changes have been observed in *Pyrrhocoris apterus*. The same origin was found by Pflugfelder (1937a) in Carausius morosus where the glands retain their original vesicular shae both in nymphs and adults. The embryogenesis of the ca as a part of the ring gland was elucidated by Poulson (1945a, b) for *Drosophila*. Their origin from ectodermal invaginations in bugs is also stated by Wells (1954).

Phylogenesis

Ca has been discovered in all groups of metabolous insects (Pterygoata) where they have so far been looked for. In Apterygota, they are fairly well developed in Japygidae and in a more primitie stage of development, in a lateral-position closely connected with the mandibulo-maxillar area of their origin, in *Ctenolepisma*. A distinct juvenilizing effect of the glands of *Ctenopisma* implanted into *Cecropia* pupae has been observed. Keeping in viw the earlier suggestions of the possible homology of the ca with the 'crops jugales' in Thysanura must be considered. It is probable that the original function of the ca was the activation of the ovarian follicles and that only gradually did they gain their function in larval development which led to the origin of Pterygote insects. The structure and development of the ca in the larvae of higher Diptera is quite complex. Here they form a part of the ring gland (1) the dorsal fusing of the originally paired ca; (2) the fusing of their nervi allati with the neighbouring pericardial glands; (3) the strengthening of the lateral arms and the dorsal part of the gland with tracheal trunks; (4) the bending of the ring with its upper part forward by a backwards shift of the brain. As pointed out by Cazai (1948) the formation of the retrocerebral endocrine system corresponds in a remarkable way to modern views on the phylogeny of metabolic insects based on the system of Martynov (1937) and Jeannel (1949).

The Direct Effect of JH

Ever since Wiggleworth (1935, 1936) discovered JH and demonstrated that it inhibited metamorphosis and activated the ovarian follicle cells, many rearcher papers have been written on various effects of the ca hormone and its different analogues on the most diverse structures and funcions of the insect organism, on its various biochemical processes and on its chemical composition. Actually taking the entire insect organism as a whole, and not its functions, structures, chemical processes or components has been found to be independent of JH.

The inhibition of metamorphosis

As it has been already stated, implantation of the active ca at the beginning of the last larval instar partially or completely prevents metamorphosis and results in adultoid forms or supernumerary larval instars. These extra instars (giant larvae) may actually change into giant adults at the next moult. The removal of the ca by decapitation or extirpation in earlier larval instars is followed by a precocious metamorphosis resulting in miniature adults, preceded (in Holomeabola) by miniature pupae. In most Hemimetabola only one instar occurs following allatectomy (e.g. Hemiptera), whilst in a number of the phylogenetically 'oldest' groups, such as Blattoptera and Phasmidae, two more moult appear following the operation, a special 'pre-adultoid' instar preceding adults.

The suppression of metamorphosis connected with the positive effect upon larval (isometric) growth is the quite obvious and important effect of JH one of the many features. Implantation of the c markedly accelerates the moulting process. For example, the intermoult period of the last larval instar in *Rhodnius* takes 24 days, whilst following the implantation of active ca at the beginning of the instar the animal moults in 17 day. On this Wigglesworth (1936) based is original concept of the JH effect, applying Goldschmidt's theory of differential velocities (1923). He finally came to the decision that morphogenesis could not be completed because of the acceleration to the moulting process.

Such an effect really appears to exist but only till a limit in the case of the so-called progressive prothetely.

The effect of the JH on the growth and morphogenesis of internal organs

Sehnal (1965, 1968) in *Galleria mellonella*, fully accepted that JH affected the surface structures. The implantation of three complexes of brain-corpora cardiaca-corpora allata at the beginning of the last larval instar completely inhibited metamorphosis. Later implantation resulted in a series of transitions between the larval and imaginal shape of the brain and other organs.

Hlinak made the same discovery (1968) in the brains of a continuous series of transitions between a supernumerary larva and a normal adult in *Periplaneta americana*. Other similar results on the internal metamorphosis of the imaginal compound eyes were confirmed by Mouze and Schaller (1971) and Mouze (1970, 1971) who distinguished it from the metamorphosis of exteral structures,in which mitotic activity (and hence differentiation) depends on the moment at which an active MH concentration is reached. Only from that moment can JH action take effect, correlated with its particular concentration in the haemolymph. On the other hand, in the MH-independent organs, the moment at which only a minimum active concentration is reached is decisive. The results are also confirmed by Riddiford (1972) in *Hyalophora cecropia*.

Depending on corcumstances JH appears also to be capable of accelerating metamorphosis. At least this would be the most probable explanation of some of the observations of Wigglesworth (1948a), according to which the removal of the ca at the beginning of the last larval instar results in somewhat nymphoid adults. From this it would follow that the presence of JH in a concentration below the morphogenetical threshold is necessary for complete imaginal differentiation. On the other hand, the positive effect of the implantation of last instar ca on imaginal differentiation in earlier nymphs appears to have another reason.

The introduction of an active ca into an adult by implantation or parabiosis can result into a partial suppression of the imaginal characters, when an artificial imaginal moulting is induced simultaneously, either by parabiosis with the nymphs, or by implantation of active pgl, or by the injection of ecdysone or by exposing a part of the imaginal epidermis to the action of the larval hormone as in the epidermal bladder method.

The effects of JH on the indirect muscles of *Leptinotarsa* corroborated with concept of all the other effects of the hormone. There is, however, an exception—the experimental results of the action of JH and JHa on the indirect flight musculature of the cricket *Gryllus domesticus*. Here the impact of the hormone is the exact opposite of observations elsewhere: the muscles ripen without J.H. and its administration causes their degeneration. The flight musculature persists in allatectomized females while the reintroduction of JH or JHa, at any time, causes its rapid degeneration. Castration of females in the last larval instar affects neither maturation nor degeneration of the muscles. In the last larval instar the action of the hormone is the same as in all other insects. This is the only one of the yet expained paradoxical effects of JH. At present, the probable theory would be that the presence of the hormone acts as a signal for nervous action causing flight muscle degeneration. Nervous inervation is evident since removal, or even mereimmobilization of the wings also causes the flight muscle to degenerate or inhibits their maturation.

Recent data by many authors, supplemented by the study of JHa effects, show that JH can unquestionably affect morphogenesis at any point from the initiation of egg segmentation up to adulthood. This is demonstrated most convincingly by the action of JHa on embryonic development (Slama and Williams, 1966b; Novak, 1969; Riddiford, 1969-197ia, Matolin, 1970; Rohdendorf and Sehnal, 1973). Some papers describe the effects of JH on the pupal instar (Riddiford, 1972; Riddiford and Ajami, 1972) and its sensitivity to the administration of JH and JHa up to the next ecdysis. Various other authors, e.g. Sehnal and Meyer (1968), arrived at similar conclusions. Gametogenesis can likewise be inhibited by JH at any of its stages.

The influence on ovarian follicular cells

The effect of the ca hormone on the development of eggs in the ovaries was demonstrated in one of the initial researcher papers of Wigglesworth papers on JH. Wigglesworth (1935, 1936) ascertaned that the presence of an active ca is absolutely indispensable in *Rhodnius* for the ripening of eggs beyond a certain development stage, at which the ovarian follicle cells become active. Following allatectomy the growth of the eggs stops at a certain stage because of the degeneration of the follicle cells. This effect may be avoided the simultaneous implantation of an active ca from another specimen. Larval and imaginal ca show the same activity in this respect. Results similar to those with *Rhodnius* were obtained by Joly (1945a, b) with the beetle *Dytiscus*, by Weed-Pfeiffer (1945) with the grasshopper *Melanoplus differentialis*, by B. Scharrer (1946) with the cockroach *Leucophaea maderae*, by E. Thomsen (1952) with the blowfly *Calliphora erythrocephala* and by other researchers with other groups of insects. The research on *Calliphora* was further developed by Possompes (1955), according to whom extripation of the ca suppresses the ripening of the eggs only when performed in the adult. But it is still unknown whether or not there is at least a partial regeneration of the ca after its earlier removal. In the higher Diptera, the adult ca are quite different from those of the larvae (the dorsal part of the ring gland) so that their development may not be prevented even by the complete removal of the larval structures. The necessity of JH for ovarian development has been found in most species examined in this respect, although there are species where no JH control exist, such as *Carausius morosus*. *Bombyx mori*, etc. This independence of the development of the eggs seems to be a phylogenetically secondary adaptation connected with a particular type of egg-laying; for example, in *Bombyx* all the eggs are deposited in a short time, not long after the imaginal moult. While in *Carausius*, they are laid one at a time at long intervals.

The presence of a special gonadotropic ca hormone has been suggested by various authors. As emphasized by Pflugfelder (1952, 1958), there is no definite proof in favour of this conclusion. There are, on the other hand, significant arguments in support of one hormone theory. One of the most important of these is the fact, already mentioned, that the effects of the imaginal glands can be reproduced by the glands of any larval instar not only of the same species but also of any other insect order. Also, conversely, the imaginal ca of any insect species can replace the larval ca in any insect (Pterygote) species. For instance, the ca of adult female P. americana in the process of ootheca formation can prevent imaginal differentiation in the last larval instar of Oncopeltus fasicatus. The histological evidence also lends no support to the theory of two or more hormones. The male gonads are in all cases independent of the presence of the JH. The fact that one and the same substance is responsible for either effect has been confirmed with juvenoids.

The influence on reproduction

The problem of the endocrine control of reproduction has been dealt with by many researcher and reviewed by Doane (1962), by Highnam (1963) and by Nayar (1964). Nervous control of egg maturation by specific neurones along the ventral nerve cord was demonstrated by

Engelmann (1964) in pregnant females of the viviparous cockroach, Leucophaea maderae. These nuerones are supposed to inhibit the ca by influencing specific regions of the brain. The researcher assumes that a hurmoral factor released by the brood sac affects these neurones, but his experimental evidence does not seem to prove this. The role of hormonal factors in the control of ovarian development in Schistocerca gregaria was investigated in a series of papers by Highnam (1962), Highnam and Lusis (1962), Lusis (1963) and Highnam, Lusis and Hill (193a, b). In Locusta migratoria castration in the last larval instar does not seem to affect the cellular volume of the ca in either males or females. Allatectomy in a highly autogenous strain of Aedes taeniorhynchus inhibited egg maturation when carried out within one hour after adult moulting; yolk deposition was stimulated by the implantation of one pair of active ca. Similar results were obtained in other aedine mosquitoes.

Some researchers have recently reported quite a different type of JH and Jha action. JH administration in the early phases of gametogenesis not only does not promote the process, but, on the contrary, inhibits it. The sensitive period for this inhibitory effect is, in general, earlier than for follicle cell stimulation. In some of the insects studied it comes at the beginning of the adult stage, in others in the pupal instar, or (in Exopterygota) in the last larval instar. It includes the whole period of gametogenesis, from the first oogonia and spermatogonia up to the mature sex cells. In principle, this is the same 'inhibition' of morphogenesis as observed in any other phase of ontogenesis. As with embryogenesis, the results of such inhibition of growth can differ considerably, depending on to the time of administration of the substance and the duration of its effect. Various kinds of deformation of the ovarioles mostly result.

Many researcher papers have been published on the ca and Jha endoerine control of oogenesis and of reproduction in general. Whatever the conclusions reached by individual researcher, the observed facts are all in agreement with the original explanation by Wigglesworth (1936) of the effect of JH on the ovarian follicles in Rhodnius prolixus.

The influence on the accessory sex glands

The necessity of JH for the functioning of the accessory glands in male Rhodnius prolixus was illustrated by Wigglesworth (1936a) in his first researcher paper on the ca hormone. Since that time a similar dependence of the accessory glands on JH has been shown in many other insect species, not only in males but in some cases also in females. For instance, the accessory glands in female Calliphora erythrocephala reach an average length of 2.58 mm, but in allatectomized females they do not surpass 1.7 mm. The effect of allatectomy in Calliphora males is less conspicuous: here the normal length of the accessory glands is 1.18 mm, but they reach only 0.91 mm following allatectomy. Their secretory activity is not completely suppressed by allatectomy. B. Scharrer (1946c) found an effect of the ca on the functioning of the accessory glands in males but not in females of Leucophaea maderae. On the other hand, no such action of JH has been found on the accessory glands of Lucilia and Sarcophaga. A clear dependence upon JH was found by Slama (1964) in Pyrrhocoris apterus, in both males and females.

The influence of polymorphism

Out of many types of polymorphism in insects the following have been found to be closely connected with JH: (a) The mechanism of caste polymorphism in social insects has been found to bear some relationship to JH activity, as for example in ants and in termites. The latter authors originally presumed the existence of a special, third ca hormone responsible for soldier production if administrated to larvae or pseudergates at the 'right time'. This possibility was ruled out by Luscher (1969), however, who showed that Roller's synthetic JH also induced the same effect. Soldier differentiation requires much larger JH does than those needed for the normal; i.e. juvenilizing effect. (b) Phase polymorphism in locusts. The influence of JH on the development of Locusta and Schistocerca, has been studied by Staal (1959, 1961), L. Joly (1960), Kennedy (1956, 1961), Loher (1960) and others. Their findings point to a complicated dependence of phase on JH activity. As confirmed by Staal, the implantation of active ca into young larval instars induces the green colouration typical of the solitary phase even in environmental conditions which would otherwise produce the gregarious phase. The adaptive value of either phase in an environment which induces this particular type of development has been emphasized by Kennedy (1961). The existence of a volatile factor acting through antennal chemoreceptors seems probably on the basis of the results of Loher (1960) who was the first entomologist to discover a substance accelerating sexual maturation in adult male locusts.

More information on this subject have been obtained by Joly and his co-workers. Seasonal polymorphism in aphids. A very complicated type of polymorphism is encountered in aphids. There is a strict seasonal alternation of forms with a pre-ponderance of parthenogenetic generations. The polymorphis forms differ with regard to the occurrence of wings, the presence or absence of vivipary and in a number of other, less marked, characters, both morphological and physialogica. Even though we are unfamiliar with the the various mechanisms involved, there are definite suggestions that the metamorphosis hormones and perhaps other chemical factors play a part in the determination of the character of the various forms. The apterous virginoparae of Megoura, for example, appear to be produced by the neotenic effect of increased JH activity. This activity, however, seems to take place during the embryonic period through the action of the maternal endocrine system on the ca of the older embryo. The still earlier activation of the embryonic ca by the maternal body appears to be the cause of differentiation of the oviparea. Similar conclusions were reached by Johnson and Birks (1960), with an Australian aphid species (Aphis craccivora Koch.), who assumed that all aphids started their development as presumptive winged forms which may be diverted from this path of development by various stimuli in the determination period between the late embryonic stage and the second larval instar. These researchers assumed that the results they obtained, could be applied to to other types of insect polymorphism.

The effect of JH on aphid morphogenesis was examined by Holman (1973). In a detailed analysis he found that the statistical regression expression the relationship between the length of individual parts of the body and total body length or the breadth of the head capsule differed relatively little in the 1st to Ivth instar, but displayed a significant shift between the Ivth instar and the adult. The regression curve corresponded to the total changes which occurs in any part of the body during metamorphosis. In most cases the absolute value of the shift was also dependent on body size. In supernumerary instars and juvenilized adults, this imaginal shift was reduced to various extends. The degree of JH effects can be determined from a comparison of these data with the average normal individuals. Mathematical formulae for the statistical evaluation of these effects are given. (d) Alary polymorphism in bugs. Alary polymorphism or the occurrence of divers degress of wing development in one and the same species is a common phenomenon in Heteropetera. A theory has been elaboratied by Southwood (1961) based in Wigglesworth's findings (1960) on the hormonal control of this type of polymorphism. Brachyptery is assumed to originate on the principle of metathetely in some species, particularly in colder, mountain conditions. It may be classified as a juvenile character in adults following the normal number of nstars (five in Heteroptera). In these insects (such as Gerris, Nabis, Bryocoris and various Tingidae) macropterous forms are found mostly in warm localities. On the other hand, brachyptery is assumed to originate in other species on the principle of prothetely (paedogenesis).

In species such as Dolichonabis limbatus and Microvelia, long-winged specimens appear in cold, mountain conditions whilst in the normal warm environment only brachypterous forms occur. This theory is yet unconfirmed as stated by its author. (e) Sexual alary dimorphism in Lampyridae. Davydova (1968) described the effects of ca implantation and extripation on wing development and metamorphosis in both males and females of Lampyris noctiluca (Coleoptera). Allatectomy often resulted in considerably prolongation of life without metamorphosis. Prothetelic males (regressive prothetely) with smaller wings were sometimes obtained. The implantation of several ca induced one or two supernumerary larval moults in the females, without inducing any wing development. The absence of wings in the females, together with other neotenic characters, is obviously of earlier phylogenetic origin and is incapable of phenotypic modification.

The effect on regeneration

A remarkable decrease in regeneration ability was noticed by Pfulgfelder (1939b) in Carausius morosus following extirpation of the ca. In some cases, particularly when an amputation was effected after allatectomy, there was a complete loss of regeneration ability. Generally the degeneration of whole groups of cells was observed in the regenerating epidermis whilst knotty proliferations appeared in other parts. But no difference was observed in Blattella germanica, either in the extent of regeneration or in its effect on the duration of the instar concerned, irrespective of whether the extirpation was carried out during the period of active JH concentration (Ist to Vth instar) or in its absence during metamorphosis. The relation between regeneration and the changes in volume of the ca were surutinizedin detail by O'Farrel et al. (1961) in Blattella germanica.

The duration of the moulting process was only affected when muscle tissue was involved and the greater the amount of tissue affected, the longer the intermoult period. But the moment the regeneration was restricted to epidermal structures, no changes could be seen in the moulting cycle. No evidence is available of any nervous influence on the regulation of the moulting process and all the effects observed appear to be the results of self-regulatory processes. A histological examination of the volume of the ca and that of the so-called headlobes carried out at twelve-hourly intervals showed a close connection between the volume changes of the two glands. Regeneration has a positive effect on the functioning of both pairs of glands, which is particularly interesting from the point of view of the experiments of Bodenstein (1953) and Engelman (1959) on the reciprocal effects of the ca and the pgl and their effect on the moulting process. Immediately after the moult following treatment which causes the regeneration, a hypertrophy of the ventral lobes and ca occurred. This hypertrophy, and the subsequent extensive regeneration seems to result in extra moults in the adult stage.

The delay caused by the regeneration observed in the moulting process in the Ist larval instar of Blatella germanica is followed by a shorterning of the subsequent instars so that the Ivth instar moult occurs at the same time as that of the controls. This can also be assumed to be as a of the positive effect of regeneration on the functioning of the gland concerned. The effect of metamorphosis hormones on the regeneration process in insects was explained in great detail by Bodenstein (1955, 1959) on the basis of his numerous experiments. He emphasized the need of MH for regeneration. It is only the degeneration of the prothoracic glands associated with metamorphosis which makes the adult metabolous insects unable to moult. All the tissues of the imago retain their full capacity for regeneration, which occurs in the extra imaginal moulting process brought about by the implantation of active prothoracic glands or by parabiosis of the adult with a nymph. It is yet unclear as to what extent the observed effect of MH on regeneration is direct or, perhaps more probably, whether it is only indirectly brought about by inducing the moulting process. According to bodenstein the results obtained by him contradict those of Pflugfelder with Carausius morosus, and attempts to find a common.

Penzline (1965a, b) has comprhersively analyzed, of regeneration and the ways in which it can be influenced hormonally. He illustrated that all three metamorphosis hormones were needed for normal regeneration and that it was inhibited by allatectomy. Another detailed survey was published by Needham (1965), who suggested that JH was necessary for regeneration processes to a limited extent in adult insects, e.g. the healing of integumental wounds and the regeneration of nerve fibres and internal organs. Thus capacity can be greatly enhanced by the transplantation of wing discs and leg fragments, etc., and by the implantation of active pgl. Rinterknecht (1964) found that allatectomy did not inhibit the reconstruction of the integument in Locusta migratoria, but caused a roughly two-fold delay and that it similarly slowed down the mitotic activity of the regenerating tissue. Normally allatectomy is supposed to speed up ageing.

The influence on tumour growth and other alterations in histogenesis

Including to the morphological effects earlier mentioned, JH has atremoendous impact on histogenesis. The first detailed analysis of the histological effects of ca extripation was carried out by Pfulgfelder with Carausius morosus. He noticed numerous changes of a pathological nature following alltectomy carried out in earlier instars. He distinguished tissue degenerations and tissue proliferations both of which often occurred in one and the same tissue. The most reoccuring pathological changes are: (a) Degeneration sometimes connected with a partial histolysis of fully developed tissue and often connected with a subsequent partial or complete phagocytosis. Simultaneously, proliferations usually appear as knotty structures scattered throughout the tissue, which sometimes contain single differentiated cells in the form of giant muscle cells. (b) Degeneration of the nervous system and the adjacent mesoderm and glial cells in which the formation of large irregular cysts often occurs. (c) Degeneration of the fat body during which vacuoles appear in fat cells, which resemble fat droplets but do not stain with osmium tetroxide and therefore probably do not contain either fat or any lipoid substance. It would be interesting to determine to what extentdefine the limit of JH deficiency syndromes agree with the alterations in the fat body of adults of other species, which are not connectred with patholotgical changes in other tissues. The prolonged effect of allatectomy is encapsulation of the pathologically changed cells by lymphocytes which often leads to the development of large cysts. Inside these cysts large concretions often appear which show concentrical layers. (d) Degeneration features are also, often observed in the Malpighian tubules in which cysts and concretions, similar to those in the fat body, appear. (e) Accumulations and proliferations of haemocytes (lymphocytes) often occur. These may give rise tonumerous knots of undifferentiated cells, which resemble those of spindle cell sarcoma in Man. (f) Proliferations of a carcinomatous nature are observed in the epidermis, gut and oviducts. Many changes occur following the implantation of extra ca (from adults into larvae), in, for example, the ovarioles, oviducts and other organs and tissues. But they are, certainly different from the changes produced by allatectomy.

Various modifications in tissue differentiation frequently occur during regeneration, particularly when a piece of embryonic tissue is implanted whilst there is disturbance in the hormonal balance, as shown by Pflugfelder (1948). He attempted verify this by the fact that young cells are more easily affected by external stimuli than older ones. Tumorous malignant growths have often been observed following the cutting of various nerves, such as the nervi corporum cardiacarum. Schneiderman and Gateff (1967) set up clones procured from a new mutant in Drosophila melanogaster, which they called lethal malignant brain tumour. The mutant imaginal discs grew considerably but did not respond to react the metamorphosis hormones. The mutant cells grew rapidly and invasively, killing the hosts, and behaved like true malignant tumours. They were cultivated for 35 transfer generations and further clones were secluded from single cells.

The Indirect Effects of JH

One must be careful to distinguish between the direct effects of the hormone as a chemical agent and its indirect effect, i.e. the consequences of change produced by the direct action of the hormone in various parts of the insect organism through the mediation of the circulatory and nervous system and other correlation mechanisms. Failure to note or respect this fact caused lot of confusion in the early days of insect hormone research. For example, JH was regarded by some researcher as a moulting hormone, because it induced extra larval moults, which MH was called the growth and differentiation hormone, etc. Since it is often difficult to identify the actual character of hormone activity, many authors are at variance as to whether specific hormones (e.g. JH) exert a direct or an indirect effect. There are, however, several cases where the indirect nature of the JH effect is confirmed.

The effect on moulting

The researchers, who were the first one to deal with the ca hormone assumed it to be a moulting hormone. Depending on the fact that, after the implantation of active ca into a last instar larva, one or several extra moults and supernumerary larval instars occurred. Further investigations have, however, shown without doubt that the moulting process is directly influenced by the prothoracic gland hormone (MH), and the remarkable effect of JH is actually the indirect effect of prolonging the action of the pgl which are normally histolysed during or immediately after metamorphosis. The seemingly contradictory observation of Bodenstein (1953a-e) that allatectomy towards the end of the last-larval instar in the cockroach Periplaneta americana may result in preservation of the pgl and thus cause supernumerary moults can also be explained on this basis. As demonstrated by O'Farrell et al. (1953, 1956), the amputation of legs or a more extensive injury to other parts of the body may have the same effect as ca extirpation. This, as noted above is an effect of the regeneration process, perhaps improved by selection into a self-regulatory mechanism, in the case of insects with a well-developed autotomy, and has nothing to do with the action or absence of ca. The indirect nature impact of JH on moulting is quite obvious from the fact, amongst other things, that neither the normal nor the experimentally induced action of the ca in adults causes any sign of moulting, whilst this may easily be induced in the adult, by the implantation of an active pgl.

The effect on the silk glands of caterpillars

It is generally accepted that in many Lepidopterous larvae the labial glands are almost inactive during the first top penultimate larval instar whereas they produce a large quantity of a silky material for cocoon spinning in the last larval instar. The situation is just the reverse in sawflies such as Cephalcia (Pamphilidae) where the silk glands are active throughout the larval instars producing the 'excrement webs' in which the larvae live gregraiously, and pupation takes place in cells in the soil without a cocoon. In each contact the action of the glands is connected with the presence of JH. The hormone, however, does not affects the silk glands directly, but affect their morphogenesis in a very general way. Whether its presence affects the function of the glands positively or negatively depends on genetic, species-specific factors.

The effect on instincts

The effect of JH on cocoon spinning in Galleria mellonella has been explained as a special case of affecting instinct. The larva of this species spins a cocoon before each moult. This cocoon is, however, different before the larval moults from that before the pupal. The larval cocoon is in the form of a tube open at each end whilst the pupal one is spindle-shaped and completely closed. It was shown by Piepho (1950b) that the implantations of ca at certain times before moulting caused the larva to produce cocoons intermediate between the larval and the pupal types, depending on the time of implantation. This effect cannot be expressed as a direct one on the instinct of cocoon spinning but, undoubtedly, as an indirect one through the morphegenesis of the brain. By preventing the morphogenesis of the imaginal brain at the required time JH determines whether, and to what extent, the material basis for a given instinct is developed, i.e., the corresponding nervous (cerebral ganglion) structure.

A detailed examination of this and similar effect will no doubt make possible an exact localization of various imaginal instincts in insects, especially when these differ from those in the larvae, as for instance in the case of insects with aquatic larvae. The question of hormonal influence on sexual behaviour in various species of the order Saltatoria has been described by Loher (1964). In some species, allatectomy influences the males, while in others the females are affected, and there are species which are not influenced by the ca at all (Gryllus campestris). No effect on sexual behaviour of either allatectomy or the implantation of additional ca was found by Roller, Piepho and Holz (1963) in Galleria mellonella. On the other hand, Beetsma, de Ruiter and de Wilde (1962) report that the injection of cecropia extract shifts the balance between the photopositive and photonegative orientation tendencies to larvae of Smerinthus ocellata towards a positive response. The implantation of active prothoracic glands resulted in the reverse effect.

Some JH effects are still debatable. However, only affects immediately connected with the morphogenetic processes of stimulation of larval growth, inhibited derepression of more advanced characters and suppressed degeneration can be regarded as direct. With the discovery of juvenoids (JHa), we have another type of effects which may not be the direct action of the hormone, but may be resulting from the side effects of molecular structure or an effect connected with experimental hormone supply (e.g. an effect on membrane conductivity, heart rhythm and also, perhaps, interruption of the diapause). More information on this group of effects is required before they can be properly difined. No difference has been discovered till now between the effects of the actual ca hormone and the various JHa, apart from their specificity form some groups of insects. This is not related to the mechanisms of action of the given JHa, but to their capacity for penetrating the integument, their resistance to various haemolymph enzymes, etc.

All JHa effects can thus also be ascribed to JH and vice versa. This is a specially significant in cases in which ca application is very difficult or even impossible, e.g. during early embryogenesis. It is often difficult to differntiate between the morphogenetical and the physiological effects of a hormone as well as between the direct and indirect effects. All morphogenetical effects are necessarily based on physiological (biochemical) ones, while most of the physiological effects can, sometimes, cause morphogenetical changes. In the earlier edition of this book (Novak, 1966) there were doublt about certain results of JH action among its indirect effects. Since then, however, we have acquired much information about the mode of action of this principleand my original assumpption on the primarily morphogenetic action of JH has been confirmed by dozens of new findings. Therefore, there is to be full justification for transferring most of the physiological and biochemical effects from the direct to the indirect group including, for example, all metabolic effects, together with oxygen consumption, the effect on fat metabolism and the fat body, etc.

The effect on the total metabolism (oxygen consumption)

It was indicated by Pflugfelder in as 1941 (on the basis of his work with ca implantations and extirpations in Carausius morosus) that JH is a factor which generally favours metabolism. This conclusion was later confirmed by the experiments of Weed-Pfeiffer (1945a). From her experiments on the effects of allatectomy in adult Melanoplus differentialis females, She discovered that the ca exert a positive effect on basal metabolism. She observed a connection between the presence of the ca and the consumption of food reserves in the fat body as well as an effect of JH on the increase in weight of the fat-free dry-matter and water. Further confirmation was obtained by E. Thomsen (1949), who found the allatectomized female adults of Calliphora erythrocephala show a 19 per cent decrease in oxygen consumption. A similar dependence on the presence of JH was discovered in males. Experiments carried out under the same conditions using castrated females. (Thomsen, E, and Hamburger, 1955) apparently suggest that this may be a direct, and not an indirect effect on metabolism and not, as earlier supposed by Pflugfelder (1952), caused by oxygen consumption of the developing ovaries. Pflugfelder himself (1952) observed an increase in oxygen consumption in Carausius morosus following the implantation of ca and a decrease following their extirpation. Both effects were, however, rather weak and only temporary. L'Helias (1956a) noticed a decrease in oxygen consumption in the same species following allatectomy. An even greater decrease was observed when the cc were removed at the same time.

L'Helias attempted to explain this as being connected with the accumulation of glycids and mineral phosphorus in the tissues and with the loss of energy caused by the breaking down of esters due to the increase in alkaline phosphatase. A progression of experiments on the oxygen consumption of normal and castrated adult females of Pyrrhocoris apterus, implanted with ca appears to indicate an indirect action of these glands on metabolism. It was noticed that castrated females do not show any increase in oxygen consumption compared with normal controls following ca implantation, very probably because of the increased share in the total metabolism of the ripening ovarian follicle cells which do not grow in the absence of the ca. Examination of the effect of ca implantation into the last larval instar of the same species gave corresponding results.

A direct connection was discovered between the extent of the induced morphological changes and therate of oxygen consumption. It was concluded from both series of experiments that the JH-effect on oxygen consumption is an indirect one, depending on the increase in the amount of metabolically active tissue. The direct effect of JH is probably concerned with protein synthesis. But, Sagessar (1960) claims to have theory an increase in oxygen consumption after ca implantation in the cockroach Leucophaea maderae even in castrated females and concludes; in agreement with E. Thomsen, that JH affects metabolism directly. His point of view based on the theory of two ac hormcne appears tibe incorrect. Results which seem to contradict to some extent the above theory, are reported by Samuels (1956). He noticed a considerable increase in the oxygen consumption of the isolated thoracic musculature of Periplaneta americana adults of both sexes 2 to 3 months after allatectomy.

Samuels tried to explain the difference observed in oxygen consumption in the whole animal and in the isolated tissues by assuming that the glycide reserves, accumulated in the tissues because of allatectomy, and thus prevented from being used inside the organism, are freed and used in the isolated muscle, thus increasing its oxygen consumption. Slama (1960) examined the correlation between oxygen consumption and postembryonic evolution in various groups of insects; he found a typical increase at the beginning of each instar and a decrease in the second half of each larval instar. The period of metamorphosis was characterized by a typical U-shaped curve in Exopterygota also.

A similar correction between oxygen consumption and ovarian development was found in adult Pyrrhocoris females, in which typical O_2 consumption cycles were observed in connection with the ovarian cycles. The cycles were controlled in a typical manner by AH and JH. The correlation was less obvious in males, in which metabolism is low and displays no clear cyclicity (Slama, 1964b). Experiments with isolated body fragments showed that JH regulated oxygen consumption indirectly, by stimulating the growth, function and metabolism of specific parts of the body (Slama, 1964c). A similar close correlation was shown to exist between oxygen consumption and the degree of morphogenetic changes in Galleria mellonella. Sehnal (1966) performed a systematic study of the effect of JH on oxygen metabolism during larval and pupal development in the same species. No effect of ca implantation was observed in the penultimate larval instar, but implantation in the last larval instar was followed by a definite increase in metabolism in specimens which transformed to supernumerary larvae.

Effects on protein and nitrogen metabolism

The first research work on the ca effect on nitrogen metabolism was published by L'Helias (1956a), who observed the accumulation of amino acids in the tissues of Dixippus morosus following allatectomy. Janda and Slama (1964) studied the effects of JH on the body nitrogen, glycogen, fat and uric acid content in Pyrrhocoris. They were all proved to be indirect and the outcome of changes produced by the hormone on the growth process in particular parts of the body. Allatectomy, inhibiting ovarian development (and hence the utilization of reserve materials) resulted in enormous accumulation of these substance in the body. If cardiacectomy was performed at the same moment the production of reserve materials likewise stopped, so that compared with the controls no important differences were observed.

In another study, Slama (1964) observed excessive accumulation of the haemolymph proteins following allatectomy or castration, which was diminshed by ca reimplantation. Minks (1965) found differences between various protein fractions and free amino acid concentrations in the haemolymph of normal and allatectomized specimens of Locusta migratoria. The two entomologists Bentz (1969), and Bentz Girardie (1969) observed changes in ovarian protein composition during egg maturation after both allatectomy and nsc electrocoagulation. Engelmann (1969) detered the de novo production of a specific protein in the serum of adult Leucophaea maderae females treated with JH.

The protein was absent from the in nymphs of either sex or in adult males, and it disappeared in allatectomized specimens. Engelmann, Hill and Wilkens (1971), using various JHa, found dependence of a similar female specific protein in Schistocerca gregaria and Sacrophaga bullata. Engelmann (1971) further analysed these effects in the original speices (Leucophaea), using different JHa, and obtained largely the same results. Since then, female specific proteins have been found in representatives of several insect orders. Engelmann (1972) later found that synthesis of this protein was blocked by actinomycin D. In agreement with this finding, JH also increased the synthesis of total sodium dodecyl sulphate phenol-extractable RNA, which stimulated specific protein synthesis in a completed cell-free system from egg-maturing females, whereas RNA from males or allatectomized females had no such effect. As stated by the researcher the information seems to support Karlson's hypothesis of hormonal control of the transcription of specific messenger RNA.

Effects on protein metabolism

The influence of JH and the ovaries on the amount of pupal and adult body far was examined in Musca domestica by Adams and Nelson (1969), who observed a decrease in the rate of disappearance of the pupal fat body following both ovariectomy and allatectomy. The size of the adult fat body was not affected by allatectomy, but in ovariectomized specimens it was larger than normal. Baumann (1969) investigated the effect of several JHa he detectedgreater stability of the membrances, which varied with the kind of juvenoids and the type of membrane.

The influence on the fat body

It has been ascertained by Weed-Pfeiffer (1945a) that alltectomy causes an increase in the fat content which is only slightly affected by castration. Similar results were obtained by E. Thomsen (1942) with Calliphora erythrocephala and by Vogt with Drosophila. The latter researcher also observed reduction in size of the nuclei and nucleoli of the fat body cells which she explained as being due to the cytoplasm of these cells becoming filled with fat droplets. Day (1943a) also mentioned conspicuous histological changes in the fat body following extirpation of the ring gland in the flies Lucilia and Sarcophaga. It is not, however, very obvious from his experiments to what extent these changes were because of other components of the ring gland. Gilbert (1966) examined the variation in the lipid content of the ovaries and fat body of Leucophaea maderae and found that lipid synthesis by a maturing ovary was enhanced by JH application in vitro. He gathered that JH might depress lipid synthesis in the fat body, with resultant mobilization of the reserve substance on behalf of the developing oocytes. There are no proof against the opposite possibility, however, which would fit everything else we know about the action of JH, i.e. that the action of the hormone consists in activiation of the ovarian follicles.

The metabolism of the ovarian follicles is much greater than that of the fat body, so they may attract reserve substances not only from the haemolymph but from the fat body also. Stephen and Gilbert (1970) also noticed that fatty acid composition in Hyalophora cecropia changed during metamorphosis and that it was affected by JH. Here again, the effects of JH are clearly indirect and caused by the different requirements of the differentiating tissue. Morophoshi and Kiguchi (1969), Morohoshi and Fugo (1972) and Morohoshi et al. (1972) studied various aspects of the effect of JH on lipid and other forms of metabolism in Bombyx (1969) and attempted to explain their findings on the lines of Morohoshi's hypothesis. Again, however, there is no information which could not be discribed on the basis of the much simpler hypothesis formulated by Wiggleworth (1936) nearly 40 years ago.

The influence on the oenocytes

Reduction in the volume of the oenocytes and in the size of their nuclei with associated pycnosis was observed by Day (1943a) in Lucilia and Sarcophaga and by Vogt (1947) in Drosophila, following ring gland extirpation. It is difficult to decide the of changes resulting from JH deficiency and the absence of the other two metamorphosis hormones (AH and MH), which also originate in the ring gland. The possibility that the effect observed may be merely the result of inhibition of the moulting process must also be considered.

The effect on the water balance

Including the neurosecretory cells of the pars intercerebralis and the cc, the ca are also said to produce an effect on water balance. Allatectomy results in a reduction of the water content of the body, including that of the haemolymph. The effect of JH is particularly evident in castrated females where retention of the ca results in a considerable increase in the quantity of haemolymph with an associated conspicuous distension of the abdomen. In such individuals the amount of dry matter also increases. However, it is not clear exactly what effect is due to JH itself and what is because of the neurosecretion stored in the ca.

The effect on colour change

Quite obvious changes have been detected in the colouration of both larval and adult insects following experimental interference with JH production. Some of these are obviously of morphogenetical origin, such as the green colouration in the solitary phase of the migratory locust. Others are, undoubtedly because of to the brain neuro-hormone. There are, however, other changes which can be assumed to be due to JH, such as those causing the colouration of allatectomized stick-insects (Carausius morosus). The colour changes are assumed to be caused by an accumulation of uric acid in the tissue together with a decrease in the purine, pterine and melanin contents. Changes in colouration were also observed in the same species following implantation of active ca. A marked morphogenetical effect of the ca hormone on the black pattern in the lygaeid bug Oncopeltus fasciatus has been written about.

Conforming the earlier findings on the colouration of Gryllus bimaculatus (1960), Roussel (1967) obtained several interesting results on the action of ca and the control of pigmentation in this species. Roussel reported very high individual variability in the response of females to the action of JH and found agreement between the control of melanin pigmentation in this species and the gregarious Locusta migratoria. The absence of the ca leads to an increase in black pigmentation, while only a small fragment of ca induced orange pigmentation. The researcher confirmed that the ca had a direct effect on the integument.

The cuticle colouration of the tobacco hornworm, Manduca sexta, was sanalyzed by Truman et al. (1973) as the basis for an ultrasensitive JH test. The presence of JH inhibited melanization of the cuticle in the larvae of this species. If the hormone was administered up to a given moment (between the 17th and 25th hour of the last larval instar), pigment formation depended on the absence of JH. This test has the same sensitivity as the wax test in Galleria, but can be experimented upon only two days after administration.

The effect on the mitochondria

An increase in the oxygen intake is stated by Clarke and Baldwin (1960) in mitochondriae preparations from Locusta migratoria- L. on a sodium succinate substrate, after the addition of ca. The increase was, however, merely detected in mitochondria prepared from Locusta migratoria adults, those from Vth instar nymphs. Of Schistocerca gregaria showing a decrease in oxygen consumption. The addition of 2 : 4 dinitrophenol in physicological amounts, on the other hand, resulted in an opposite effect : it decreased the oxygen consumtion of the mitochondria of adult Locusta and increased that of mitochondria of Vth instar Schistocerca. The addition of both dinitrophenol and ca at the same time increased the oxygen consumption of the preparations from both species. The researcher discuss the possibility of the ca controlling metabolism by their action on the mitochondria and the possible relationship between the effect of JH and those of the dinitrophenol. But it is possible that these are indirect effects, and that they depend on the type of mitochondria on which JH acts. JH has no effect on imaginal (JH-independent) cells, but may act on the structures of larval (i.e. JH-dependent) cells. This also seems to be indicated by the lack of differences between the alary muscle mitochondria of normal and allatectomized Locusta migratoria (Joly et al., 1969), findings by Novak and Slama (1962).

The effect on pheromone production

It has been confirmed by Loher (1961, 1962) that JH controls sexual maturation and the corresponding changes in colouration and mating behaviour in the locust Schistocerca gregaria males in a very interesting way. The maturation of the adult males of the gregarious phase in this species is marked by three distinctive features : colour change from the light-brownish grey and white of the freshly moulted adult to the yellow of the mature specimen; a specific sexual behaviour pattern; and the production by the epidermal cells of pheromone which acclerates the maturation process in young adult males in the immediate vicinity of the mature specimen by an olfactory effect on their chemoreceptors. None of these indications is demonstrated allatectomized males. They are however, induced by the implantation of active ca from a mature specimen. The same type of pheromone was discovered by Barth (1962) in the cockroach Byrsotria furnigata.

The virgin females of this species produces a volatile sex attractant which stimulates the courtship behaviour of the male. The production of this pheromone is also controlled by JH. It is not manufactured by allatectomized specimens, but its production can be induced by the implantation of active ca from another specimen. In this species no other role of JH in the reproductive behaviour of either of the sex was found. More particular JH (or ca implantation) effects in various indectswere observed by researchers. All of them were definitely indirect, or were perhaps, in some cases, caused by extrinsic neurosecretion contained in the ca. For example Baumann (1968) found that JH clearly increased cell membrane conductivity in the salivary glands of Galleria mellonella, causing strong depolarization of the membrane. The source of the hormone was cecropia oil. The same oil extracted from allatectomized males was inactive.

Here again, the action of JH can be taken to affect prolongation of the activity of similar larval (JH-dependent) cells about to degenerate, or which have already actually started to degenerate, and that this also indirectly influences membrane potential. This theory is yet to be tested in both larval and imaginal cells. Shaaya and Sekeris (1970) examined the effects of JH on protocatechuic acid-4-0, β-glucoside synthesis in Periplaneta americana. They found that JH acted in the biosynthetic chain leading from tyrosine to protocatechuic acid and that the primary site of biosynthesis of the substance might be the integument. Specific local effects of JH on the integument were described by Levinson et al. (1966). Roussel (1970) described the influence of the ca on the heart beat of last instar larvae and adults of Locusta migratoria. Allatectomy in the last larval instar slowed down the heart rate, while ca implantation accelerated it. In adults, alltectomy caused a remarkable decrease only if performed after the 8th day. The implantation of a pair of ca did not have any discernible effect on heart-beat. The researcher suggests two possible methods to bring about this effect - the induced release of neurohormones from the cc, or a direct JH action. A third possibility would be that the ca themselves contain an active neurosecretion from the brain. Extirpation of the ca in the last larval instar of the red locust, Nomadacris septemfasciata, produces a state comparable to adult diapause, which can be reversed by reimplanting several pairs of ca from normal adults.

Is There More Than One Corpus Allatum Hormone?

The seemingly opposing effects of JH, such as the negative effect on metamorphosis and the positive effect on growth, the activation of larval tissues and, on the other hand, its necessity for the development of such typically imaginal structures as the ovarian follicle cells, led to the hypothesis of two or more ca hormones. A number of indications in favor of this theory have recently been collected. It was debated by Luscher and Springhetti (1960) that the ca in Kalotermes flavicollis F. produce two different hormones which have different effects in caste differentiation, a 'juvenile hormone' and 'gonadotropic hormone.' The first of these is supposed to increase the capacity for supplementary reproductive differentiation and the other to initiate pre-soldier differentiation.

There is, however, no indication in other insects for the absence of one of the two effects in a gland which both would be present. The practice of this theory, therefore, results in complex contradictions. Sagessar supported the theory of two ca hormones in his finding that there is no qualitative difference in the oxygen consumption curve between the penultimate and the last larval instar in the cockroach

Leucophaea madera. Actually, there is only a quantitative difference in the supposed production of JH, and the qualitative changes in its effects on morphogenesis can be completely explained on this basis.

Moreover, there are many other processes acting during each instar, e.g. all those connected with moulting which are practically equivalent in both the penultimate and the last larval instar and which more or less mask the changes because of differences in endocrine balance. All other experimental findings indicate towards the fact that there is only one ca hormone. According to Schneiderman uses the fact that the extract of the cecropia male abdomens does not induce the development of ovaries favours the theory of two ca hormones. This, however, is much more an argument against the identifying the extract with the ca hormone. This however, is much more an argument against the identifying the extract with the ca hormone. On the other hand, Wigglesworth (1961) found another morphogenetically active lipoid substance, farnesol, with a positive effect in preventing metamorphosis and in activating overian development.

Wigglesworth interprets his results as a 'support for the belief that the yolk formation hormone and the juvenile hormone are likely to prove identical.' There are various arguments in favour of this identity: (1) As confirmed by the histological investigation of the ca during the secretory cycle, there is no evidence for more than one type of secretion, there being only one type of secretory cell in the ca. (2) There is no difference in the endocrine activity of the ca in either direction during the whole life cycle: the ca of larvae from the first to the penultimate instar affect both morphegenesis and ovarian development in the same way as those of the adults. (3) As illustrated elsewhere, there is no major difference in the process of on growth and function of the larval parts of the body and that on the ovarian follicle cells, and both of these effects necessarily result in an increase in oxygen consumption. Even if none of these arguments can be taken as definite evidence of the existence of only one ca hormone, they are, nevertheless, much more relevant than the opposing arguments. There is therefore no evidence for the hypothesis of two ca hormones.

We get the same result from the fact that all these effects can be reporduced by a single juvenoid (JHa), i.e. by a single chemical substance. Pener (1967) arrived at the same conclusion regarding the identical nature of the ca secretion of both male and female Schistocerca gregaria adults. The minute differences which to have been noticed by this author were related to differences between the target organs in the two sexes rather than to differences in actual ca activity.

The Control of JH Production

Criteria of corpora allata activity

The activity of the ca may be understood either indirectly, by the size and the histological appearance of the gland, or directly, by its effects on the recipient after transplantation. Each of these methods has both advantages as well as disadvantages. There has been a great deal of discussion regarding the reliability of estimating the function of the ca by the increase in the volume of the gland. It must be considered, that there are various instances of increase in volume, both normal and pathological, which are not connected with a corresponding increase in hormone production. For example, the gland increase in size from one instar to the next and this increase is not related with an increase in activity per volume unit. It may be estimated by comparing the size of the gland at two consecutive ecdyses. Similarly, the increase in volume observed following castration in adult females is probably not because of an increase in the secretory activity of the gland but due to restriction in the removal of the secretion by the haemolymph.

The increase in volume owing to the secretory activity of the ca can be calcuted approximately as the difference between the total increase in the volume of the gland and the increase due to growth of the ca up to a given moment. It may be accepted that under normal conditions in comparable developmental periods or during short periods, an increase in volume is a definite sign of an increase in the secretory activity of the gland.

Histologically, the active stage is characterized by the swollen cytoplasm which stains more deeply with acidophil (nuclear) stains. The amount of cytoplasm increases in proportion to the nucleus even if the nuclei also become swollen. The local or general inhibitory effect on metamophosis following the implantation of a ca into the last larval instar is usually taken as experimental evidence of JH production. The defect of to this method is that the effect cannot be calculated until after the next ecdysis. During this time the activity of the implanted gland may change drastically. The probability of a quantitative test for ca activity based on its metabolic effects has been stated by Slama (1963). The implantation of an active gland into an allatectomized adult female of Pyrrhocoris apterus causes a specific increase in the oxygen consumption of the recipient. This increase appears to be directly proportional to the activity of the gland. Before it can be adopted for general use, however, detailed elaboration of the criteria for the test is required, together with the determination of the time limits of its applicability and the experimental conditions for its full acceptability.

The changes in corpora allata activity during the instar

The confirmation all the experimental (Wigglesworth, 1936, 1940a), histological (Mendes, 1948) and morphological (Novak, 1951b) information agree that the ca are inactive at the beginning of each instar and that the haemolymph does not contain an effective quantity of JH at that time. The amount of hormone gradually increases and the maximum volume of the ca in usually observed in the second half of the intermoult period. The production of the hormone ceases as the moulting process progresses and, as a result, its concentration in the haemolymph decreases. The main reason for this decrease is, undoubtedly, the excretion of JH by the Malpighian tubules as shown by the experiments of Bounhiol (1953). He discovered that ligaturing the Malpighian tubules of the silkworm at the beginning of the last larval instar has an effect similar to ca implantation, i.e. is suppresses metamorphosis to a greater or lesser extent. The operation is effective during a period which corresponds to that for an effective ca implantation. It is longer than the critical periods for AH and MH. The length of this period may, however, vary noticeably in different species.

In Lepidoptera, for example, the critical periods for all these hormones seem to be greatly prolonged as a result of the long feeding period and short moulting process. This may explain the discovery by Legay (1948) that the ca of silkworm larvae reach their maximum volume at about the time of each ecdysis. The selective value of the deviation, together with extreme abbreviation of the actual moulting process in many holometabolous insects, evidently depends on the prolongation of the feeding period; this is particularly significantin the last larval instar, where it enables the accumulation of reserve substances for metamorphosis and eventually for ovarian development.

Corpora allata activity during postembryonic development

The production of JH begins in the final stag of embroyogensis after the ca have been fully formed. The secretory activity of the gland continues during larval development with a temporary interruption at the time of each ecdysis, as mentioned above. A particular time is therefore necessary at the beginning of each instar before the active concentration of the hormone is again reached. It is not reached at all in the last larval instar in Hemimetabola, or in the larval and pupal instars in Holometabola. The production of the hormone continues in adult insects. It is clearly cylical in the females of those species which lay eggs in several separate batches, such as cockroaches and many beetles and bugs. No such cycles are noticed in males and in females of those species which lay only one batch of eggs (e.g. Bombyx mori) or which lay one egg at a time with wide gapof time in between the process. (e.g. Carausius morosus).

The secretory activity of the corpora allata in the last larval instar

Wigglesworth (1940, 1948) initially believed that JH production ceases completely in the last larval instar or even that the ca actively eliminate JH from the haemolymph during this period. While Piepho (1951), condidered the hypothesis that the presence of some JH in the last larval instar of Galleria (Holometabola) was a necessary condition for the development of the pupa : in the complete absence of JH, adult differentiation should occur. This theory was later confirmed by Wigglesworth (1954), Karlson (1956), Schneiderman (1960) and a number of other workers. That the ca continue to produce JH in the last larval instar was shown by Novak and Cervenkova (1959). They implanted ca from last instar nymphs of Pyrrhocoris apterus into other last instar nymphs immediately after moulting and at different times during the intermoult period and in most cases they obtained juvenilizing effects of varying extent. This does not prove, however, that the ca were producing a morphogenetically active concentration of JH as assumed by the above hypothesis. There is definite evidence against thisassumption : when the ca are removed from the last larval instar immediately after moulting no morphological change is observed. While, the implantation of another last instar larva ca which itself is inactive in the donor, is sufficient to produce a juvenilizing effect.

Hypetrophy of the corpora allata

Castration of Calliphora erythrocephala females immediately after the imaginal moult results in hypertrophy of their ca (Thomsen, E., 1946). The most possible elucidation this phenomenon is that in the absence of the ovaries JH is not removed from the haemolymph. When the concentration of the hormone here, exceeds that inside the gland cells, further diffusion of the hormone into the haemolymph is inhibited. It Therefore, it collect in the gland causing its excess growth and a subsequent decrease in secretory activity. This seems to be a more acceptable explanation than the theory suggested by E. Thomsen (1946, 1947) of a specific hormonal influence of the ovaries which suppresses the activity of the ca. The hypertrophy of the ca observed in functional termite females appears, however, to be of quite a different type, being connected with hyperfunction of the glands.

Effect of an implanted corpus allatum on that of the recipient

The first person to demonstrate at was Pflugfelder (1939a) that the implantation of an active ca causes a more or less complete degeneration of the recipient's gland. A detailed biometrical analysis of this relationship was carried out by Novak and Rohdendorf (1961) using adult females of Pyrrhocoris apterus. They fixed active ca into freshly moulted adults and measured the volume of the recipient's glands on the 3rd, 6th and 9th days after implantation and compared it with that of controls implanted with a piece of muscle of similar size. The ca of the recipients were distinctly smaller than those of the controls. No such effect on the recipient's gland was observed if the implantation was made later in the instar (6th day after moulting), when the recipient's gland had reached its full activity. The explanation of the effect of the implanted gland followes. The follow of the hormone from the gland into the haemolymph can take occur only by diffusion through the surface of the gland as there is no particular mechanism (e.g. muscular) for this.

But diffusion can only take place particular when the concentration of the hormone inside the gland is higher than that in the haemolymph. This condition is not maintained when an active gland is implanted before the recipient;s gland has reached its full activity. Under such circumstances the concentration of the hormone in the haemolymph, produced by the implanted gland, exceeds, that inside the recipient's gland and so diffusion and thus also the secretion of the hormone is avoided of course, if the implantation is effected later in the instar, when the ca of therecipient has reached its full activity, there is already a high concentration of hormone in the haemolymph. This goes beyonds the concentration in the implanted gland so that the position is now reversed and the secretion of the implanted gland is stopped.

As mentioned by Engelmann (1965) the ca was first activated by an increase in the blood protein level (to a small degree) and then by maturation. The decrease in ca activity associated with egg maturation can be attributed to a decrease in the haemolymph JH level caused by reduced food intake at this time. In other insects, however, such as the viviparous cockroach Leucophaea maderae, it was shown to be due to a hormonal factor produced by the brood sac. JH secretion in Dermaptera was probnable first examined by Ozeki (1965).

External factors controlling the activity of the corpora allata

The direct influence of various factors on JH productio is yet unknown. The only paper analysis the effects of external factors on development is that by Wigglesworth (1952a) on Rhodnius prolixus. He discovered that subjecting Ivth instar nymphs to a high temperature approaching the living maximum (c. 35°C in Rhodnius) causes an extension of the intermoult period and the development of somewhat adultoid Vth instars (regressive prothetely). A low temperature approaching the developmental minimum (below 20°C in Rhodnius) prolongs the intermoult period more than the high temperature, but the resulting Vth instar nymphs are somewhat juvenile (progressive metathetely). The results may be explained as effects on the action of JH in the tissues. A reduced oxygen concentration (below 5°C) produces exactly the same effect even at of high temperature.

A comprehensive morphological investivgation of the shape and size of the ca in the bug Eurygaster intergriceps during postembryonic development and in adults in the different seasons of the year was made by Teplakova (1947). She noticed an increase in the size of the gland in adult females in the spring, associated with the development of the ovaries. The ca of bugs parasitized by larvae of flies of the family Tachinidae were very small. In the summer period of increased activity, the ca of some of the females with functional ovaries were extraodinarily large, whereas in most others they were of medium size. Variations in the size of the gland were also found among females from different heights above sea-level.

Internal factors controlling the activity of the corpora allata

Various explanations have been given for the what has been called by Wigglesworth (1948) the 'counting of instars'. Wigglesworth stated on the basis of his experiments with Rhodnius prolixus first that the decrease in JH activity in the last larval instar is not connected with the age of the gland. This is obvious from the fact that the ca of the IVth (penultimate) larval instar when implanted into the Vth (last) instar is able to induce juvenilizing effects, i.e. it retains its activity even when its age is equivalent to that of the last larval instar when it would normally cease to function. From this, Wigglesworth deduced that 'it is not the corpus allatum itself that counts the instars' and he suggested the possibility of a nervous stimulus being the cause of this decline in ca activity in the last larval instar. The good innervation of the gland by the nervus allatus appears to indicate nervous control of the ca.

The theory of nervous control was confirmed by Engelmann and Luscher (1956, 1957) on the basis of their experiments with adult females of Leucophaea maderae and further developed by Engelmann (1965). They examined the stimulating effects of an interruption of the nervi corporum cardiacarum on the swelling of the ca previously observed by B. Scharrer (1952). They discovered that a similar result could be obtained by destroying specific part of the protocerebrum. On the while, no changes were observed after electrocoagulation or surgical removal of the neurosecretory cells of the pars intercerebralis. The treatment induced egg formation in adult females and several extra larval instars in last instar nymphs. The researchers assumed their results as evidence for nervous control of the ca. They believed that the absence of JH both in females carrying oothecae and in last instar larvae is the result of by nervous inhibition. But this is not the only explanation of their results. Deducing from with other known facts, it seems probably that the observed swelling and activation of the ca is because stimulation by the nervous irritation and regeneration process induced by the operation.

The fact that no effect was noticed after the removal of the neurosecretory cells may be due to neurosecretory material accumulated in the cc, as ca are incapable of either secretory activity or growth in the absence of AH. A nervous stimulus inhibiting AH production and in this way, undoubtedly, JH production might, however, be the mediator of the inhibitory effect of the oothecae on the activity of the ca, in the same way as the distension of the abdomen has a positive effect on AH production in Rhodnius. A nervous control of the ca was even accepted by Johansson (1958) in his work with adult Oncopeltus fasciatus. He realized that interruption of ca innervation in a starving female can induce egg-laying in the same way as the implantation of an active ca. This treatment, however, was infective when a brain with unbroken nervicorporum cardiacarum was implanted at the same time as the ca.further evidence nervous control of ca activity was provided in the research papers by joly (1945a, b) on Dytiscus marginalis, by Detinova (1954) on Anopheles messae, by de Wilde (1958) on Leptinatarsa decemlineata, etc.

The control of the function of the corpora allata

Over the years there has be lot of research on the question of the innervation of the ca and the form of the supposed nervous and neurosecretory control of their secretory activity. The relevant papers were reviewed by Highnam (1963). The existence of afine nerve connecting the ca with the suboesophageal gang lion, demonstrated by Engelmann and Luscher (1956) in Leucophaea maderae, was also found in periplaneta in schistocerca and in Locusta. Neurosecretory granules were detected in this branch of the nerves allatus. When this nerve is ligatured, neursecretory material accumulates on the ca side of the ligature, neurosecretory material accumulates on the ca side of the ligature. The different forms of control of the activity of the ca have been put forward.

Nervous control, was considered to be of an inhibitory character in some insects, e.g. Leucophaea while possibly having a stimulatory effect in others, e.g. in Oncopeltus or Schistocerca in which removal of the brain neurosecretory cells by cautery or extirpation prevents the ca from attaining their full size. The possible cause of this contradiction was pointed out by strong (see below). The non-specific effect of regeneration of the operation injury, must also be taken in to account. It was discovered, for instance, in Schistocera, that the development of the eggs can be greatly advanced by wounding, as shown by the experiments of Norris (1954) and Highnaman (1961b, 1962a, b). The function of the nervous system in the activation of the ca was examined by removing parts of the brain and by sectioning the nerves of the glands infemale adults of the Central American locust? (Schistocerca paranensis Burm), and by a histological investigation of neurosecretory activity with the following results striking changes in the volume and appearance of the ca were found during maturation, where as only slight changes occurred in the amounts of neurosecretion at the same time; no material from the median neuroseretory cells was found in the nerves which innervate the ca, but granules from the lateral cells with different histochemical properties from those of the median cells were found jin these nerves; extirpation of the cerebral neurosecretory cells resulted In reduced volume and inhibited the activation of the ca. Unlike other experiments, the implantation of ca did not restore oogenesis in allatectomized females.

The implanted glands could only become and remain active when they retained intact connections with the central nervous system. Never sections and cautery of the neurosecretory centres of the brain showed that the activation of the ca was mediated through the lateral neurosecretory cells in the ipsilateral half of the brain. Transsection of the nervi corporum cardiacarum II resulted in inactivity and reduction of the volume of the ca on the corresponding side. The contradiction between the supposed inhibitory effect of the neurosecretory cells of the brain and their stimulatory effects seems to be due to a disregard of the possible breaking of this connection. It is yet unknown whether the effects of the lateral neurosecretory complex upon the ca are mediated through nervous or neurosecretory stimuli, or, perhaps more probably, through the humoral action of AH., No function could be attributed to the nervous connections between the ca and the suboesophagela ganglioin. Local effect of neurosecretory granules from the neurosecretory cells of the brain reaching the ca via the nervous allatus.

Granules of Gomori-positive materials were found in the ca of various species by a number of authors. B.scharrer (1958) assumed that 'the restraining influence on the ca takes place by nervous activity, whereas the stimulatory affect is brought about by a substance present in the neuro secretory material'. Khan and Fraser (1962), who found neurosecretory granules in the ca of newly moulted adults of Periplaneta americana, believed, contrary of B.scharrer (1958) that they were responsible for the straining influence there. Mordue(1963) observed neuro secretory material in the ca of female pupae of Tenebrio moliitor before imaginal moulting; after mating, it disappeared from the ca, remaining, however, in the allatic nerves in close proximity to the gland. The ca were found to grow more rapidly in mated than in unmated females. Here again there is no unequivocal evidence of a local physiological function of the neurosecretion in the gland, and on based on the information of neuro hormones and their way of reaching the target its explanation is improbable.

Humoral activation by the AH of the brain, corresponding to the activation of the pgl. It seems that activation or inhibition of the production of AH by nervous stimuli from other parts of the body (e.g. the ovaries, ootheca etc.) or by various external factors (e.g. photoperiodism) might be the only way to control ca activity. Among other findings , this is shown by the fac that the imagical dispause in Leptinotarsa, aphids and Pyrrhocoris which is an AH-deficiency syndrome, can be completely abolished by implanting active brains and/or cc, whereas the implantation of active as is not effective in the absence of AH (in brain and cardiacectomized specimen), as demonstrated by Slama (1964) for Pyrrhocoris.

The following result from the available experimental evidence seems to be the most probable. The brain controls the activity of the ca both by neurosecretion (AH) and by nervous activity. The latter may produce either a stimulatory (distension of the abdomen in blood sucking insects) or inhibitory effect (carrying of oothecae) depending on to the biolohy of the species concerned. Neither of these factors, however, seems to be responsible for the decrease in activity of the ca in the last larval instar which is the immediate cause of metamorphosis. The nervous system here is merely a mediator of the external stimuli which on average remain unchanged in each instar.

The dependence of JH production on gland volume

A different view towards the problem of the control of metamorphosis was stated by Kaiser (1949) who stressed on the increasing disproportion between the size of the ca and the pgl and concluded that the amount of JH produced, rapidly decreases compared with that of MH, and the surplus of MH causes metamorphosis. This idea was further developed by L'Helias (1956). He based his views on to suppositions which have been disproved : that of a direct positive effect of MH on metamoirphosis; and that of an antogonism between MH and JH. In addition, the actual amount of MH as estimated by the volume of the pgl in releation to the volume of the body, and thus also the concentration of MH in the haemolymph, remains static.

Another solution which takes into account the surface area and volume of the gland in relation to the volume of the body as responsible for the time of reaching the active concentration of the hormone, was used by the author. Thwe concentration of the hormone in the haemolymph depends on the following factors: (1) the quantity of hormone produced in a unit of time; (2) the volume of the haemolymph and, in direct proportion to this, the volume of the body; (3) the quantity of hormone consumed in the body; (4) the quantity of hormore removed from the haemolymph by the Malpighian tubules (and perhaps other organs) in the unit of time; (5) Another factor controling the rate of release of the hormone is the surface-volume relationship of the gland.

Observation of the changes which these factors undergo during postembryonic development confirms that JH concentration in the haemolymh necessarily decreases with each subsequent instar. Sooner or later, therefore, there comes an instar (which varies with the species) in which an effective minimum JH concentration is not attained before the advancing moulting process inhibits further growth and differentiation. This decides the last larval instar and the beginning of metamorphosis.

The relationship between the surface area and volume of the ca and that of the body, during development

To investigate the above theory, measurements were made of the surface area and volume of the ca and the volume of the body in freshly hatched larvae (the beginning of the first larval instar) and freshly moulted larvae int he last instar. The result obtained in 18 differnt speices of both Hemimetabola and Holometabola show that there is a marked decrease in tghe volume of a ca relative to that of the body, in addition to a very great decreaase in the surface area of the ca compared with the volume of the body. This comparative decrease in the volume of the ca is because of the very slow growth of the gland, which during the whole larval periods is slightly greater than that of various parts of the nervous system and much less than the growth of the most other parts of the body, such as the breadth of the head, breadth of the pronotum, length of the tibia, etc. It is confirmed that the relative decrease in the surface area and volume of the ca in all the species studies is too great not to affect the production of JH, and is large enough to explain the lack of an effective concentration of JH in the last larval instar and thus to induce the onset of metamorphosis.

The Mode of Action of the Juenile Hormone

Many theories have been forwarded to explain the mode of action of JH, some obviously contradictory. According to Wigglesworth's original idea (1936), JH affects morphogenesis on the principle of Goldschmidt's hypothesis of differnt reaction velocities. Wigglesworth believe that the degree of differentiation attained by a given specimen after ca implantation is the product of competition between two simultaneous processes : the differentiation of the imaginal structures, and moulting. Variation is only possible in the period between the moment of detachment of the old cuticle and that of the deposition of the new. The effect of JH was assumed to depend on accelerating the moulting process so that the permaturely deposited new cuticle would inhibit further differentiation. On this basis, Wigglesworth initially used the term 'inhibitory hormone' for the secretion of the ca. Wigglesworth (1940) himself replaced this theory with the hypothesis of two alternative enzyme systems inside each epidermal cell, larval and imaginal, The larval system was supposed to be dependent upon the presence of JH and the imaginal system in order to operate in the absence of JH. The terms 'inhibitory hormone' was therefore changed to 'juvenile hormone' or to 'neotenin'. The hypothesis of polymorphism which has been described earlier was based on this concept.

While Pflughfelder (1939, 1941) and Weed-Pfeiffer (1945a, b), stressed on the positive effects of JH on total metabolism as shown by their experiments. To avoid the obvious contradictions, the hypothesis of two differnt ca hormones was created; one of them being identified with the JH of Wigglesworth, and supposed to affect only larval development. The other, a 'gonadotropic' hormone, was assumed to activate the ovarian follicles and affect metabolism. However no cogent evidence against the first concept Wiggles-worth on a single a hormone has been produced.

On the basis of the gradient-factor as a factor conditioning the JH-independent growth of the imaginal parts of the body, a theory which is discussed later, a hypothesis on the mode of action of JH has been suggested by the author (Novak, 1951a,b, 1956; cf.p.258). This seems to resolve the above-mentioned contradictions in the experimental results and to synthesize earlier views. According to this theory, JH produces its effect by taking the place of the GF in those parts of the body which lose it in the course of development. Such parts are the lateral parts of the body during larval development, the ovarian follicles in adult females, and a number of other tissues at particular times during development and Slama (1963), the effect of JH seems to depend on the moditioning of protein synthesis as well as other functions in the larval parts of the body. Its effect on oxygen consumption is therefore marely indirect, arising from an increase in the amount of metabolizing tissue (larval structures, ovarian follicles). The negative result on oxygen consumption produced by the implantation of can into castrated females seems to show that JH has no direct effect on the immaginal parts of the body. This question, however, needs further experimental evidence.

Put it in nutshell JH is principally a morphogenetic hormone. Its action consists in restitution of the capability for growth of parts of the body which have lost it in the course of morphogenesis. All other changes observed in the body after the administration of JH are indirect consequences of this action and are due to the altered biochemical balance produced in the body by the changes proportion of metabolically active tissue.

Chemical Characteristics of JH

Previously nothing was known of the chemical character of JH, largely because all attempts to extract it from ca failed. The situation changes when Williams (1966a) found that a lipid extract from the adbomen of cecropia males produced JH effects in Coleoptera (Tenebrio molitor, Schmialek and Wigglesworth, 1958, 1961) and in bugs and cockroaches, etc., as well as in cecropia and other moths. This discovered facilitated chemical identificaion of the active principle. Since then many papers on the principle of action of substances of this type have been published and many different chemicals, both natural and synthetic, have been found to produce similar effects. Of the many worker investigated the nature of this 'cecropia oil', the most successful were Roller and his co-workers who succeeded in isolating the active principle by means of gas chromatography. Later they identified the chemical formula of male cercopia extract, which was found to be methyl- 10- epoxy-7ethyl-3, 11-methyl-236-tridecadienoate. A number of related substances are also active.

Table 3.1. Juvenile hormone content of cecropia during development

Stage of development	*Hormone content per gram fresh weight compared with adult male (%)*
Unfertilized eggs	4.3
7-day-old embryos with yolk	3.7
Ist instar larvae (freshly hatched)	6.4
Vth instar larvae (mixed ages)	0.50
Freshly moulted pupae	0.75
Diapausing paupae (1 month old)	0.55
Chilled pupae (6 months old)	0.0
Pupae 2 days of adult development	0.0
Pupae 11 days of adult development	0.0
Pupae 17 days of adult development	0.0
Pupae 20 days of adult development	0.5
Pupae 22 days of adult development (males)	50.0
Adult males, 2 days	100.0
Adult females, 2 days	3.2

The occurrence of the active principle in the body of the cecropia silkworm (Hyalophora cecropia) during ontogeny was studied in detail by Gilbert and Schneiderman (1960). As illustrated in the following table, the cecropia JH content is fairly high in unfertilized eggs, during the embryonic period and in the freshly hatched larvae. It decreases at the end of larval development and in the pupa, diasppears when development recommences after the pupal diapause and does not reappers until just before adult emergence, when it increases slowly in the female, but at faster rate in the male.

Roller and Dahm (1970) demonstrated that the cecropia factor was identical with the actual JH produced by the ca of this species. A number of brain-cc-ca complexes dissected out from pupae two to three days before adult emergence were cultivated in a tissue culture medium for six to seven days. The ability of the complexes to produce JH after this time was illustrated by reimplanting them into last instar larvae of Galleria mellonella. The culture medium was then extracted three times with diethyl ether. With 50 brain-cc-ca complexes, an extract containing 10 000 to 15 000 Tenebrio units was obtained. This is evidence that the active principle of cecropia oil is actually JH produced by the ca is quite convincing.

4

Social Insects

Living together of insects develop different types of inter-relationship of which the social life is the highest development of intra specific inter relationship. From simple animal aggregations there may evolve complex animal societies composed of specialized types of individuals such as the colonies of bees, ants and termites. Although ants (like termites) live solely in organized societies, bees and wasps may have either a solitary or a social way of life. Some times the boundary between the two is hazy. Various simple forms of family life are shown by a few mining bees of the genus Halictus, as described earlier, and by a few vespid wasps. Vespid wasps of the genus *Stenogaster* include both solitary species and species whose members live in united families. Among the latter, the females reared by the mother remain in the nest for a short time giving help. After their departure, they are replaced by younger sisters. Frequently such offspring, instead of establishing their own nests at a distance, attach them directly to the maternal structure, forming a home that contains a loose federation of families.

Honeybee as a Social Insect

Honeybee is a social insects and exhibits a clear cut phenomenon of Division of Labour. For the first two three days of their life, honeybees take care of cleaning tasks and the regulation of warmth. By means of community co-operation, they maintain a continuous brood temperature of about 95°F to 97°F (35°C to 36°C). Ventilation and cooling are accomplished by mean of whirring the wings, in case of need even by fetching water and pouring it over the combs. On the other hand, heating result from vibratory activity of the flight muscles,

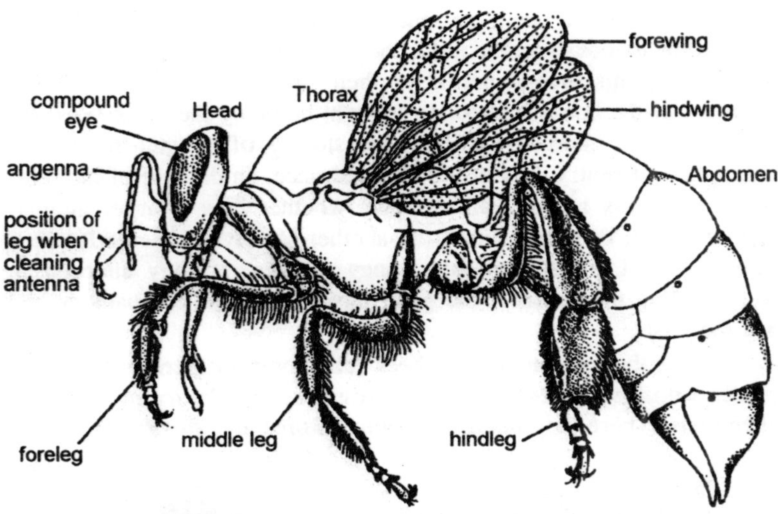

Fig. 4.1. Worker honeybee (lateral view).

from which the wings have been uncoupled, and thus from burning the carbohydrate that has been derived from the honey. From the third to the sixth day of their adult life the bees use pollen and honey to feed the medium-sized to full- grown larvae, which are four to six days old, Both of these foods are stored, carefully separated, in different groups of cells in the combs. From the seventh day onward, for about a week, two large glands develop in the bee's head. These are vitally important to the colony in that they secrete an extremely growth - promoting, predominantly protein - containing fluid, the royal jelly. This flows from the worker's mouth and is fed to the young larvae and the queen. Unlike them, she is given royal jelly continuously, for to ripen about 100 eggs every hour of the day and night her body must have an enormous metabolic turnover. The "royal court," whose members feed and lick their queen, consists of individuals of various ages.

The queen does nothing but deposits an egg at the bottom of each cell prepared for it. Every day she lays over 2,000 eggs, with a total weight more than her own. At any one time the bees in a colony have perhaps 10,000 larvae to feed, and each of these may receive several thousand feeding visits during the six days it takes to mature. In a strong hive with 10,000 to 80,000 individuals, the 2,000 to 3,000 males (drones) that develop and also fed with royal jelly. But gradually the males come to be regarded as useless parasites, are starved to death or killed by stinging, and are thrown out. During the twelfth to eighteenth days of a bee's existence the wax glands begin to function; in addition to construction tasks the workers have much to occupy them during this second period of their life. Above all there is the reception of the honey and pollen brought home by the other bees, the feeding of the larvae with it or the storing of it, and finally the assumption of sentry duty at the hive entrance. In defense of the nest, the bee is ready to sting an intruder, whether it is a larger animal, another insect, or even a bee from another society. Particularly when times are bad, bees turn into honey thieves. If they succeed in overpowering the guard at the gate, they fetch their companions and raid the nest.

Among the sting less honeybees (meliponines) there is even one species whose societies specialize in organized predatory invasions of other nests. There they pillage the honey stores and also steal wax for

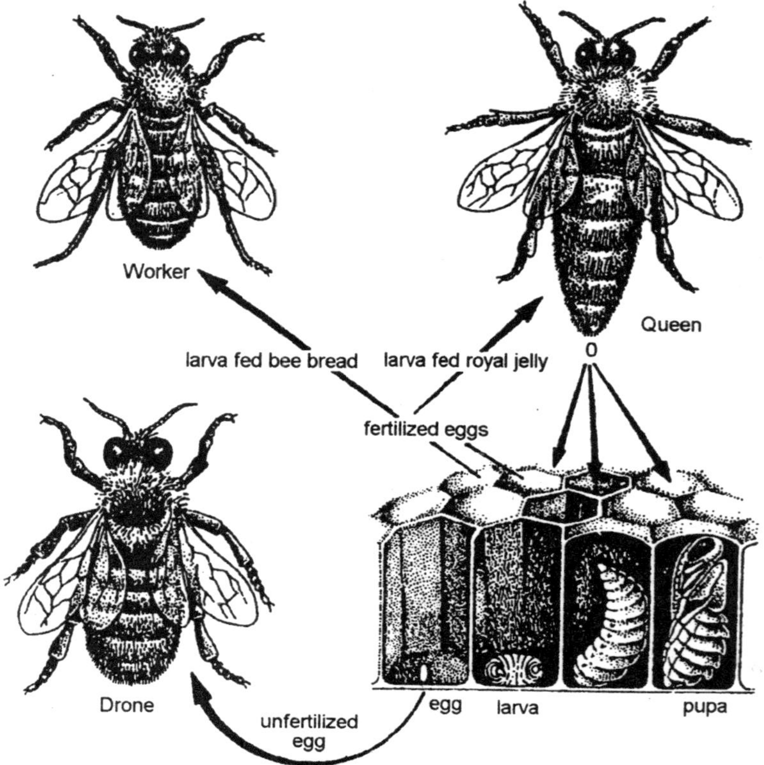

Fig. 4.2. Honeybee. Life cycle.

use as building material. The third phase of a bee's life, beginning at an age of about three weeks, is one of active service in the field. The worker prepares herself by undertaking several preliminary orientation flights in which she impresses upon herself the location of the nest and its immediate surroundings. Now she begins her food collections, and when she finds sources of food learns how to pass on information about them to her comrades. After perhaps ten days of this, the bee's life is at an end; the average age is 30 to 35 days, with a probable maximum of 55 days.

One might be tempted to regard the division of labour within the bee society as merely an automatic consequence of various glandular functions. But experiments have shown that such conduct can be repressed or can be forced ahead of schedule if conditions so demand. In other words, the requirements of the social organization are the moving force of all behaviour. Thus, if a hive happens to lose its normal building workers, old bees whose wax glands have regressed with experience a redevelopment of these glands and will begin to produce wax again. Or suppose a hive is divided into two nests, one with only young bees and the other with only older ones. In the first of these the catastrophe of starvation seems about to take place. And yet many of these young bees go flying out betimes to gather food, although this is not their job, and they do this in spite of their active royal-jelly glands, which then degenerate prematurely. On the other hand, in the older bee's nest, these glands continue to function in a number of bees longer than normally, simply because, for lack of young nurse bees, this has become a vital necessity.

The Workers

In other social bees and wasps, the female progeny do not leave the home, but continue to live there. They participate, as workers, in the care of the brood, in building activities, in the preservation of hygienic conditions in the dwelling, and in sentry duty. Domestic tasks include cleaning out of empty cells, removal of walling off of foreign bodies, and temperature regulation an renewal of air by fanning the wings in the vicinity of the nest entrance and air vents. This last is done in concert and mostly silently, except by bumblebees. Early in the morning a single bumblebee, dubbed a "trumpeter," takes a position immediately beneath the roof of the nest and ventilates it, with a loud buzzing.

Another chore of the workers is to receive food and building material fetched home by their foraging sisters. The food is eaten by

Fig. 4.3. Castes of honebee.

the workers and fed to the larvae right away or is first stored in special provision cells. The building material is used at once. An especially important job is the care of the brood and, for honeybees and ants, care of the nest mother or queen. With out proper attention the eggs may die. The larvae, which live in groups within the structure, have to be shifted again and again from chamber to chamber, according to the temperature and the humidity. Even newly matured insects may need to have the cells opened for them when they emerge, or may have to be freed from the pupal skin, as happens with ants.

Aside from these chores, nursing is confined to distributing food adapted to the age of the brood. In general, wasps and ants proffer food prepared from captured insects, spiders, or even fresh meat; the wasps provide it in the form of little balls saturated with saliva, the ants usually as droplets of predigested juice. Bees, on the contrary, feed the brood only with pollen and honey. Just as young songbirds, apparently by means of their strikingly bright-coloured and widely opened gullets, keep alive and augment the urge of the old birds to continue feeding them, the larva of wasps and many ants (but not of bees again and again offer a drop of secretion on their extended mouthparts. This is freedily licked off by the older insects and thus indirectly stimulates them to feed the larvae. Such exchange of food, called trophallaxis, may also induce a swarm to rear as many offspring as possible. Moreover, many ant larvae do not produce this delicacy from the mouth, but instead exude it from peglike processes or even from the entire body surface. In the vespids of the genus *Belonogaster* the assisting daughters are just like the mother. Since their urge toward nest construction and search for food awakens only with sexual maturation, they are occupied during their first week solely with cleaning operations and with distribution of food in the nest. Then they begin to lay eggs, gather food and enlarge the nest as needed. Nevertheless the family still needs the original mother, for if she were lost the daughters would finally devour their own brood, and the entire society would gradually die off. But otherwise the family gets bigger and bigger, and males, too, are developed. For four to five days the males are fed in the nest with stored up honey by their sisters or else, pilfering on the sly, lead a parasitic life, as is also the case with various species of the wasps *Polistes*. If the members of the community become so numerous that for technical reasons the structure no longer can be enlarged sufficiently, some of the wasps depart and found a new home. But they do this in a most indelicate manner. They not only bite off material from the old home for use in the new one, but even pillage it of the larvae, feeding them to the new brood, with the result that old nest soon is destroyed. *Polistes* wasps are of a rather predatory nature anyway, and occasionally they plunder one another's nests.

In other wasps, such as *Ropalidia* and some species of *Polistes* the worker caste begins to show some of the differences that among the rest of the social bees and wasps make it a clearly distinct form. The offspring of the nest mother are smaller females and have reduced fertility or lack it entirely. This results from a restricted supply of food. Except in honeybees and ants, the later offspring, as the family grows in size and the amount of feeding becomes larger, develop into bigger and bigger individuals. With wasps such as Polistes and occasionally even with bumblebees (*Bombus*), the later-emerging females scarcely can be distinguished from the mother any longer, but with the other wasps and hornets (*Vespa*) they always are clearly smaller than the queen. The tasks of the workers in the nest frequently are determined by their age-that is, by the degree of maturity of definite internal organs- and thus the life cycle of the individual unfolds in a predetermined sequence. Such a sequence of duties has been best studied in honeybees and, with some exceptions, follows the schedule described here.

Ant Workers

Ants are also social insects. In an ant society, though brood care and division of labour are more highly organized, the type of work seems to be less definitely regulated than it is with bees, and its individual facets are only slightly dependent on the age of the individuals, or not at all. So the society is much more adaptable. In place off males that merely are reduced in size and more or less sterile, there are workers with distinctly different body structures specially suited to the tasks at hand. The workers of a given ant species may be more or less uniform in size and shape (monomorphic), may be variable (polymorphic), or may appear as two distinct types (dimorphic). These last, known usually as "*workers*" and "*soldiers*" may represent the extremes of an originally continuous chain of variations, the intervening links of which no longer exist.

In genera with a single type of worker, only one variant, usually a smaller one, has remained, with ants low in the evolutionary scale such monomorphism may be a primary manifestation. Where a species has different types of worker, the large ones are designated as "*soldiers*," yet by no means are they always the defenders or the warriors of the society. Big heads and mandibles of extra ordinary shape may be based upon other biological pre-requisites, for instance, they may be necessary for mincing up seeds or animal prey. In some species the nest entrance is closed off by the head of a "porter" stationed there; in certain genera the porter's head is so constructed that the door is not only barricaded but is concealed besides. Various worker forms, ranging from about 0.08to 0.6 inch in length, are found among the leaf cutting ants (*Atta*). As in other ant societies, the different tasks appear to be parceled our to the types best fitted for them. The

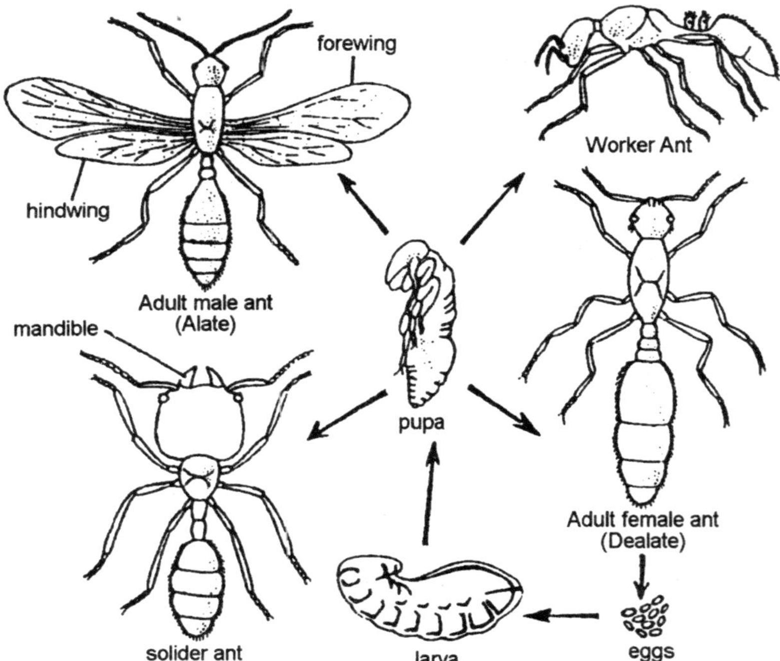

Fig. 4.4. Ants. Castes and life cycle.

dwarfs take care of the mushroom garden, the middle-sized individuals gather leaves and work them over in the nest, and the big workers guard and defend the nest.

Different tasks in the nests of ants all can be carried out by any worker, and usually a single worker executes a particular job only for a limited length of time. Nevertheless, the small individuals seem to devote themselves preferentially to the actual care of the brood. Both the eggs and the larvae are licked over and over and put in piles, arranged according to their age, and the pupal coccons (often mistakenly called "*ant eggs*") are also separated and heaped up in different chambers. Ants mostly feed their larvae, as they do their queen, their fellows, and their guests, from mouth to mouth with drops of nutritive juice from a "crop" or "social stomach" that, like the honey stomach of bees, lies in the fore part of the abdomen. But more primitive species, such as many ponerines, proffer chopped-up insects; and a few genera offer balls of minced tissue as food, in some cases from a special oral pocket.

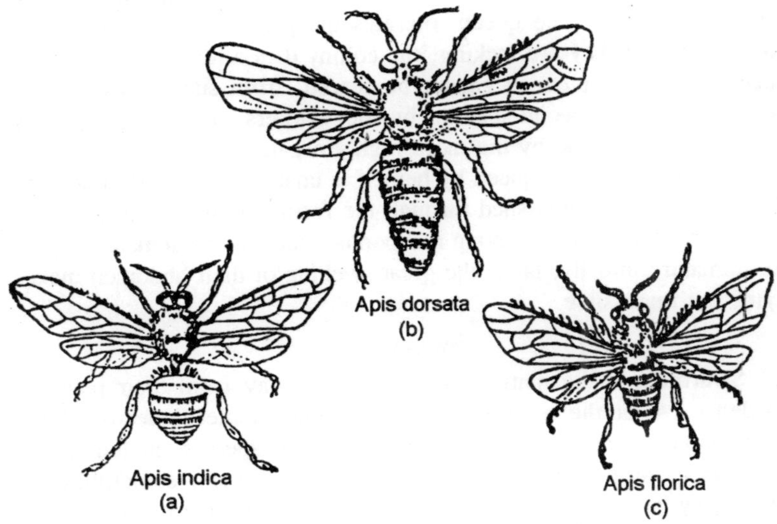

Fig. 4.5. Workers of three species of honeybee.

Forms of Social Life

After this consideration of the worker caste and its duties, we return to the forms taken by family and civic life. In the tropics the life span of families of vespid wasps and bumblebees (Bombus) does not depend on portion of the hive together with one or several queen-

or else, as in other wasps, simply by individual fertilized females. These *Polybia* wasps, moreover, have the strange peculiarity of flying soundlessly, and are particularly prone to attack intruders within a circumference of about two years of their nest. The bleeding puncture made by their sting is very painful and makes the affected part of the body whiten at once, yet causes only a slight swelling. The frequently large societies of the "stingless" honeybees (*Melipona*, *Trigona*), which do have a sting, though an ineffective one, seem more primitive than the societies of *Polybia*. For these honey-bees close off their brood cells, as do the solitary bees, after providing them with honey paste and an egg. Further feeding, with consequent relations between larvae and workers, does not occur at all. But again, the queen of the nest is more highly specialized than the queens of bumblebees and wasps. With shortened wings, small aggregations of pollen on the legs, a little head, and weak mouthparts as well as reduced brain, she is greatly dependent on the assistance of workers.

With a single queen, stingless honeybees form societies that endure for several years. The queen tolerates the presence of other, virgin queens. Where they are lacking, the colony dies after the death of the queen, for another queen cannot be reared, from such larvae as are present, through special feeding by the workers, as can be done by honeybees (Apis) and by the more highly organized among the wasps. On the other hand, the queen by herself is unable to establish a colony. This has to be accomplished through the formation of a swarm led by a young female. In fact, among meliponines such young females usually are immature initially; later, the great swelling of their abdomen makes flying out impossible.

Swarming

Swarming is the mating of individuals away from their place to residence. After the sexual forms-males and females (queens) have been reared, swarming occurs. Since in many species, including the honeybees, these forms are larger than the workers, sometimes far larger they have to be reared in special large cells. One might imagine that honeybees would construct their cells of a size proportional to their own body mass. But where do honeybee workers get the proper measure for their fatter brothers, the future males (or drones), which they have never seen before in their lives? What signal suddenly causes these bees to build larger cells in the comb, and impels the queen to deposit eggs that are unfertilized and that therefore produce males? Truly they have great power over the life of their folk, these workers.

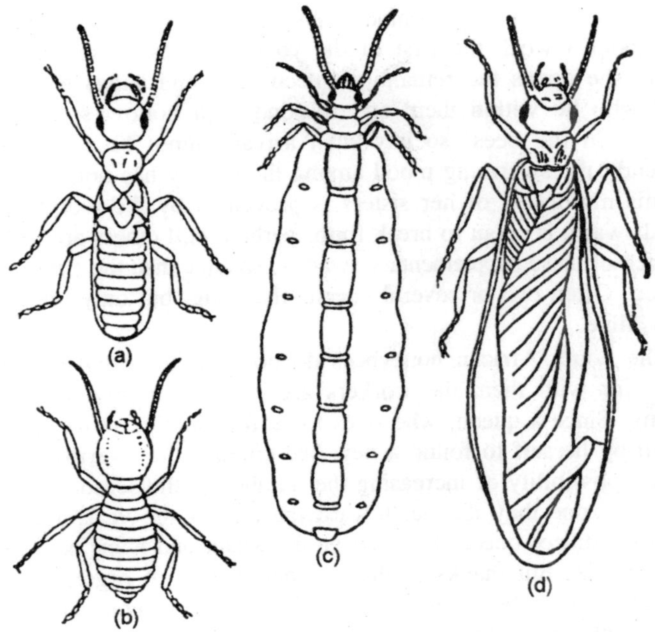

Fig. 4.6. (a) Queen of termites without wings after nuptial; (b) Wingless sterile workers; (c) Mature queen; (d) The winged male before nuptial.

Ultimately, too, they determine the number of the young queens. These are reared with special food in the few big, pendant, peg-shaped queen cells, which strangely enough bear on their surface the size—cornered pattern of the comb. Inside the cells the young larvae hang head down ward. When they are ready for pupation, their cradles are closed over with waxen covers. The worker caste is then gripped by that feverish, pervasive agitation, the swarming mood. This leds to the departure, as if on order, of about half the population, after each of the individuals has filled up on honey from the plentiful stores in the nest. In the course of only a few minutes, they fly off from the exit, perhaps some 30,000 of them, together with the old queen. Not being an especially good flier, she soon settles some where on a branch, and the whole billowing swarm of bees gathers around her like a bunch of grapes. From this place, usually a few hours later, they make off to a new nesting place that has been explored mean while by scouts. Or perhaps the swarm is packaged up by a beekeeper and given a home in his apiary.

It is precisely this situation that has enabled man to culture bees a practice known to the Egyptians over 5,000 years ago. Among the ancestral stock that stays behind when the swarm leaves, there emerges after about a week the first of the young queens. Excitedly singing "tu-tu" she opens the remaining queen cells and stings to death her rivals who lie within them and respond with hollow sounds to her call, for in the bees' society such arival cannot be tolerated. But frequently the swarming mood among the colony has not yet subside, and this murdering of her sisters is prevented by the workers, for a second swarm is about to break forth, perhaps still others are to follow, and each of these supplementary swarms, smaller than the first, requires a queen. Often two or several queens fly along, but only one of them is left alive.

The South African honeybees do not absolutely have to have a queen, for with them the workers are capable of producing female progeny. Since a queen, who lives to be four to five years old, is not able all by herself to found a new bee colony, the swarm constitutes the only possibility of increasing the number of the season, but in the temperate zone they die out toward the end of the year. Here, only the new, fertilized queens live through the winter and found new societies in the spring. But thanks to their higher degree of organization, the societies of honeybees and ants are able to persist continuously even in the north; ant colonies have been observed to last for over eight years.

Many wasps and bumblebees have simpler and more similar forms of social life, and an example will describe their normal course. A bumblebees queen, after over wintering, finds a site that appeals to her and prepares it for nesting. After plastering a spot on the ground with wax, she mixes a lump of pollen and honey, builds a circular wall of wax on top of it, lays some 7 to 16 eggs inside, and closes the cell with a wax roof. From the same material she builds a storage vessel for pollen and honey to one side, placing it where she can reach it without leaving the waxen cell. Sitting on the cell, she keeps the eggs inside war, and they soon hatch into larvae. These are fed from time to time be the mother, who bites a hole through the lid of wax and then closes it afterward. The larvae also eat the mass of pollen and honey on top of which they were born in the cell, which the mother enlarges, as they grow bigger, until finally they spin cocoons that hang together; in these they pupate. The queen then attaches more waxen egg chambers here and there to the old cell, using in part material from it, so that the lighter-coloured cocoons now lie bare. After the first bumblebee workers emerge, these cocoons are used as storage vessels. The workers help the mother, and the more numerous they become the less she works herself, until she perhaps has become nothing but a producer of eggs. Thus the number of queen in two ways embodies the transition from the solitary to the social bees. First, she gathers food like the solitary bees and closes off the cell after laying eggs, but on the other hand constantly feeds the young through openings. Second, in the beginning she carries out all the tasks, but later merely lays eggs like a queen bee.

Somewhere in the midsummer, when the family has become numerous and food is plentiful, new queens are reared. Bigger and bigger workers develop, and some lay unfertilized eggs. If there are excess eggs, the bumblebees frequently feed on them. The unfertilized eggs always develop into males, as is everywhere the case among social Hymenoptera. Thus males may be descended from workers. But more often they are produced by the queen, who is able voluntarily to prevent access of semen to the egg she is laying. This semen, which suffices for the entire duration of her life, was received from the male with whom she mated in the preceding year. The emerging males, fed by their sisters, remain in the nest for a few days until their hairy covering, at first a whitish-gray, has gained its full colour. Then they fly off, visit flowers and mate. In the tropics bumblebee societics may last for years; then, instead of a single queen, a whole group of them is active in the nest. In the far north, there frequently is another variant. During the very short period of the year when flight is possible, the mother's offspring may not develop into workers but instead directly into males and new queens, as is customary in the solitary bees.

The family life of many vespid wasps corresponds in great measure with that of the bumblebees, apart from the wholly different manner of constructing the nest and feeding the young. In both groups, the mature insects imbibe nectar and sweet of fermenting plant substances, as well as honeydew. But the vespids, with the apparent exception of the nectarines, feed their with a paste of chewed-up insects. In the spring two or several vespid queens occasionally found a nest in common. In *Polistes* wasps, this urges for sociability may even extend to where a female will leave her own nest and her still-young brood in the lurch and will attempt to join a distinct family. Whether she succeeds is another matter.

For the majority of species drive away intruders, even those of the same line of descent; the cause of this is nothing else than that very possessiveness toward the brood that to some degree has been lost by the worker caste in correlation with their degenerate sexuality, a loss that has made possible the selfless cooperation of the individuals in a society. Through this loss a certain loosening of the rigidly established instinct for the care of the brood also entered and brought with it space for other, new tasks. The most highly developed social organizations among the wasps are those of certain species of *Polybia*. They are said to surpass in this respect even the social organizations of honeybees, which are very similar. With one and the same *Polybia* species the establishment of new colonies can take place in very different manners according to the circumstances. It can take place by swarming, as in honeybees- that is, through the departure in flight of a bee population ad thereby preserving the species. Lack of food or any kind of disturbance or menace an also provoke bees into swarming; such migratory swarms then represent merely a change of dwelling place and not a fractionation of the population.

The honeybee Apis mellifica, originally from the Old World, has been Spread by man over the entire earth. Until recently its races were not products of intentional breeding, but were primal geographic forms. Today, at all event, crosses can be produced by artificial insemination of queens. Of the remaining three *Apis* species (dorsata, flora and indica), all of which live in Asia, only the last has been domesticated.

In ants a quite different kind of swarming is seen among ants. Excited workers appear before the best entrance and get more and more numerous. Finally, there appear the winged, sexually mature individuals, which usually are reared in large numbers only years after the society is founded. First come the males, and then the females. They rise into the air, perhaps like clouds of smoke, later to cover the earth with a layer of ants. This is no swarming in the sense of the bees, for these are mating flights.

In many genera the males and females, which as a rule are equipped with huge wings, are almost equal in size, but in other genera they are very different. One may even see a tiny male riding about on his partner's abdomen as if on a dirigible. Verging on the incredible are the differences in size between sexually mature forms and works, especially in legionary ants (dorylines). Here, a male may measure 1.2 inches, and the smallest worker a mere 0.08 inch; or a worker 0.12 to 0.16 inch long may have a queen more than 2 inches long.

Role of Males

In hymenopterous colony the males are of no use. After mating, they die. The fertilized females then found their purely female societies, as queens with a final cohort of many myriads of workers. There are ant socities with over a million individuals, and one population my hold sway over an area of hundred, even thousands, of square yards. Certain *Farmica* ants pile on top of their nests heaps of earth more than a yard high and perhaps four yards in diameter. Such digging by ants is of great benefit to man in loosening and aerating the soil.

Life Span

On the average the workers live to be about three years old. Thus they have much more time than bees, of instance, to gather impressions and gain experience, and their adaptability is decidedly greater than one might credit insects with. The queens may live for as long as three decades, not as rulers, it is true, but merely as producers of eggs, for among ants even more exclusively than among bees it is the colony that does the deciding. In the simplest case, a fertilized female founds a colony by enclosing herself in a hole she has dug or in some other hiding place, after shedding her wings, which is facilitated by a performed line of rupture. Here the ant remains closed off from the world at large until the emergence of her first workers, which may take a year. She ingests nothing but moisture, living on her bodily reserves. A good part of which come the breaking down of the thoracic flight musculature.

This source of nourishment also enables the queen to rear her first larvae on glandular secretions in spite of the long period without food. The queen may eat some of the eggs she lays, but they of course, are products of her own body. When the larvae are fully grown, the mother helps them in the spinning of their cocoons by providing them with bits of dirt, which they incorporate in the structures. Later, she helps the workers emerge from the workers emerge from the cocoons. At this point her instincts for the care of the brood usually disappear, since subsequent brood care is taken over by the workers; only in the smaller societies of certain ponerines and *Leptothorax* does the queen continue to help.

Societies of a few of the primitive poneriness may be no larger than ten individuals. In founding them the queen does not enclose herself, but goes on the hunt, fetching food for her brood in the manner of the solitary wasps. Frequently several females work together in establishing the nest, but when the first brood of workers appears, the mothers battle one another until only a single one survives. But several or even many of them are tolerated when spatially more extensive societies are concerned. The young queens then settle down in outbuildings after their mating flight. This procedure is also followed by species in which queens have lost their capacity for digging or for some other reason are unable to found a society by themselves. By remaining within the existing society, they have ant helpers at their immediate disposal.

In the genus *Carebara* a young queen solves this problem by taking along with her on her mating flight a number of tiny workers from the nest; these cling fast fast to her en route. Or a female of Formica sanguinea, a species that is particularly many-sided in its methods of establishing colonies and its ways of life, first joins forces for founding a common nest with a female from the company of one of her slave species; then, as soon as the latter's first workers appear, she kills her. When females that are not self-sufficient but are dependent on worker.

Ants establish branch settlements of a flourishing colony, there frequently are schisms and estrangement between the original population and it offshoots. On the other hand, the adoption of young queens is by no means limited to populations of their own species, but strangers of the same genus are accepted, resulting in mixed communities. These occur frequently, especially in the genus *Formica*, and also in *Lasius* and others. They are only temporary, however, is the queen of the original population dies, and often enough she is beheaded or murdered by the adopted stranger or the latter's workers. In this case a pure stock of the adopted species gradually arises. Communities that have lasted a long time frequently consist of several mixtures and a confusing diversity of interrelationships. For instance, a female of a third species may establish her nest in a *Formica* colony that already consists of two species; and, in addition, the workers may sally forth to capture slaves, bring home pupae of still another species, and let the ants that hatch from them likewise become servants of the state. Such mutual associations of different species stretch through countless possibility, including mere neighbourly activity, the mutual benefits of symbiosis, hospitality, and larceny. They range in one direction to slavery (in which only the ruling ants produce males and new colonies, while the helping ants are allowed to produce only workers), and in the other direction to parasitism.

Slavery

The dependence on slaves varies from one society to another. The custom of capturing slaves seems to have developed from the dependence of those queens that are able to establish a nest only with the help of workers. The pupae of foreign societies are stolen, and a permanent supply of the enslaved working a population that merges from them is assured to the society by forays repeated annually. But occasionally the slaveholders cradicate all the potential sources of slaves in the area and have to return gradually to self-sufficiency. Slave hunts may be extended over such great distances that the soldiers establish guarded camps at intervals of a day's march from home. The saberlike mandibles of the Amazon ants (Polyergus) make them fine fighters, but also make them so incapable as workers at home that they are fully dependent on their slaves, which usually far outnumber them. On the other hand, many *Formica* community and the greater its own resources, the fewer slaves it maintains. The species of *Formica* that steal slaves, although they are less capable soldiers than *Polyergus*, thus remain good workers.

The Amazon ants must depend on their slaves even to feed them, for their salves even to feed them, for their dagger like jaws, tailored for piercing the head of adversaries, do not permit them to chew their food. In addition, these implements are no good for caring for the brood or for construction, and apparently this formal specialization is accompanied by a degeneration of the corresponding instincts. Thus, the true masters in the Polyergus community are the slaves, which happen to be Formica species. These would by no means exchange their life of slavery for one in the maternal nest, for among the warrior-folk they are better protected, quit apart from the fact that ants feel at home in the place where they were born. And more than ample food is at hand, for during the summer the daily slave-hunting expeditions yield an abundance of captured larvae and pupae that can be eaten.

The predatory attacks of *Amazon* ants on *Formica* communities are carried out with refined tactics adapted to prevailing conditions, and are uncannily abrupt. After the return of scouts sent out to ascertain the route to be followed for attack and the proper initial disposition of the forces, the soldiers assemble before their nest, excitedly drumming each other's heads with their antennae. As if responding to an order, part of the population, numbering from many hundred to perhaps 2,000 warriors, sets out in a close column, marching directly at full speed and perhaps straight through heavy grass to the gathering place near the selected nest. When the attackers have consolidated here, they carry out an assault so suddenly and with such concentrated force that the Formica community is caught by surprise and in a few minutes is sacked. Nevertheless, in the first confusion of battle, many Formica ants succeed in fleeing to a place of safety, carrying part of the brood in their mandibles. A *Polyergus* colony has been observed to undertake 44 expeditions in 33 days, collecting a booty estimated at 40,000 slave pupae. In addition to various species of *Formica* and *Polyergus*, the myrmicine genera *Strongy lognathus* and *Harpagoxenus* take slaves.

Stronglylognathus ants have saber-like mandibles, but unlike the Amazons are capable of feeding themselves. And they have different tactics or battle methods. Thus certain *Stronglylognathus* communities maintain throughout the duration of their attack, a constant liaison between their own nest and the *Tetramorium* nest they have assaulted; their aim is to intimidate the inhabitants of the latter, namely to render their more or less defenseless by pinching them suddenly from behind with the mandibles. But if battle is joined, nonetheless, fights break out between the robbers slaves, participating in this expedition, and the similar workers of the besieged community, from which it is possible that salves themselves have descended, having been captured in an earlier raid. Consolingly, however, such an occasion may end in reconciliation, with attacker and attacked giving up the battle and joining forces. The methods of slaveholders vary. `Colonies of *Harpagoxenus sublmevis* drive *leptothorax* societies out of their own nest, set themselves up in it, and obtain the indispensable helping ants from the Leptothorax pupae left behind. For these slaveholders, which even try to make workers out of already-hatched queens by biting their wings off, are unable rear their own brood.

Composite Nests

Composite nests also are shared by very different groups of ants and even by ants and termites. Frequently societies of this sort are brought together by such things as the presence of favourable dwelling places—for instance, hollow branches or the hours made by termites. Each of the species, more or less forced by circumstances into a neighbourly existence, has its own dwelling, but with a common entrance to the nest. This may result in various forms of symbiosis. A defenseless species may live with a species capable of defending itself and may bear a deceptive resemblance to it. The combined defenses of two or more populations living in a given habitat may enhance the safety of each. In the outer parts of the ant gardens of a *Componotus* colony, for example, a population of *Cremastogaster* may establish itself. Under moderate menace only the latter takes the appropriate measures of defese, but when seriously threatening attack occur, the fundamental owners appear from the depths of their spherical nest.

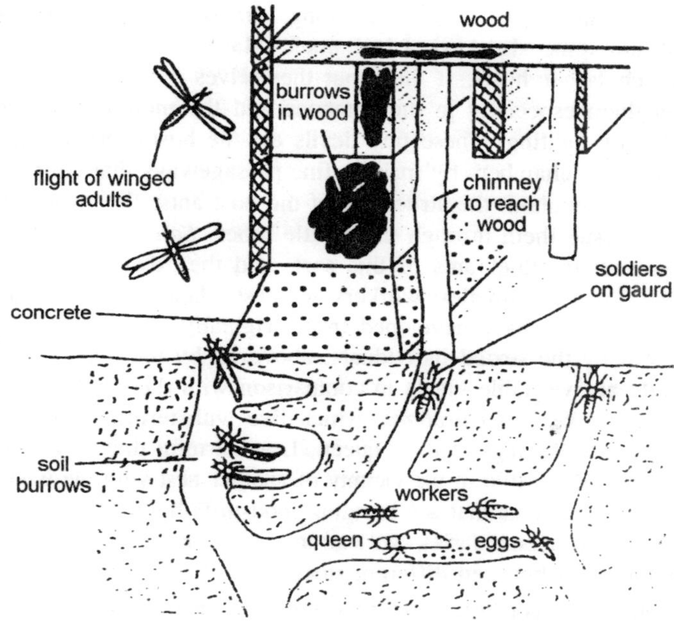

Fig. 4.7. Diagram showing the termite colony.

Guests

A few genera of tiny ants live as *obligatory lodgers*; they build their little cities and dwellings in the middle of the nests of much larger species, being tolerated a friends of the latter and fed by them on solicitation. Certain *Leptothorax* species are ant guests of this sort. They often ride on the back of their host ants, licking them. Other such guests include a few *Formicoxenus* and *Symmyrmica* ants, the males of which, as a rare exception, are wingless. Except in human society, hospitality probably is nowwhere manifested in so varied and interesting ways as among ants (and to a much lesser extent, among termites). In addition to welcomed and uninvited ants guests include certain mites and isopods (soe bugs). One may estimate at several thousand species the number of living forms that, by craft or by friendly means, overcome the natural standoffishness of ants and profit directly from them in one way or another.

Thieves

There are a great many thieving ant species whose workers settle among other ants as bad, antagonistic neighbours or even as lodgers, frequently in numbers nearing the hundreds of thousands. They are dwarfs only about 0.04 to 0.08 inch long, almost too small to be taken hold of and some also have a bad smell. Masses of them will attack their much larger hosts, if these put themselves in the way, and by stinging them especially in the antennae and the mouth, make them incapable of fighting. These tiny devils eat the host's brood and get into the hosts' chambers by making fine passageways that course like a network throughout the structures of the host ants; the hosts cannot possibly pursue them through these little tubes. *Soienopsis* and many other genera are thief ants of this sort, and they dwell also among termites. Since the smallest workers are best adapted to this away of life, these species have developed from originally larger forms, as is still shown by the sexual individuals, above all by the queens, which frequently are veritable giants in comparison with their tiny workers. Thus, the workers of *Carebara*, which live with termites, seem like lice hanging to the legs of their queen. Less harmful thieves are those highway men that settle in the vicinity of another nest with the intention of plundering the ants that are bringing foodstuffs home. Both robbery and quarrels over boundaries can lead to open war or perhaps to feuds lasting for years between different stocks.

Contest of a more ritual nature also occur among ants. *Camponotus* ants, for example, run headlong against one another, with the abdomen audibly beating the ground. But ants in general engage at once in mortal combat. The principal weapons are the mandibles and, among the more primitive ants (ponerines, myrmicines), the toxic sting. In lieu of stinging. *Formica* curls the abdomen forward between the legs and sprays out its poison, formic acid. Certain species disseminate repugnant odours (for example,.many *Lasius*), and dolichoderines plaster the enemy's antennae with a sticky, ill-smelling secretion from the rear end. And finally, the soldiers of many species are able to blockade the nest entrance with their head.

Harvesting and Hunting

Ants usually feed upon plant juices, nectar and honeydew, as well as on insects (ants not excepted), worms, and other small animals. Insect-eating ants can be highly useful to man. A large community of wood ants of the genus *Formica*, for example, destroys about 100,000 insects daily. In specializing on certain types of food, some ants have developed unusual organizations and ways of life. Ants that feed on honeydew, a sweet excretion of aphids, are good example. These ants in their own nest raise aphids, feeding them leaves or roots and even building special chambers-stalls, so to speak- for such "milk cows."

The ants comprehend to an astounding degree the essentials of aphid care, a capacity that has probably developed in great part from the instincts for the care of their own brood.

Walking Honey Jars

Certain ants (such as Mymecocystu; Prenolepic, Leptomyrmex, Melophorus, a few Componotus and Plagiolepis) of warm and dry Countries store juice in the bodies of their own workers. A number of such workers are fed by their sisters until their expandable abdomen inflates into a gigantic, amber-coloured, shimmering sphere, upon which the normally contiguous segmental sclerites are widely separated from one another. These "walking honey jars" can creep forward only slowly and with difficulty, and usually remain hanging calmly in groups on the ceiling of a cavity in the nest; particularly when food is scare they dole out their store from the mouth drop by drop to their enterating sibs.

Man gathers and eats the bodies of such honey ants, or at least their contents, rating them as a delicacy that surpasses the honey of bees. Incidentally, the honeydew of aphids is not the sole constituent, for these ants frequently also gather nectar and exudations from plant galls, in America particularly those from oaks. Quite different specialists are the harvest ants, especially *Messor*, *Pheidole*, *Holeomyrmex*, and *Pogonomyrmex*, the last of which strings energetically. These collect plant seeds, preferentially those of grains and grasses, and accumulate them in their nests. The seeds, later to be broken up and eaten, are laid out in the sum to dry; if they nevertheless sprout, the sprout is often bitten off. Certain species throw such unwanted seeds out of the nest, and surrounding the nest entrance there soon grows a dense fringe of grass, a phenomenon that has led to the erroneous designation of these ants as 'farmer ants." The harvest ants are not only eaters of seeds, but also hunters of insects; in addition, a hard-stinging American species (Solenopsis geminata) eats fruit and cultures aphids.

Ant Farming

In contrast, the well-known leafcutting ants (Atta) of tropical and subtropical America are one-sided specialists that grow "vegetables." There vegetables are the protein-containing bodies that grow on the mycelia of certain fungi. And the majority of these fungal species will flourish only in the gardens of certain leafcutting ants. The culture medium in which the fungi grow is prepared by these ants and consists of a fermenting mass of chewed-up leaves carefully manured by the ant's own excrement. The amount of plants that an *Atta* community will cut up in a very short time and carry into the fungal chambers is so great that the result is frequently as destructive as a catastrophe of nature. The nest workings, which may be many yards below ground and spread over a wide area, may perhaps have a volume of several hundred cubic yards.

Rocking and reeling, the ants hustle nervously along their routes, stumbling past their fellows coming the other way, who draw aside for them, and, falling suddenly into the aperture of a nest entrance, are swallowed up by the earth, one after the other, or even two at a time, over and over again. On her flight into new county, a young queen, as founder of a colony-to-be, brings with her not instruct for tending a fungus garden but, in a special pouch of her head in back of the mouth, a tiny was of fungal mycelia from the maternal home. In the new little brood chamber the ant puts this down, deposits five or six eggs on it, and manures the fungus by pushing little bits of it beneath her anus and then returning them to the pile. The wad grows, and yet the young queen takes none of it; instead she eats some of her own eggs and also feeds her first hatching larvae exclusively on these.

The tiny workers that after a few weeks have developed from these larvae are the first to consume the "vegetables" that have grown on the fungus plant, which now is about 0.8 inch in diameter. The mother and brood still partake only of eggs. The workers also manure the precious fungus, open the brood chamber perhaps some ten days later, and gather and chew up pieces of leaves. This now permits the vitally important vegetable to develop adequately.

There are certain species of insects which are not leaf cutters, but culture their fungus on insect feces, especially that of caterpillars, or on all sorts of decaying plant tissues. In great contract to these vegetarians are two groups of warlike carnivores. First, there are the primitive Australian ponerines of the genus *Myrmecia*, known as bulldog ants and notorious for their stings and furious attack. Like certain other ponerines (*Odontomachus*, *Harpegnathos*), they are able to leap upon their adversary by snapping themselves backward from the ground with their powerful mandibles. Otherwise, ability to jump has been seem among Hymenoptera only in a few chalcids, and here is done with the legs. In their organized predatory raids, the bulldog ants hunt mostly termites. The second group of carnivores are the driver, or migratory, ants (dorylines). They also prey on termites, as well as on other ants. In tropical America the driver ants are represented especially by the genus *Eciton*, and in Africa by *Anomma*. Their sexual individuals are gigantic, the males resembling wasps more than ants, the females wingless from the beginning. The workers, remarkably unequal in size, have saber-shaped mandibles. The workers, remarkably unequal in size, have saber-shaped mandibles.

Army March

The driver ants travel in well-organized formations that differ according to the species of ant. Some columns may be more than 100 yards long and several yards wide. At times, the workers with the most formidable mandibles march on each side of a column. Biting, stinging, murdering, and pillaging, these armies move over the countryside, some travelling more through open territory, others through woods and thickets. Whatever is unable to flee from them is torn to pieces and carried forward in little bits. The ground, vegetation and trees, holes caves, nests, even houses are in a short time completely cleared of every living thing, and even larger animals and people take flight. Clamoring antbirds accompany these horrifying processions and eat their fill of the insects before them. Many driver ants migrate below ground, and others under the protection of leaves and fallen debris. In case of need they may even build uninterrupted tunnels above ground over open stretches. The driver ants build no nest, but in a sheltered place crowd together in a heap, leaving cavities that constitute the brood chambers. These satisfy all requirements for the care of the brood. In the nomadic life of these ants, resting intervals approximately twenty days long alternate with slightly shorter migratory periods, a rule that is determined by the rhythm of egg-ripening in the queen, for she hardly could take part in the march with her abdomen swollen with ripe eggs.

In a circle of as much as a mile around the resting places, everything edible is soon eradicated, yet the signal for a new migration apparently does not stem from a shortage of food, but rather from the 25,000 or so larvae produced from the previous egg-laying period. In fact, this signal seems to emanate from their secretions, which are licked up by the attending workers and stimulate them. Because of their social effects, these secretions are called "*social hormones*." Certain other insects go along as undisturbed guests on the expeditions of migratory ants. Rove beetles (Staphylinidae) frequently accompany the ants. That they are allowed to do so may seem odd, for they rarely resemble the hosts in appearance. But with the exception of the males, the ants have no compound eyes, and at most have ocelli; the majority are blind and gauge their environment exclusively by the senses of smell, test and touch. Evidently these guests meet the specifications covered by these senses deceptively well. Along with such adaptation, the guests have developed to an astonishing degree the ability to respond satisfactorily to antennal "fingering," the so-called "antennal language," despite the difference in the structure of the host's and guest's antennae. When the migratory ants raid termite houses, frequently some of the accompanying guest get into the termite chambers, perhaps lose touch with the ants when these withdraw, and are left behind, usually to starve to death or to killed. But certain of these guest are able to make themselves acceptable to the new hosts and live henceforth as guests of the termites.

Insect Communication

The ability of insects to live together in communities rests on the means of communication among them. Such communication is based first of all on the species and nest, which are perceived through the antennal sense organs, Sentries at a nest entrance admit no individuals contaminated with the odour of strange nest, and populations refuse to accept them. But if strangers somehow succeed in taking on the community odour, they are adopted without further ado. Some ants regularly use this "fifth column" system to gain entrance to a nest, and certain others are rendered inconspicuous naturally by the possession of deceptive body odour. Mother and children (or workers) recognize one another by smell, even after long separation, at least up to a certain age; after recognition may fade, for older workers often change. Odours serve not only for recognition, but also for marking pathways. By their odours, for example, honeybees become direction signs for their companions returning homeward. In front of the hive entrance, bees stand and do a waggle dance, during which fragrance is emitted from an exsertile, membranous organs between the last two segments. The odour is disseminated of the winds by means of continual fanning movements of the wings. In the same way bees that have discovered a source of food call in their fellow. Or the fragrance of the flowers, clinging to a bee that returns home successful, can induce her nest-fellows likewise to scour the vicinity in search of such blossoms. The simple urge for imitation may play an especially large role in ant communities; it seems that there the older workers, leading the way in all actions, set an example for the younger ones.

When a large prize is being transported by several individuals, for instance, there is a constant participation of new individuals, that attracted by what is going on, hasten up and crowd into the work. Thus, there is a constant succession of more or less compulsory relief of those already on duty, and the object passes through many "hands" before it is finally fetched home. Such community transport takes much longer than that engaged in by individual ants, parincipally because the different temperaments and intentions of the participants are likely to be a mutual hindrance. Quite similarly, any moods or conditions of excitement may be disseminated, via imitation, throughout the population, yet ants also have at their disposal concrete means of expressing themselves. As one instance, there is capability of many species of producing high-pitched chirping notes by rubbing two abdominal segments together. And, of course, there is the "antennal language," which is rooted in the sense of touch. Definite signs and signals are given by antennal movements, which include stroking, tapping, and drumming; with these may be included similar motions of the foreleges, as well as blows with the head. Added precision frequently is given the communications not only by the varying intensity of the movements, but also by the production or transmission of odorous materials. Within the community, the most frequent signs of this sort no doubt are connected with the mutual solicitation of food.

Language of Bees

The honeybees (*Apis mellifica*), and quite similarly a few advanced wasps, have developed an incredibly highly differentiated sign language. For if a honeybee has found a source of food in the immediate vicinity of the nest, say up to 50 or even 100 yards distant, once home she executes a small circle on a comb, first to the left and then to the right, and so on. This is done in the midst of her sisters, both busy, and idle, and they carried away by the display of so much zeal in their immediate vicinity, are infected by her spirit and join the leader in her circular dance. Thereafter the bees fly out and round about the hive until, by means of the particular flower odour on the leader of the dance, they have located the source of honey or pollen she has announced so ingratiatingly by her dancing. But this is only part of the story. For if the flowers are farther away, then the discoverer, still dancing alternately to the left and to the right, changes from a single circle to a broad figure eight; the longer middle portion of which is a straight line. Every time the dancing bee runs over this stretch, she waggles her abdomen, surrounded and followed by her sisters, who are not just looking on but are talking the message in. For the more slowly the dancer runs, the more her waggling movements over the middle part of the course are accentuated, and the further is the distance to the source of food.

Not just an approximate distance but the precise one is communicated and comprehended, measured by the bees no in units of length but in terms of the required expenditure of flight energy which has to be greater in head wind, for example. Yet the most incredible feature of the dance is that it communicates the exact compass direction. As is well-known, the sun serves as a compass for bees, even when hidden behind clouds, as long as clear patch of blue sky is visible. If the dance takes place on the "flight board" in front of the hive entrance, then the mid-stretch of the dance points straight as an arrow to the food source. But this direct indication of direction is impossible in a dance within the nest on the comb, which hangs vertically. Here the straight part of the path is run through from below to above when the location to be indicated is in a compass direction exactly toward the sun, and from above to below when it is away from the sun. But if the source of food, as viewed from the hive, lies to the right or left of the sun, the bee dances the straight stretch upward at the same angle to the right or left of the vertical by which the direction of the discovery site departs from that of the sun. Hence, vertical direction on the comb is equivalent for the bees to the position of the sun, which signifies a translation of the direction learned through orientation in sunlight into perception of the force of gravity.

Astonishingly, the dancing bee also indicates then precise compass direction when the place she has found cannot be reached in a straight line of flight, but only by means of side excursions, such as flight around a mountain. But since the bees reading her message are oriented by the dance as to the distance, they now produce the required amount of exertion in flight and, after flying around the obstacles but always holding to the proper compass direction, they arrive at the goal. The dance expresses, in addition to distance and direction, the productivity of the source of food, which is communicated by the degree of liveliness.

After a rich discovery the dance is a correspondingly excited one, and even an arousing one. The dance is also used by swarming bees in searching for a new home. Each bee returning from the search dances on the hanging mass of the swarm. She communicates not only the location of the place she has found but also how good it is. The discoverers of the best situation prove to be the most infectious dancers and induce more and of their comrades to fly out and inspect the place. These return, dance with equal elan, and soon the other scouts, which are behaving less convincingly, are ignored and either give up or simply join with the rest. Thus finally all are doing the same dance, the decision is made, the swarm breaks forth, and, led the many dancers, they fly off to the new home. In this way, the swarms is likely to select the best hollow tree even though it may be miles away.

Winter Swarming

Bees swarm densely not only when seeking a new home, also when passing the winter in the hive. A single inactive bee, like all resting insects, has inappreciable body warmth of its own and is therefore at the mercy of the environmental temperature. When the temperature drops to about 46°F (8°C), the bee is incapable of moving and soon dies. Nevertheless, together in the hive, bees are able to survive a rigorous winter. They cluster in a thick spherical mass around the empty inner combs, and individual bees inside the cluster produce heat by vibrating their wing muscles. In this way they are able to maintain a temperature in the center of the swarm of 68°F to 86°F (20°C to 30°C) even when it is as cold as-4°F to –22°F (20°C to 30°C) outside. The temperature of the outer layer of the swarm is about 45°F to 46°F (7°C to 8°C). When raising a winter brood, the raise the cluster temperature to 95°F (35°C). Ants spend the cold season of the year in deeper chambers or in special winter nests. A few house ants—such as the notorious yellow pharaoh's ants (Manomorium pharaonis), which have been spread all over the world and which establish their colinies everywhere, even in the handle of a table knife-live in comfort in the heated houses of humans and avoid the problem of overwintering altogether.

Social Parasites

Among certain bumblebees, wasps, and ants, individuals social species have become parasites. They differ from their hosts in that they have lost their worker casts, and in part, their own instincts for brood care. Whereas a few *Polistes* and *Vespa*, which resemble their wasp hosts almost to the hair, live quite peaceably in the latters nests and lay their eggs in the comb, the female parasites of bumblebees and ants are compelled to gain possession of foreign workers for rearing their own brood. With some exceptions, this done usually with naked force, but among certain ants with treacherous guile.

A parasitic bumblebee (*Psithyrus*) forces her way into the nest of her host (*Bombus*), intimidates the workers with her bites slaughters the queen, and destroys eggs and larvae. In some cases, *Psithyrus* will spare the life of the bumblebee queen, which then no longer pays any attention to her own progeny. In either the bumblebee colony perishes. In rare instances a genuine bumblebee (*Bombus*) adopts the despotic method of a *Psithyrus*, kills as relative, and with the latter's workers found her own community. But Psithyrus found no new community, merely rearing a few males and females. In ants, social parasitism may transpire in a significantly more peculiar manner.

Parasitic queens frequently have arisen from former slaveholders, whose own workers, instincts for brood care, and even ability to feed themselves independently become superfluous and gradually were lost altogether. Some of these of these ants are pitifully reduced creatures, unable to rear their offspring, yet are endowed with the refined cunning needed to compel others to carry out these tasks. The female of a *Bothriomyrmex* for instance, climbs onto the back of the much larger queen of a *Tapinoma* community, bites her head off, and now usurps her place in the society, to which she gained access thanks to her similar body odour. Other parasitic females are not in so simple a position; first they must patiently make friends in the outlying quarters with the workers, and attempt to accustom these to themselves, until their own body odour has become imbued sufficiently with the family smell.

Still another procedure is used by the female of *Epimyrmex stumperi*; she takes a firm seat, from which she cannot be dislodged, on the

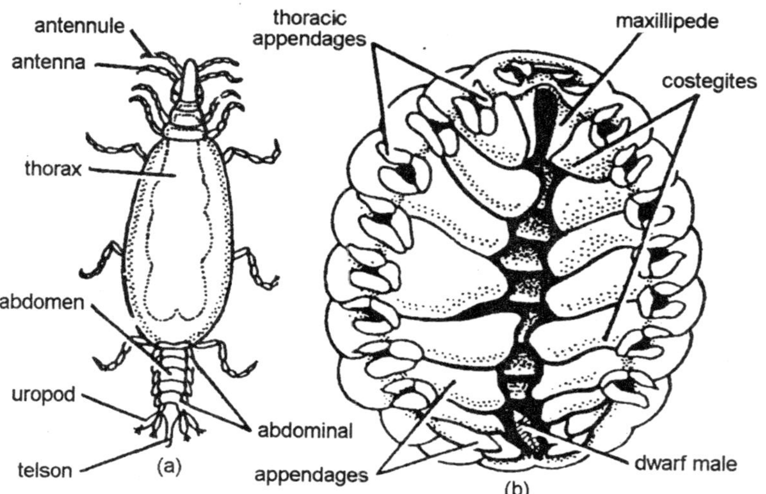

Fig. 4.8. Parasitic isopoda. (a) Pranzia larva of Gnathia; (b) Female Bopyrus.

back of a worker of the selected *Leptothorax* colony, and for a long time brushes the worker as well as herself with her bristly forelegs. Thus the crafty *Epimyrmex* perfumes herself with the *Leptothorax* odour. And then finally, being unrecognized in the darkness of the passageways and chambers of the nest, she comes upon the queen, rolls her over, and bites a firm hold in the throat of her victim, who is much larger but does not defend herself. There the Epimyrmex hangs for days, until the queen dies or is completely exhausted.

Workers are in the vicinity no doubt, but none offers help against the perfumed strangler; no one knows of the drama that is being played out silently in the middle of the community. Finally their own queen, perhaps still alive though paralyzed, is not heeded any further and must slowly starve to death, while the monster sets out after another queen, if there is one, but leaves her unmolested if she is young and has not been fertilized. What a horrible triumph of bare natural instincts. How friendly, in contrast, seems the relationship of *Teleutomyrmex schneideri* with their hosts ants *Tetramorium caespitum*. For their these little invaders are tolerated, and hence, as a rare exception in ant communities, the queens of two different species dwell in peace side by side. Not literally side by side, it is true, for the parasites, themselves pitiably incapable figures, ride upon the host queen and are tended by her royal court.

Termites

Termite populations run into the millions. They forge the complexion of landscape like no other organism except man. Their secretive activities result in almost uncanny accomplishment; these grow out of their remarkable social organization, species although they make inroads not on the brood but on the food stores. As remarked earlier, termites and ants sometimes live together, and even certain bees (*Trigona*, *Hemisia*, *Acanthopus*) set up their societies unmolested in the arboreal nests of termites. The same is done by a few parrots; and in Africa the great Nile lizard (*Varanus noiloticus*) lays its eggs in a hold it has clawed out of a termite hill, where the constant temperature and humidity are favourable for brooding.

Termite guests include members of the most varied insect orders, as well as mites and spiders. Many insect guests that are greatly desired and well-tended by the termites have strongly swollen, often grossly misshapen abdomens (physogastry), as in the wingless flies *Termitoxenia*, which live in the fungus gardens, and in numerous beetles of the family Staphylinidae, among which there is a viviparous species from Brazil (Corotoca-melantho) that deposits first-stage larvae instead of eggs.

Broader Relationships

Indeed all insect-eating organisms are active as enemies of the sexual individuals, which swarm in huge numbers, and during such swarming even man traps termites as food. But the pursuit termites within their structures requires specialists. Among vertebrates these include the armadillos, anteaters, and a few others usually adept at digging. Among birds the woodpeckers take precedence. But none of these is an destructive as bulldog ants, migratory ants, and a few others. These decimate the termites, emptyping an entire house at a time and then frequently talking possession of it, or even eradicating the termites from whole localities. In general termites cannot cope with these enemies. But the so-called "nasute" forms are exceptions. The sticky excretions rom their "nose" gum up beyond repair the antennae and mouth of the attacking ants thus are much more effective than implements for biting.

Humans that inhabit the warm Countries must be on guard against the termite horders, for these insects devour almost anything and everything, even (possibly with the help of their solvent saliva) metal and ivory. Coming from the ground, they penetrate into the wooden parts of the houses, hollowing everything out yet leaving the outer layers deceptively untouched, and in the same way get into furniture and utensils of all sorts. In addition to nonliving materials, they attack trees and other plants. As proof of their destructiveness, man has been forced to establish institutions for testing termite-proof materials, of which there are incredibly few. A problem everywhere in the tropics, termites also threaten to become one in the temperate zone. Notorious pests there are *Kalotermes minor* and Reticulitermes in North America, and *Kalotermes flavicollis* in Europe. Although considered destructive by man the armies of termites have inestimable value in the natural economy of the tropics. Their dissolution of dead materials, above all fallen timber, and their loosening up of the soil enable new generations of plants to grow. On the other hand, in certain districts the secretions give off by termites during the activities of building can harden the ground to such an extent that it becomes unsuitable for cultivation.

Classification

About 2,000 tropical and subtropical species of termites are known. They range from about 0.08 to 0.9 inch long (large queens up to 4 inches). Of the three to five families of termites, the kalotermitids and the mastotermitids seem the more primitive; the latter are represented only the single Australian species *Mastotermes darwiniensis*, which has been designated the most primitive of all. Rich in species and more advanced in their organization are the termitids and rhinotermitids. Only among them is there, strictly speaking, a true worker caste, namely fully developed workers and soldiers no longer capable of transformation. Those of the other groups are merely nymphs of different stages. Such nymphs can be developed into workers, soldiers, or sexual individuals by special feeding applied at a definite period of their development. At first glance, insect societies may seem strangely similar, or at least comparable, in some ways to human societies. But it is well to keep in mind the infinite distance that separates the two. Our social existence is based on the family as a unit, has been developed by human intellect, and depends upon the free will of the participants. Insect societies, on the other hand recognize no unit other than the population as a whole, are governed by instinct, and have compulsory membership. There is really no need to make comparisons; the life of social insects is marvelous enough in itself.

5

Pheromones in Social Insects

Taste and smell, which can be also termed as chemical sences are much more important to insects than to humans. Insects conntinuously receive with meaningful chemical signal that tell of food, of nest sites, of prey, of predators, or of a suitable mate. Many of these signals produce a rapid and particular behavioral response. The study of the chemical signals that act between organisms constitutes a new disciplines called chemical ecology. Much study in this field has resulted into several recent books and innumerable research articles, and is one of the most exciting frontiers in entomology. Chemical signals include hormones, pheromones, and allomones.

Hormones are internal signals that are produced by an endocrine gland and that are carried through the blood to act on target tissues within the animal. Chemical ecologists study external signals that are secreted into the environment by exocrine glands and that act between different organisms. Pheromones act between individuals of the same species. Pheromones attracts the sexes, aid in courtship, announce the presence of danger, mark trails and home territories, and control many other intraspecific-interactions. E. O. Wilson and W. H. Bossert, of Harvard University, divided pheromones into two general subclasses: (1) *releasers*, which act rapidly to produce a behavioral response, and (2) *primers*, which act to modify the physiological state. *Allomones* act between different species to benefit the producer. Allomones repel predators, confuse prey, and mediate symbiotic interactions. Inspite of the fact that the terms *pheromone* and *allomone* are of recently coined, our knowledge of chemical signals began in the nineteenth century for pheromones and even earlier for defensive allomones. Yet it is only since 1950 that the analytical power of organic chemistry has developed to the point that identification of the molecules can be accomplish repeatedly from very small samples. With the resurgence of information on the molecules themselves, is now possible to ask precise questions about their impact on behavior and physiology.

The molecular structures are interesting in themselves but the most fascinating aspect of chemical ecology is the delicate intercourse among organisms that is mediated by the chemical signals. The factor which are necessary for chemical communication are a source of signal molecules, the chemical signal itself, and a receiver. The source is most often an *exocrine gland*, which opens on the surface of an insect. Glands may be associated with mouthparts, intersegmental membranes, sclerites, legs, and other body parts. Somehow the products of these glands must cross through the impervious insect exoskeleton. Type I secretory cells lie well below the epidermis itself and pass their secretions to the outside through elongate ductules that are invaginations of the epicuticle. Type II secretory cells have cuticles that are filled tiny perforations so that the secretion percolates up from the secretory cell to the outside. The molecular signals and their receptors will be discussed below.

PHEROMONES

From among a noisy jumble of competing chemical stimuli, a pheromone must become conspicuous and convey a clear message. Not all molecules make good pheromones. The signal must be unique and in a terrestrial environment it must be volatile. Glucose and glycogen, for example, are poor candidates on both counts. In theory, the larger the molecule, the greater are the possibilities for a unique structure that conveys a clear message. However, in actuality the requirement for volatility restricts pheromones to molecular skeletons of 20 carbons or less. Most are derived from common biochemicals like fatty acids or amino acids. In the entire class of Insecta, the molecular structure of most known hormones varies little.

In contrast, many thousands of different molecules have been identified as pheromone components. Their diversity among species helps to avoid signal confusion. To increase the information content even more, most pheromones are mixtures of several chemical substances so that in the actual signal several components are mixed together.

Active Spaces

Pheromone signals occupy space and persist for some time. Many of the special features of the communication systems are related to the expansion of the signal and to its fade-out in space and time. Imagine an insect sitting in the middle of an empty basketball court and emitting pheromone. Assume that there are no breezes. As the pheromone begins to evaporate from the insect, its concentration rises in the air nearby. As times goes on, the molecules diffuse outward to the edge of the center circle and beyond. There will be a hemispheric could of pheromone, with its concentration highest at the source and spreading around it. To begin a chain of reaction the pheromone must be perceived by another insect.

In most cases the response is fairly stereotyped, when the pheromone has exceeded a threshold concentration. Far away from the source, pheromone concentrations are extremely low, and no insects respond. As one moves closer to the pheromone source, concentration in the hemispheric cloud reaches threshold. And at lesser distances, concentration is above threshold. Thus there is a zone within which other insects respond and a surrounding environment where they do not. Wilson and Bossert called this zone of response the active space. In Wilson's words, "The signal is the active space." The size of an active space is determined by both sender and receiver. When an animal has given off pheromone for a long period, the active space is likely to be at its maximum size.

An active space will be larger for a receiver with a low threshold of response than for one with a high threshold. When the emission frequency of the pheromone is high the active, space will expand faster and achieve a larger volume than when the emission rate is low. By altering the rate of emission, the threshold, and the chemical signal itself, the communication system can be tuned to optimum performance for a particular task. The active space for a sex pheromone may be large, but that for an alarm substance should be quite small. A basketball court is an artificially created environment, but the lessons learned there about out hypothetical insect can be applied to the natural world. The major differences between the gymnasium and the forest are winds and obstacles. In nature, breezes act on odor clouds. The cloud of pheromone drifts downwind and forms a plume, or aerial odor trial. The plume will rarely be perfectly symmetrical, since breezes increase and decrease unpredictably. The resulting turbulence gives the plume ragged edges, and these are further distorted by flowing over hills and into valleys , by going around and through trees, and by changes in temperature and air pressure. Every plume is a little different from all the others. Although there remains much debate about the detailed shapes of these clouds and about the effects of wind and weather, the general concept of an odor plume serves well to give us some clue for the chemical signal.

Sex Pheromones

Some pheromones attract one sex to the other over long distances, while others act at close range to actival a sequence of the elements is a chain of courtship behaviors. These Pheromones are not produced continuously by all members of a species, but only by animals that are reproductively mature and often only at a particular time of day. Because disruption of reproduction is an appealing strategy for control of insect pests, a large number of species have been studied in university laboratories and agricultural research stations throughout the world.

The research began as a descriptive search for the molecules, and as more and more have been identified, the field is shifting toward precise study ofhow pheromones function. During the period around 1960, it was thought that one sex of each species might produce a single, unique substance that automatically triggered a rigid sequence of stereotyped behavior in the opposite sex. That view of pheromones was attractive because of its simplicity. However, natural selection has left much more interesting and complex phenomena for the entomologist to analyze. Most pheromones are complex medleys of scents, and these odor blends have many subtle shades of meaning. Rather than list all the detail the biology and chemistry of the sex pheromones which affect the reproduction of two groups of insects: moths and queen butterflies.

Moth Attractants

The French naturalist J. H. Fiber a nineth century entomologist demonstrated that male peacock moths are attracted to a paper that previously held a female but pay no attention to a female under a glass bell jar. He correctly concluded that a residual sex scent on the paper was responsible. But it was not until 1959 that Adolf Butenandt, of the University of Munich, Germany, became the first to determine the chemical structure of a pheromone. That was the culmination of a 30 year effort with Bombyx mori, the oriental silkworm moth, which has been cultured for many centuries in Japan and China. As soon as a female Bombyx moth comes out from her pupal cocoon, nearby male become highly excited and will copulate immediately with her if given the opportunity. Males do not even need to see the female. They exhibit characteristic excitement (wing fluttering, flying upwind, abdominal movements, attempting copulation) when exposed to female scent, generally, the scent is emanated from two glands that open in the intersegmental membrane at the tip of the female abdomen. Butenandt and his colleagues painstakingly snipped off the tips of the abdomens of 500,000 virgin female *Bombyx*. These tips were then extracted with ethanol-ether, and the active components were carefully purified from the crude mixture and then characterized by brilliant microchemical techniques that were devised by Butenandt for this project. He obtained about 12 mg of pure attract ant which is termed as bombykol and identified it as a 16-carbon alcohol.

Male *Bombyx* are extremely responsive to female scent, and the male fluttering response is used as a bioassay to compare different concentrations of pheromone. For each test a sample of dilute pheromone is placed on a filter paper cartridge and blown by an air-stream to resting males. A typical does-response curve for pure bombykol is shown in Figure. If carefully examined such curves may seem dry an abstract but they nearly summarize the data and show how they fall into a pattern. Below 10^{-6} μg the response is no different from background activity. Between 10^{-6} μg and 10^{-4} μg there is an abrupt rise as an increasing percentage of males respond. Above 10^{-4} μg nearly all the males respond. For a considerably period of time, it was wrongly presumed that bombykol is the only component emitted by female Bombyx. But in 1978 K.E. Kaissling and his colleagues at the Max Planck Institute in Sewiesen, Germany, found that the female also produces the corresponding aldehyde in the ratio (bombykol: bombykol) of 10: 1. When males are stimulated with pure bombykol at moderate doses, there is no behavioral effect. But when the alcohol and aldehyde are blended together as in the natural pheromone, bombykol partially inhibits the response of the males to bombykol. The dose-response curve for this mixture (which presumably approximates the natural pheromone). Note that relative to pure bombykol, the curve is shifted to right, and its rise is much more gradual as the dose increases. The curve shown for the blend reflects that the use of pheromone concentration results in a more graded response.

The part of male Bombyx which detects pheromone are the antennae. The large, feathery antennae are studded with about 64,000 sense hairs, of which about 80% are specialists that respond only to sex pheromone. One sensitive to bombykol, the other to bombykol, and each sends a signal independently to the brain. The brain computes the relative concentrations and decides the physical responses. On any summer evening many different species of moths emit pheromones simultaneously, and somehow males must distinguish conspecific female scents from all the others. Wendell Roelofs and his co-workers at Cornell University have shown that the pheromone blends emitted by some species are carefully times. For example, two species of the genus *Archips* (leafroller moths) share the four components in their pheromones. But the two species differ in the relative proportions of the components-60:40:4:200 for *A. argyrospilus* and 90:10:1:200 for *A. mortuanus*. Males of each of species have specialist receptors for each of these molecules. Their sensilla have at least four neurons, and apparently each neurons recognizes one of the components. Signals from these sense cells are sent in parallel to the brain, where the relative proportion of the components must be computed. In this way males are able to discern the particular type of scent emitted by females of their own species.

Now the question arise that how well can moths discriminate between a signal, such as bombykol, and other molecules of similar size and with similar properties? The fluttering response can be used to compare similar molecules, such as geometric isomers of bombykol. Bombykol has two double bonds and thus four geometric isomers, Butenandt synthesized all four isomers, and experiments showed that the threshold for bombykol itself is 100 to 1000 times lower than for any other isomer. Most of the experiments have resulted into the belief that precise molecular shape is recognized by pheromone receptors.

Another question which is arised about the number of male each female can attract. The pheromone gland of each female *Bombyx* contains about 164 mg of bombykol, in theory enough molecules to excite 10^{12} males if each received a threshold dose. But such calculations ignore the impact of bombykol in the blend, and even if so adjusted, they would not be biologically meaningful. The important advertising signal is the active space. Production of large amounts of pheromone is largely an adaptation to maximize the volume of active space. By spreading wide the area of active space, the female increases the likelihood of attracting males. As soon as attracted by pheromone within the active space, a male requires orientation cues in order to find the female. For long-range orientation, he appears to use mostly *anemotaxis*-movement based on the direction of wind. The upwind movement of insects takes them toward the source of wind-borne scent.

It is difficult to observe sustained oriented movement in free-flying moths, but Iise Schwinck, of the University of Munich, Germany, found a clever alternative system. She used a flightless *Bombyx* mutant. When excited by female pheromone, flightless males wave their useless wings and try to run to the female. Schwinck watched the males as they ran along a table in the laboratory. When exposed to female scent, males ran upwind in a *zig-zag* course irrespective of the actual origin of scent. Further research has proved that the "Schwinck effect" also applies to free-flying moths. J. S. Kennedy and D. S. Marsh, of Imperial College, London, confined male moths to a special wind tunnel and placed a "calling" female upwind. Males flew upwind in a series of diminishing, irregular *zig-zags*. But at the sudden removal of the calling female, males flew back and forth across the wind with frequent left-right reversals. In some other field experiments, the males actually blundered out of the plume. To find the female, they flew black downwind to reenter the odor trail and began anemotaxis again.

Orientation of moths can also be done by *chemotaxis*, detecting slight differen- ces in pheromone concentration that would permit them to follow a chemical gradient upstream toward its highest concentration where the female sits. Some evidence suggests that moths can detect and follow chemical gradients in still air, but it has yet to be shown that they can tell upstream from downstream. Chemotaxis certainly is used for short-range orientation in insects, but as an important mechanism in long-distance orientation of moths is yet to be experimented. As soon as males get close to a female, their behavior changes. They hover or land and run until collide with her. There are many ways by which males locate the calling female. In some species it appears that the male sees a silhouette against the sky or sees the female herself and then alights.

In other species a hierarchy of active spaces may be involved. In the redbanded leafroller moth, Roelofs and his colleagues found that the pheromonal blend included components that acted at a distance as well as one, dodecyl acetate, that served in close-range orientation and woke her responses of landing. Unique feature of some arctiid moth is their use of pulselike signal William Conner and Thomas Eisner, of Cornell University, have recently shown that a female arctiid moth emits its sex lure in short pluses (100 times per minute). At distances of several meters from the source, the pulses would probably be smeared together by turbulence, but close in, they remain discrete. When males are but a few centimeters downwind, their antennal receptors show corresponding pulsations in electrical potential. It is possible that when the continuous signal transforms into an on/off pattern of pulses, males switch from upwind anemotaxis to the research in shorter range.

Queen Butterfly

Visual signal attract males to females in most diurnal butterflies, such as the queen (*Danaus gilippus berenice*). The courtship sequence, a chain of signals between male and female, was described in detail by Lincoln and Jane Van Zandt Brower while they were at Amherst College in Massachusett.

Male queen butterflies have a pair of large organs, called *Hair pencils*, which are really tufts of fine hairlike processes of cuticle. As soon as a male queen sees a female in flight, he quickly overtakes her. As he passes over her dorsal surface, he extrudes large hair pencils from the tip of his abdomen and bobs up and down over the female's head and antennae. As demonstrated by Thomas Pliske and Thomas Eisner, of Cornell University, the hairs are covered with dustlike particles that are brushed onto the antennae of the female. These numerous particles are actually microsopic spherical bits of epicuticle that were formed during pupal development. The particles clump together and stick to the hairs and to the antennae because they are coated with an oil film. The oil contains the real signal, and the dust is just a means for its transfer. The oil is apparently produced by a secretory cell at the base of each hair. Jerrold Meinwald and his team at Cornell identified two major components in the secretion: a ketone and a viscous terpenoid diol. As a result of to the hair penciling, the female lands. The male continues to hair pencil vigorously, and the female folds her wings. The male then alights and attempts copulation. Pliske and Eisner deprived males of their hair pencils and found that, although they courted actively, the courtship very rarely led to copulation. Males raised indoors on cuttings of the major food plants failed to mate successfully, although males raised outdoors, mostly on the same intact food plant, courted successfully.

Comparison of these two male groups showed that the indoor males had the diol but lacked the ketone, whereas the outdoor males had both components. The outdoor males obtained the chemical precursor of the ketone from another plant on which they spend very little time and of which they eat little. Therefore it is clear that the intake of ketone by males is vital in the courtship procedure. When artificial ketone in an oily vector was applied to the deficient males they began to court successfully. The behaviorally active aphrodisiac is clearly the ketone. Which the diol provides the oily vehicle for its spread and adherence.

Aggregation Pheromones

Many types of insects form aggregations for feeding, mating, shelter, oviposition, or some other purpose. The gregarious behavior may wax and wane with the passing seasons, perhaps occurring only in the cold months for hibernation. It may be limited to a certain time of the day, as in sleeping aggregations of butterflies. It may be associated with a developmental stage, as in tent caterpillars, which are gregarious as larvae but not as adults. It is generally understood as opportunistic behavior, as when lycid beetles form large aggregations to feed on sweet clover and to mate. Once the temporary or restricted resource is exhausted, the aggregation breaks up. The majoirty of insect group form in response to pheromones. The most intensively studied species are bark beetles and ambrosia beetles, both members of the family Scolytidae.

Bark beetles and ambrosia beetles join together for feeding, mating, and oviposition. Because the beetles and the fungi that they carry from tree to tree do tremendous damage to pines of our forests and to the American elm, his aggregation behavior has been exhaustively studied. Bark beetles' pheromones explain the complexity and natural history of a chemical signal system. Immediately after their final ecdysis, the adult bark beetles emerge from their host tree and begin a dispersal flight to find a new host. Depending on the species of beetle, either males or females (but never both) are the pioneers. When they reach a new host, the pioneers burrow through the bark and construct galleries. As they chew their way along, the beetles seed the galleries with fungal spores, which germinate to form the myeelial mass which become the food of beetle larvae. A new tree is colonized by pioneer beetles that orient to a particular host tree by several cues, such as its shape and odor. In the genus *Dendroctonus*, females are pioneers; in the genus *Ips*, males are the first to colonize. As soon as they land, the beetles burrow into the bark.

Healthy trees produce a flow of resin or pitch that may eject the burrowing beetle from its partially completed hold or may drown it. But when a few pioneers manage to complete the entrance tunnel and to establish themselves, they release aggregation pheromone. More beetles are attracted, and the balance begins to shift toward the intruders. The mass attack causes damage to the phloem, and fungal hyphae block the conducting vessels of the tree. The tree gradually begins to die. Since dead timber dries rapidly, a quick explaitations of the resources become must for beetles. The bark beetle pheromones which have been till now identified are monoterpenes, structurally similar to those produced by the host tree. The aggregation is a response to a medley of molecules that enhance each other's effectiveness. Such molecular components are described as *synergistic*. Each of the synergists may be somewhat attractive when tested alone, but the attractiveness of the blend is much greater than we would expect the sum of all the contributions from each separate component is combined.

The very first bark beetle pheromone to be identified was that of *Ips confusus*, the California five-spined ips. As the beetles chew their way into ponderosa pine trees, the wood passes through their gut and is deposited pellets called frass. David Wood, an entomologist at the University of California, Berkeley, and Robert Silverstein, and organic chemist now at Syracuse University, isolated three terpene alcohols, ipsensol ipsdienol, and cis-verbenol, from the frass of attacking male engraver beetles. In a laboratory bioassy with walking female beetles, ipsenol alone was excitatory as a single component, and it was synergized by mixing with one or both of the others. In field tests only the combination of all three synthetic compounds proved attractive. When the synthetic mixture was later compared with "natural" scent from male infested pines, that natural sent was much more effective than a synthetic blend of ipsenol, ipsdienol, and cis-verbenol. Clearly the natural pheromone contains components which are till now unidentified.

Near the male-infested tree, a female runs about, searching for entrances to a male burrow. Outside its entrance, she stridulates by rubbing the serrated edge of her elytra against a scraper on her abdomen. In response to the female chirps, the male may allow her to enter his burrow; mating follows. One male will allow only three females to enter the nuptial chamber, and after mating, each female constructs an egg gallery in the phloem and outer xylem. When more females stridulate or try to intrude the burrow, the male uses his body to block the intrance.

Many species of Ips have common geographic distributions. Both *Ips paraconfusus* and *Ips pini* attack ponderosa pine in California, but the two species rarely attack exactly the same tree. Martin Birch and his colleagues at the University of California, Davis, have shown that resource competition is minimized because each species responds only to its own attractant blend and because each blend includes an inhibitory scent for the other species. The pheromones emitted by males is composed:

Ips paraconfusus	Isp pini
100% S-ipsenol	
90% S-/10% R-ipsdienol	100% R-ipsdienol
IS,4S,-5S-cis-verbenol	

S-ipsenol inhibits females of I. pini, and pure R- ipsdienol inhibits the attack by females of *I. paraconfusus*. One component of the aggregation pheromone of each species is an inhibitory allomone.

As it has been stated earlier, in bark beetles of the genus *Dendroctonus* it is the females that are pioneers. This genus includes many serious pests of coniferous forests, such as the southern pine beetle and the western pine beetle. From the frass of the latter species, Wood and Silverstein isolated and identified the unusual terpene *exo*-brevicomin. Pure exo-brevicomin is quite attractive to males, but many more males were attracted to synthetic mixture of exo-brevicomin and myrcene, a terpene produced by wounding the host ponderosa pine. After responding to these substances, males at a tree laden with female and gove off a pheromone of their own, frontalin, which attracts other females. Like exo-brevicomin, frontalin is synergized by myrcene. The combination of these substances is cause for the mass attack of both sexes. After mating, both sexes begin to release two additional pheromones, *trans*-verbenol and verbenone, which stops the arrival of other beetles.

Inhibitory pheromones are one device to control density so that the tree is not overcrowded with larvae. Another device to control density is high attractant concentration, as demonstrated in trapping experiments. Beetles are readily caught in artificial traps baited with a mixture of brevicomin, frontalin, and myrcene. However, when pheromone concentration gets extremely high, the catch in the trap declines. At very high concentrations of pheromone the incoming beetles turn away from the baited tree and shift their attention to the nearby trees. Consequently a cluster of infected trees is produced, which becomes a "hot spot" of brown foliage in the forest.

It is remarkable that the of beetle receptors to recognize the differences between the pheromone components and similar molecules. Geometric isomers are easily discriminated; thus transverbenol is an effective inhibitor for attacking western pine beetles, while cis-verbenol is not. It is yet unknown that how the chemicals in bark beetle aggregation pheromones are manufactured. No discrete exocrine glands seem to be responsible. The aggregation scent is not produced if the host terpenes are absent, and many of the pheromones seem to be oxidized derivatives of the terpenes.

For instance, male Ips paraconfuses convert myrcene to ipsenol and ipsdienol. David Wood and his accociate, John Byers, have proved that the oxidation of Myrcene to ipsenol and ipsdienol is inhibited by the antibiotic streptomycin. Then suggest that symbiotic bacteria, restricted to the male, are responsible for myrcene oxidation. If bacteria play such a role, the beetle depends on the tree for the starting material and bacteria for the metabolic conversion in order to produce an insect aggregation pheromone. Initiating an aggregation carries a risk. Pradaceous carabid beetles, which feed on the bark beetles, are also attracted by the bark beetle aggregation pheromones. When the carabid arrives at the infested tree, it feeds on the beetle which emits the signals.

Alarm Pheromones

The major disadvantage of such mass gathering of insects is that a predator has an easier time finding prey. Upon approach of the predator, The group may disperse or it may invoke a colonial defence response. In some hemipterans and in social insects, an alarm pheromone results in the similar behavioural pattern.

Aphids gather in large number to feed. When a predator attacks, the victim secretes a droplet of fluid from a large gland opening called a *cornicle*. Within that fluid is an alarm pheromone, of which the most common is farnesene. Nearby aphids respond by falling to the ground, walking away, or leaping from the plant.

Alarm Pheromones of Ants

Anyone who has stepped on an ant nest has seen the outrush of defenders, which open their mandibles and swarm over the attacking foot. Such alarm behavior is caused by a pheromone emitted by the mandibular glands and by glands associated with the sting. It has been predicted by Wilson and Bossert that an alarm signal should be localized, blatant, and brief. The pheromone should spread rapidly to coordinate colonial defense and should fade out rapidly to avoid prolonged false alarms. For the site of the disturbance to be easily located, the active space should remain small. To prove there findings they investigated the size and time-course of the active space for alarm in a harvester ant.

The major component of this ant's Mandibular secretion is 4-methy1-3- heptanone. Wilson and bossert qucikly crushed the head of a worker ant to release all of its mandibular gland contents at once. They watched the response of nearby workers, and by repeating the experiment many times, they were able to deduce the properties of the active space. The active space expanded to its maximum radius (6 cm) in 13 seconds, and as the pheromone continued to diffuse, the active zone shrank to nothing by 35 seconds. In a natural situation with a serious invasion of the nest, other workers, excited by the first secretion, would have encountered intruders and added their own pheromone. This added pheromone would amplify and prolong the signal and expand the defensive frenzy. The extent of the group response decide to the severity of the attack.

Alarm pheromones are very volatile, and most are of low molecular weight, containing 12 carbons or fewer. Those from the mandibular glands usually have a keto-or aldehyde group, while those from Dufour's gland (near the sting) are most often hydrocarbons. Many are mildly toxic and distasteful; thus they function as alarm defensive devices. In case of the african weaver ants construct an elaborate nest by sewing leaves to one another with thread from the silk glands of their own larvae. Both the nest itself and the territory surrounding it are vigorously defended by the major workers. Any insect that happens to wander into the territory is rapidly attacked and often becomes prey for the ants. The defensive behavior consists of four stages. (1) alerting, in which ants raise their heads and open their jaws; (2) attraction toward the disturbance; (3) stopping near the source of the disturbance; and (4) bitting and holding for several minutes.

The distress signal follows the expulsion of the fragrant but musty of or from the mandibular glands of a major worker. J.W.S. Bradshow, R. Baker, and P. E. Howse of the university of Southhampton, England, have shown that at least 33 volatile components are released by crushing the head of a major worker. Four of these seem to trigger specific aspects of the alarm response. These molecules differ in their volatility from hexanal (most volatile) to 2-buty11-2-octenal (least volatile). Their defense mechanism can be easily apprehended if we imagine that an excited ant deposits a droplet of mandibular gland secretion at the center of the concentric circles. In still air each component diffuses away from the site of deposition. Because they differ in volatility, some fill space more quickly than others. After a short time, hexenal defines the largest active space, indicated by the outermost circle.

When ants enter this space, the alerting behaviour is characteristic. As they move into the next zone (hexanol), the major workers are attracted toward the site of its deposition. In the innermost zone 2-butyl-2-octenal is the biting marker, and 3-undecanone contributes to biting and is also a short-range orientation signal. When the secretion has been allowed to evaporate for some minutes, no effects of hexanal or hexanal are detectable, but as soon as the workers come close to the deposition site, biting behavior is released. Consequently a chain of behaviors is sequentially triggered by the discrete chemical components of the alarm pheromone that are localized on the intruding object.

As opposed to the major worker and the minor workers of the weaver ant emit a considerably different set of components from their mandibular glands when they are disturbed. If other minors are exposed to that secretion, they do not defend the nest but instead flee from the site of disturbance. In contrast, major workers are attracted and arrested by high concentration of atleast one component in the minor's perfume. Thus major workers come to the defense of the distressed minor worker and of the nest.

How much different is one alarm signal to the other? Are they specific for each ant species or for each colony? Apparently they are much less so than for sex pheromones. Often the same components are shared along many species within a genus and among genera within a subfamily. However, there are subtle and biologically important differences between similar species, as shown by studies in the ant genus Myrmica. Four species of this genus that coexist in similar habitats in England and Belgium have been found to have similar blends of several components in their alarm pheromones, but their relative proportions differ. In effect, each species has a different form of communication.

Trail and Territorial Markings

Both food and habitat are restricted. Territorial and trail pheromones enable efficient resource exploitation. Many foods, like blue-berries or caterpillars, act as the container of insects eggs. Female insects that place their eggs in such organic matter and mark them with a pheromone that is like a sign that says "Occupied." Some parasitic wasps include such a territorial phero- mone with each egg injected into an insect host. Female apple maggots mark each fruit after ovipositing in it. After she has laid an egg, the female drags her ovipositor over the surface of the fruit and thus deposits pheromone. The size of the mark depends on the future needs of one larva from hatching to maturity. A small fruit, like a cherry, is completely coated after one oviposition and thus will be avoided by subsequent females. A large fruit, like an apple, is marked on only a portion of its surface. It will go on collecting several eggs until its surface coat of pheromone reveals that it is loaded to capacity with larvae. Effective utilization of the resource material requires choices among alternative.

Many social insects use the guiding trails to direct nestmates to the richest food sources. Recruitment trails are continuous directional signals that precisely guide the foragers to abundant goodies. Stingless bees use secretions of their mandibular glands to lay recruitment trails. Like honey bees, stingless bees have large colonies with many workers, use wax as a building material, and collect and store honey and pollen. Martin Lindauer, of the University of Munich, showed that stingless bee foragers guide each other efficiently. When a pioneer forager finds a new nectar source, she makes a direct flight back to the nest with food. If the nectar source is still copious after several trips, she returns to the nest in a different manner. Instead of a quick flight home, she flies in small stages-stopping repeatedly to wipe secretions from her mandibular glands onto rocks, plants, or dirt along the route. At the nest she arouses a number of her sisters, and with the pioneer as leader and pilot, the group flies along the trail of scent marks. The foragers stop along the way to check the marks, and finally arrive at the source of rich nectar.

The tactics which guide these stingless bees require both the pilot bee and an odor trail. In many ant and termite species an odor trail alone is sufficient to guide a follower to a food source. In the case of ants and termites the trails are laid by dragging their abdomens along the substrate and depositing streaks of chemicals. The active space of the trail is a narrow corridor and extends upwards in a half-cylinder. Followers run along with in this active space, weaving from side to side and turning back to the center as their waving antennae break out of the active space into an usecented zone. The followers do not have to touch the odor marks but have only to stay within the active space. They can even follow a trial drawn overhead on the ceiling if ceiling and floor are not too much apart.

The function of these trails differ according to the biology of each species. In harvester ants, Bert Holldobler, of Harvard University, has shown that some are fixed trunk trails while others are more temporary recruitment trails. Trunk trails, laid by secretions of Dufour's gland, are like spokes radiating out from the nest entrance into foraging territory. Trunk trails are fairly permanent and are used mostly by ants that are returning to the nest. Recruitment trails, laid by the poison gland, branch off from trunk trails. They are laid by a successful forager as she returns from a food source. These trails entice other foragers to follow the chemical signals and thus to help retrieve the food. As food-laden foragers come back to the nest, they lay recruitment trails that reinforce the previous signals. Foragers that do not find food and return hungry to the nest lay no recruitment trails. The recruitment trail has a short half-life, it gradually fades away along with the finishing of the source of the food.

It is apparently easy to say that a given species of ant or termite follows a chemical trail, but the molecular components of the trail signal have proved difficult to isolate in most cases. To demonstrate the presence of a chemical trail, a researcher need only paints line of a chemical sample along a substrate, such as a piece of filter paper, and then watch to see whether foraging ants or termites follow the line when they chance to cross it . Extracts of whole insects or their glands or chemical composition can be easily analysid.

Not more than a dozen trail Pheromone components have been identified, and most of these are not very volatile. Neocambrene A, a complex terpene, is the trail substance of nasute termites, and a pyrazine is the trail substance is several species of myrmicine ants. At least in these ants, the true trail pheromone is not species specific. This pyrazine is found in all seven species of British *Myrmica*, and it is found also in related genus of tropical leaf-cutter ants. The power of modern microchemistry is such that only 50 Myrmica workers provided enough pyrazine for the adequate examination of the chemical composition of the trail.

The poison gland produce trail pheromone of Myrmica. When Marie-Claire Cammaerts, of the University of Brussels, Belgium, drew lines with poison gland extracts, workers that blundered across the lines turned to follow them, but foragers were not attracted to the trail. Further experiments revealed that the volatile components of Dufour's gland are attractants and excitants. Both glands have a common opening at the tip of the abdomen and both are responsible in forming the trail.

In the natural habit trails are not public thoroughfares that are shared commonly by all species. Each species, and usually each nest, follows its own trails. An artificial trail, laid with a combination of the pyrazine and volatile components of the Dufour's secretion of one species of Myrmica, is followed quite well by foragers of the same species and equally well by foragers of other species. There is no obvious species indentification. However, when Cammaerts added the remainder of the Dufour's secretion, which is an oily mixture of low volatility, she produced trails that were species specific. The oils differ among the species and allow each species to identify its own unique trail. The secreted oils from Dufour's glands are *territorial pheromone* for Myrmica. Pioneer foragers, seeking new food sources far from the nest, move forward slowly and tentatively. As they walk ahead onto new ground, each forager repeatedly marks the ground with her Dufour's gland. The oils persits for long periods, and subsequent foragers will run quickly across the same ground that was so slowly crossed the first time. The oily marks differentiate home territory from foreign territory.

The natural foraging trails of *Myrmica* have three kinds of components, each of which serves a distinct purpose. Secretions of the poison gland define the trail itself. Volatile substances from Dufour's gland attract and increase running speed, but these signals are short-lived. Finally, oily substances from Dufour's gland survive and mark the home territory.

Pheromones and Individual Differences

A few members of species do not emit the same complement of pheromones. Differences between life stages (Larva versus adult) and between sexes are obvious. Also, genetic strain, physiological state, and nest and caste identification is social insects are identified by pheromones. For example, for successful mating of the fruitfly, *Drosophila melanogaster*, the female must emit pheromones that stimulate the male to court, and the male must emit pheromones that cause the female to accept. Both sexes prefer to mate with "strangers" that are genetically quite different, and they recognize the differences between related and unrelated partners by their differing pheromones. Males and females from the same inbreed line do not spontaneously court one another unless they are foled by being brought together in an environment filled with alien pheromones.

Most of the halictine bees are colonial and nest together in burrows. At the entrance to each colony, a guard bee inspects newcomers and rejects those that are not residents. Charles Michener and his associates at the University of Kansas wondered how the guards of one species. *Dialictus zephyrus*, distinguished nestmates from aliens. Bees were brought to the nest opening in a piece of plastic tubing and porked out into the entrance with a pipe cleaner. Of the 228 nonresident bees introduced into other nests, 50% were attacked by guards before contact and 92% on contact. Fifty residents were introduced in a similar manner and all were accepted. Then, Michener killed 20 residents and 32 nonresidents by freezing and presented them, after thawing, one by one to the guard bees. Merely 10% of the frozen residents were rejected, while 97% of the nonresidents were attacked. Nonresidents were also rejected when the room was illuminated with a red light to which bees are blind. Thus, it is confirmed by these experiments and others established that touch, vision, and hearing are not used for rejection. Odor must account for recognition of nestmates. Does the guard remember the scent of a nestmate, or does she merely compare the odor of the entering bee with that of the nest? When guards are placed on duty in a foreign nest, they accept only their own nestmates and reject the owners of that nest. It is the odor of the bee, not that of the nest, that these guards use to detect the identity of the intruder.

The scent if the colony has a strong genetic component. Nestmates are sisters and are genetically similar. Guard bees of *Dialictus zephyrus* can recognize individual odors that are variants on the pheromone blends produced by their nestmates, and they will thus admit genetically related bees that were reared in other nests. In honey bees. Even in the case of honey bees, colony odor has a genetic basis. When a honey bee queen is suddenly presented to workers of another colony, she is usually attacked. But she is much more likely to be accepted by these workers if she is closely related to their own queen. Michael Breed, of the University of Colorado, Boulder has shown that inbred sister queens are accepted in 35% of the exchanges, while the queen bees which are unrelated are definitely not accepted.

Conditioning plays a part in the recognition of individuals or classes of indviduals. In laboratory colonies, an halictine guard bee, placed in a "foreign" nest, can habituate to the odors of that nest and then will accept a bee with that odor, but such a situation is probably rare in nature. The response of male halictines to females is more meaningful. Upon first scent, a male responds avidly to the novel odor of a new female. But if she is unwilling to be courted, he becomes unresponsive to her now familiar odor. When exposed to a second new female with yet a different odor, he immediately attempts to court again. It is clear that the male bee learns and accepts quickly the response of the female bee. Probably the glands which emit a distinct body odour are many. The surface layers of insect cuticles contain odorous molecules that originate in the underlying epidermal cells. Other contaminants are left on the surfaces around the exocrine glands and dissolve into the coat of epicuticular and surface lipids. Each exocrine gland has many components, and their proportions vary slightly between individuals. The rich mixture of secretions from many glands, the scents from the environment, food odors, and the intrinsic odors of the epicuticle combine to produce pheromonal polymorphism enable each insect with unique identity.

Primer Pheromones

Primers bring changes in the physiological state of the recipients. It is usually assumed that primer effects are mediated by the endocrine system. Thus pheromones influence endocrine glands, and hormones directly regulate the physiological state. Primers are found in subsocial and social situations where the persistence of the group allow stimuli produced by one insect to influence another. Most primers influence reproductive capacity. A pheromone from male desert locusts increases the rate of development in immatures of both sexes. The scents of male and female mealworm beetles accelerate egg development and other aspects of reproductive maturation in young female beetles. Honey bee queen substances stop the growth of ovaries in the workers.

It is quiet interesting to note that the primers that switch development from one end point to an alternative. Desert locusts have such a primer, locustol, which triggers development from the solitary form to the gregarious form. The gregarious form is responsible for the devastating locust swarms that have revaged crops in Africa since biblical days. As more and more locusts feed together, they begin to produce locustol in their faeces. Locustol causes a gradual transformation of the light-colored solitary types into much darker gregarious forms that gather together and migrate, devuring all that comes in their way.

Generally the colonies of termite have only two functionally reproductive indviduals at any time. If an accident or an entomologist removes the male or female reproductive, one of the nymphs of the same sex develops into a "supplementary reproductive". The same sex develops into a "supplementary reproductive". The presence of a functional reproductive inhibits the maturation of others of the same sex. It is presumebly each reproductive insect produces an inhibitory pheromone, which is passes around by mutual grooming to workers and among workers. In at least one species by termite, it is possible to inhibit sexual maturation of nymphs by introducing extracts of the functional reproductives. Soldier production is probably limited by an analogous soldier-specific inhibitory primer, but there is no experimental evidence to confirm this.

Defensive Allomones

Most insects prevent the possibility of being eaten because they are distasteful. Fóul-smelling and eviltasting chemicals in both animals and plants confer protection from attackers. Probably the easiest chemical defensive behavior for an insect in regurgitation-vomiting gut contents at an attacker, as do many grasshoppers when they "spit tobacco." The regurgitate may be enriched an unplesant chemicals derived from the host plant. For instance, sawfly larvae that feed on pine trees sequester terpenoid resins from the needles in two pouches of the foregut. The purches, like the rest of the foregut, are lined with a cuticle that prevents diffusion of the resins into the blood of the insect. When attacked by an ant or a spider, the sawfly rears up its front, regurgitates a droplet of resinous fluid, and wipes the regurgitate onto the attacker. The regurgitate is rich in the resins of the ine tree, which are repulsive and distasteful to birds and so arthropod predators such as ants and spiders.

Blood-Borne Allomones

Most of the unpalatable insects are aposematic (warnings colored), such that predators can readily associate the appearance of the animal with its unpalatability. For example, monarch butterflies are strongly aposematic with their striking orange and black wing patterns. When blue jays try to eat monarch butterflies, they vomit up the partially eaten carcass. Linclon and Jane Van Zandt Brower, them at Amherst College, Massachusetts, offered moncarch and other butterflies to captive jays. The birds quickly learned to associate the monarch color pattern with vomiting and thereafter the butterflies were refuced. Monarch butterflies are unpalatable because their blood and tissues contain cardiac glycosides. These glycosides were sequestered by the caterpillars from milkweed in their diet.

The insects which carry the defensive allomone in blood, it may be liberate a special localized bleeding, or autohemorrhage. Blood bubbles out from weakened sites at leg joints, antennal joints, or between sclerites when the insect is attacked. Rough handling of a blister beetle will cause it to release clear droplets of blood from its leg joints, and lady beetles exude yellow drops of blood from similar sites. The blood of the blister beetle contains cantharidin (Spanish fly), While that of the lady beetle carries the alkaloid precoccinelline. In contrast to the cardiac glycosides used by monarch butterflies, both cantharidin and precoccinelline are manufactures by the insect itself rather than extracted from a food plant. The substance such as, cantharidin and precoccinelline are distasteful to many vertebrate predators, but not to all. Thomas Eisner reports that toads will repeatedly consume blister beetles. Small predators are further deterred by the mechanical properties of the droplet. Ants become entangled in the sticky blood. As it coagulates, one leg is glued to another, antennae become stuck to mouthparts and ants adhere to each other. Ants retreat off the attack and go through a frenzy of cleaning activity.

Allomones Produced in Exocrine Glands

Defensive allomones are frequently found in cuticular reservoirs (as were those of the sawfly). Most often the reservoirs are storage compartments for secretions made within exocrine glands. The reservoir may be eoerted and wipied on the site of attack, as seen in the larvae of sawllowtail butterflies. The defensive fluid of these caterpillars cantains isobutyric and 2-methyl butyric acid. Alternatively, the secretion may ooze from the gland opening, much as blood oozes from the knee joints of blister beetles. Finally, the secretion may be ejected in a jet of acrid mist. The spraying glands are dramatic in their impact and often extremely accurate. The defensive fluid of a Florida walking stick is aimed directly at the site of attack and effectively repels both vertebrate and invertebrate predators.

Cuticular reservoirs increases the chances of potential defensive allomones, for within a sac of impervious cuticle, small reactive molecules can be safely stored. In contrast, these toxins cannot be stored in the blood. The chemical substances which protect the insects operate in a very different context from that of chemical signals such as pheromones. The potential prey is at an advantage when the signal is so blatant that the predator cannot ignore it. The predator is at an advantage when it can ignore the blatant insult or when it is insensitive to the defensive substance. Any insect has a host of potential predators, and one might expect natural selection to favour defensive allomones that are nonspecific. That, are the repellant to many species. The ofensive secretions of many insects contain small reaction molecules that are broadly toxic to most living tissue. Small acids, like formic or acetic, readily coagulate protein and irritate excitable tissues.

The chemical substances, such as formic and acetic acids are used in the defenses mechanism of ants, beetles, and caterpillars.

Reactive aldehydes, which bind readily to proteins, are painful tear gases to humans and animals. Molecules such as these are found in the defensive secretion of walking sticks (*Anisomorpha*), cockroaches, beetles, and true bugs. Benzoquinos, which also bind proteins and "tan" them, are common in the defensive secretions of beeltes and millipedes. Many predators find such small reactive substances unpleasent and split out the prey that sprays them. In addition, these defensive allomones are effective as antibiotics against microorganisms. Storing the reactive defensive molecules could pose a problem for the insect producing them.

There are two ways by which an animal avoids being poisoned by its own secretions. First, the secretions are stored within reservoirs of cuticle that have special permeability properties that prevent any leakage of the toxic fluids. Effectively the secretion is stored outside the animal. Second, many of the secretions are produced in a stepwise manner, such that the cytoplasm of the secretory cells contains only harmless precursors rather than the toxic end product. The progression of the production machinery is best seen in the adaptation for quinone production by bombardier beetles and darkling beetles. The term bombardier beetles is in keeping with the defensive behaviour of this particular species of beetles. When disturbed, they emit a jet of host secretion at the site of attack, and the ejection of secretion is accompanied by a clearly audible explosion. The glands open at the tip of the abdomen into movable turrets that are aimed when the tip of the abdomen is bent toward the attack. The spray contains a mixture of p-benzoquinones.

Quinones are not stored in the reservoir of the gland but rather are produced at the instant of attack. The gland reservoir consists of two compartments. It has been demonstrated as shown by H. Schildknecht, of the University of Heidelberg, Germany, that the inner compartment contains hydroquinones (diphenols), which are precursors to the final defensive quinones. The hydroquinones are mixed with hydrogen peroxide. As soon as a beetle is attacked, some of the content of the inner compartment is passed into the outer compartment. The author compartment is a reaction chamber. The small outer compartment contains a mixture of enzymes (catalases and peroxidases) that oxidizes and decomposes the hydrogen peroxides to give water and oxygen and also oxidizes the hydroquinones to yield benzoquinone. The oxidation reactions liberate heat and gaseous oxygen. Thomoas Eisner has shown that the temperature of the mixture reaches 100°C. The gaseous oxygen provides that propellent that accounts for the audible "pop" when the glands discharge. Eisner has also used high-speed photography to analyze encounters between attacking ants and bombardier bettles. Thus entire sequence, from ant bite to beetle discharge to ant release, takes place within a second.

Another species known as Darkling beetles, or flour beetles and mealworm beetles, also use quinones in their defensive secretions. Unlike the bombardier beetles, these darkling beetles store the actual quinones, not a precursor, in the reservoir of the glands. But like the bombardier beetles, the darkling beetles protect their secretory cells from quinoid products by a strategic sequence of localized reaction steps. The cells which secrete these defensive glands pass their products to the reservoir via fine cuticular ductules. In these defensive glands there are two distinct populations of secretory cells. The secretory cells that are of particular interest to quinone production are types 2a and 2b, which are arranged in pairs. Each pair is strung in series on a cuticular ductule that leads to the reservoir. Each cell of the pair surrounds a central cavity within which lies the effective ductule.

The secretory product of the cell passes into that extracellular cavity and then through the wall of the different ductule and finally to the reservoir. The structure of cell cytoplasm, cavity, and ductule can be viewed as reaction compartments, arranged in series. The two final steps in quinone production are hydrolysis and oxidation. The reaction steps for quinone production apparently occur in the sequential compartments. In the cytoplasm of the distal cell, phenyl glucoside is stored. The glucoside is hydrolyzed as it is secreted into the cavity of the distal cell, and the resulting diphenol can then pass along through the wall of the ductule. Within the ductule, the diphenol is oxidized to produce the final quionone. The oxidative enzymes (phenolases and peroxidases) are apparently built into the wall of the cuticular ductule. Thus the secretory cells avoid being poisoned by quinones because they are never actually exposed directly to poison.

Another mechanism, to which protects insects from the poison they develop themselves gradually to secrete appropriate enzymatic catalysts and precursors of the allomones into the reservoir. Reservoir fluids often consist of a watery phase and an oily phase. Murray Blum and his colleagues report that in the two-phase secretion of a true bug (*Leptoglossus*), the enzymes are in the aqueous phase and the reactive defensive allomones are in the oily phase.

Like pheromones, defensive secretions are generally mixtures of several substances. The components often differ sharply in their polarity, as in the darkling beeltes, where several relatively polar quinones are mixed with nonpolar hydrocarbons. The mixture is adequate to dissolve and disrupt a variety of permeability barriers at the surface of the integument. Furthermore, a mixture of several different repellent substance contributes to the nonspecific effectiveness of defensive secretions in repulsing many different kinds of predators. In some species, 20 to 40 distinct components make up the defensive fluid.

Defensive Allomones of Social Insects

As soon as the nest of a social insect is invaded, members of the colony pool their resources for collective defense. In some ants and many termites, a special caste, the soliders, is dedicated to nest defense. The secretions they use are doubtly effective repellent to the attackers and alarm signals for nestmates. As shown by endure Quennedey at the University for Dijon in France, the major source in ligher termites is the frontal gland of the soliders in *Cubitermes*, the strong mandibles pinch the attacker (usually an ant), and the frontal gland secretions pour out on the sties of contact. The secretions are somewhat repellent, but in this genus the mandibular pinch is the major defensive act. In *Rhinotermes* there are two caterogries of soldier: a major caste with enormous mandibles and a minor caste with tiny mandibles. The minor caste has a huge frontal gland reservoir that extends back from the head all the way to the abdomen. It secretes a mixture of ketones toxic to many insect species. In the third style of defense, found in nasute termites, the chemical weapon is paramount. The soliders are remarkable in that their mandibles are reduced in size and their heads are huge, elongated snouts-squirting devices for the frontal gland secretion. When the nest wall is broken, soliders advance outward in parallel or in staggered line, precisely spaced by just the distance of antennal contact. A soldier sprays the frontal gland contents when it is bumped. The secretion contains volatile terpenes that are alarm releasers for termites and irritants to the attacker, and other unusual terpenes that are sticky and entangle ants, which are major threat to the lives of termites.

Allomones Promoting Associations

Defensive allomones terminate interspecific associations between potential prey and a diverse array of predator species. In contrast, there are a host of less-studied allomones that promote intimate associations between two paticular species. Many of these associations are mutually aggreable with both partners deriving substantial benefits, while others are parasitic, when only one of the partner gets the benefits. A strange phenomenon is seen in the behaviour of specialized insect speices are "guests" in the nest of social insects.

The alien insect enters an ant nest or termite nest without being attacked and is sometimes even assisted by a host worker. In some cases the volatile body lipids of guest and host are very similar in composition so that the intruder appears to mimic the body odor of the hosts. In other cases the guests placate their hosts with allomones that E. O. Wilson has termed *appeasement substances*. Jacques pasteels, of the University of Brussels,Belgium, has shown that that appeasement substances of rove beetles are produced in a battery of special exocrine glands that are found only in species that are social parasites. Bert Holldobler has observed the process by which a rove beetle guest grains entry to an ant nest. The ant feeds on the appeasement secretion. Next, the ant licks an "adoption gland" on the lateral margins of the beelte's abdomen, and finally it welcomes the beetle into its nest. When inside the nest some guest species behave generously enough, merely soliciting regurgitated food from the workers and exploiting the shelter and stable habitat provided by the hosts. But some guests return the hospitality with a bad turn. Kenneth Hagen, of the University of California, Berkeley, has reported that some lacewing larvae eat their termite hosts.

The lacewing larvae move about freely in the termite galleries. At certain developmental stages the lacewings approach termite workers and wave the tips of their abdomens in front of the termites. The termites are paralyzed by a gaseous allomone, and the larvae then devours the paralysed workers. Allomones also promote symbiosis between insects and micro-organisms. Leaf-cutter ants, certain termites, and bark beetles transport and consume symbiotic fungi. The association are quite specific-each insect species cultures only one or two fungal species. As the fungus grows in ant nests, the cultures remain clear of any contaminating species.

Leaf-cutter ants weed their crops by mechanical and chemical means. Alien spores are picked out of the fungus garden, and allomones, produced by the metapleural gland of the thorax, regulate the species composition of the growing fungi. H. Schildknecht and V. Maschwitz reported that the metapleural secretion has three components: phenylacetic acid, β-indolyacetic acid, and myrmicacin β-hydroxy-decanoic acid). Phenylacetic acid is a general-purpose antibiotic that prevents the growth of bacteria and of some weed fungi. Indolyacetic acid, a plant hormone, is thought to promote growth of the symbiotic crop. Myrmicacin inhibits the generation of fungal spores. The speices of fungus normally cultured in colonies of these ants is unaffected by phenylacetic acid or myrmicacin. Both, Bark beetles and ambrosia beetles transport symbiotic fungi within cuticular pits or in large cuticular pockets. The pits and pockets are associated with secretory cells, and cycles of secretary activity correlate with the peak of fungal proliferation in the pocket.

In the case of the southern pine beetles, the secretions apparently nourish the symbiotic fungi and restrict their growth to yeastlike budding rather than production of long hyphal strands. The yeastlike progeny drop out of the transport pocket as the beetle asses through the tunnels and galleries and start new growths of symbiotic fungi that will nourish beetle larvae. In addition, the microflora in the pocket is restricted to two species. Presumably the secretions favor the growth of these two symbionts while excluding all the weed fungi that contaminate the integument of the beetle. Unfortunately we unfamiliar of the chemical nature of allomones that play these role.

The microsymbiont of the association may be an aid to digestion. As shown by the late L. R. Cleveland, of Harvard University, intestinal protozoans (flagellates) help the primitive cockroach *Cryptocercus* to digest its meal of cellulose. The protozoans live in the anaerobic hindgut of the cockroach. In the presence of high concentrations of oxygen, the protozoans die. To grow, the cockroach must molt its entire cuticle, including that of the hingut. If abruptly exposed to the atmosphere together with the shed skin, the protozoans would not survive, and the cockroach would lose its vital digestive parts.

The protozoans save themselves from a molt because they "anticipate" it and enclose themselves in a cyst that excludes poisonous oxygen. The allomone that triggers encystment is the insect the hormone that induces apolysis: the familiar steroid ecdysone. The newly molted cockroach consumes its old cuticle, and once within the anaerobic gut environment, the protozoans become activates again.

An Overview of Chemical Signals

There has been chemical signals between cells since the moment first cells appeared on the earth, and intercellular chemical communication is a distinguishing feature of all major phyla. Hormones, pheromones, and allomones share some common origins in the earlier phase of the evolution of life form on the earth. Exocrine secretions for diverse purposes often have metabolic byproducts, and is not surprising that these products, which are produced only in certain physiological states, might communicate that state to animals of the same species. For phermoones are those allomones that promote associations, selective pressures act on both the producer and the target organism to evolve an unambiguous signal that is detected at low levels. For defensive allomones, slection operates on the producer to make the singal blatant and insulting, while the potential target organisms are at an advantage if they can ignore the insult. Thus many defensive allomones are broad-spectrum reactive protein poisons that are vitally important in the oves of insects.

Same kind carbon chains form the skeletons of the molecules that function as allomones, pheromones, or hormones. The same molecule may play more than one role, as does ecdysone in *Cryptocercus*. Theoretically structural diversity of chemical signals should be almost unlimited, but in practice only a few families of carbon skeletons predominate among pheromones, hormones, and allomones. These include aliphatic straight chains, terpenes and the related steroids, and small ring compounds. There are also a significant number of hormones are peptides, and a significant number of defensive, allomones are complex and unusual poisons. The explantaion of the widespread molecular conservatism is probably rooted deep in the metabolic pathways common to living systems. All cells have the enzymes for manufacturing certain skeletons. To develop a singal molecule, it is simpler to amend a preexisiting biosynethetic pathway and a preexisting receptor surface than to evolve totally new ones.

All the categories of molecular signals have common characteristics. Mostly pheromones and many allomones are volatile. Many defensive allomones are simple poisons with quite reactive functional groups, while others are complex and unusual toxins (such as percoccinelline) that irritate sensory receptors and are difficult to degrade metabolically. Most pheromone components are scarcely reactive, but their skeletons tend to be rigid and to have asymmetric carbons that determine precisely the shape of the molecule. Both defensive allomones and pheromones are medley, but for very different reasons. Defensive allomones are medley, but for very different reasons. Defensive allomones are medleys so that the producing animal covers as many bets as possible when it plays its defensive hand. Pheromone medleys allows the construction off fever number of indifiniteness signals.

6

SOCIAL WILD ANIMALS

About 37 million yeas ago, when primitive fissiped carnivores were evolving from their miascid predecessors, some of them forsook their way of life on land and went to sea to find food. This move must have worked out well, because many of the descendants of these pioneers—the seals, sea lions and walruses—are still making a living in the sea today. No doubt, descendants of the carnivores that remained on land fared well, too. Because land and sea are such different environments, we might expect the two animals groups to have diverged in many ways since their separation. Indeed, it should be interesting they have in common, an ancestor, and because of the obvious difference in the environments in which they live. In this chapter, some differences and similarities aquatic carnivores like between seals and their terrestrial counterparts with respect to gross morphology, feeding and predatory behaviour, reproductive behaviour, and various aspects of social life have been discussed. The similarities between aquatic carnivores and land carnivores were obvious to Laymen and early scientists. Fifteenth and 16th century sailors described the seals they saw as marine carnivores and called them "*sea lions*," "*sea bears*," or "*lobos de mar*" (*sea wolves*).

Linnaeus and several later taxonomists classified seals, sea lions, and walruses in the order Carnivora. However, as early as 1811, there was an attempt to place these animals in a separate order (Illiger, 1811). No a days, many investigators class these marine mammals in their own separate order, the Pinnipedia. However, classification of these animals if far from settled. Some maintain that the new order is inappropriate and that pinnipeds should be considered merely a suborder of the Carnivora. Other argue that the pinnipeds can be classified into two living families (walruses are placed in the same family as sea lions) under the super family, Canoidea, in the order Carnivora. The earliest pinnipeds probably entered the sea in one of the following areas: the Arctic Basin, the northwest coast of North America, or the Tethyan-Mediterranean area. It is not clear whether seals and sea lion (including the walrus) were already differentiated at this time. The oldest remains of pinnipeds are from the Miocene era, a time when they were already well adopted for aquatic life and when seals and sea lions were already distinguishable.

One opinion holds that all pinnipeds derived from canoid or dog-bear stock. An opposing view is that sea lions and the walrus derived from *ursine stock*, and seals derived from *lutrine stock*. This question is far from settled. Nevertheless, there is agreement that all pinnipeds are more closely related phylogenetically to members of the superfamily Canoidea of the order Carnivora the dogs, raccoons bears weasels and the like—than to the feloid carnivores. After their entrance in to the water, the distribution and diversity of pinnipeds were influenced by geomorphic and climatic barriers, distance between land falls, ocean currents, and water temperature. *Davis* hypothesizes that pinnipeds are, and always have been tied to a cold water environment. He argues very persuasively that the distribution and differentiation of present day northern pinnipeds reflects periods of expansion and contraction of sea and glacial ice that occurred during the Pleistocene era.

Scheffer points out that "as local shore lines and islands rose and fell and glacial barriers came and went, pinnipeds moved back and forth in order to maintain favourable breeding grounds along the edge of the sea." On land, many contemporary carnivores were being forced into extinction by periodic fires, floods, droughts and dust storms. *Scheffer* suggests that the early pinnipeds moved out along the shores of continents from island to island and later to polar ice fields. The rate or evolution was faster at the frontiers of advancing lines where immigration was unidirectional. The primitive populations were small and the members did not wander far, had no well developed homing instinct, were not migratory, and did not need to rendezvous in special places in order to find mates. Polygyny had not developed in any of the pinnipeds stocks. The animals were perhaps smaller than most recent pinnipeds had less fat, lived in more temperate waters, and perhaps made crude hens in beach grasses and among boulders.

The advancing pinnipeds met, from sea birds, and cetaceans, little competition for food. They met no competition for breeding room nor do they often today. Pinnipeds early lost the habit of feeding on beach organisms such as crabs, mussels, periwinkles and blennies. As each generic stock became isolated, it was transformed in response to the local physical and biotic environment and adapted to its peculiar niche. Among species frequenting isolated bays, gulfs, islands, inland seas, or lakes, there has been and still is, rapid evolution in the last few centuries, the effect of commercial sealing upon island and continental pinniped population has been catastrophic. Millions of seals have been slaughtered and entire breeding populations wiped out.

Fig. 6.1. Lutra (Otter).

It is doubtful that some species will ever recover genetically from undergoing such a severe population "*bottleneck*". In one recently exploited species, absolutely no polymorphic variation was found in 21 blood proteins fro 159 seals representing five rookeries! In this regard some land carnivore populations have suffered a similar fate at the hands of man-for example, the wolf. The increasing range of human populations is changing the distribution of both land carnivores and pinnipeds. Inspite of the fact the pinnipeds are marine, all species have retained an attachment to land.

In this respect, they differ from the Cetacea, descendants of primitive ungulate ancestor, who have become completely aquatic. All pinnipeds give birth to their youngs on land or ice, and some species spend the majority of their time resign out of the water. Pinnipeds have exploited the marine niche for food. A few carnivores, such as the sea otter, *Enhydra lutris*, and the polar bear, *Thalarctos maritimus*, also obtain food from the sea, but they have not undergone the same degree of morphological transformation to aquatic living as the pinnipeds. The anatomical and behavioural adaptations of the pinnipeds to life in the sa these adaptations influence their behaviour on land, it is worth reviewing some of the major ones.

Morphological Comparisons

Size

There are several morphological difference between pinnipeds and land carnivores simply reflecting the different environments in which they live. The average body size of pinnipeds is greater than that of carnivores. The smallest pinniped, the ringed seal, *Pusa hispida*, weighs 90 kg. The largest pinniped, the southern elephant sea, *Mirounga leonina*, weighs 771 kg, about 3629 kg and is 650 cm long. Several seal species are larger than the largest land carnivore, the grizzly bear, *Ursus arctos*, which weighs. Many land carnivores such as the foxes of Africa weigh less than 4 kg.. The Fennec fox, *Fennecus zerda*, weighs less than 1 kg! The greater body size of environment. Large body size is better fro hear retention in the cold sea. Moreover, since the water medium gives more support than air, large size could more easily evolve in the sea than on land.

Shape

Aquatic adaptation favouring swimming and diving have been observed in pinnipeds Body shape is streamlined to reduce drag. The

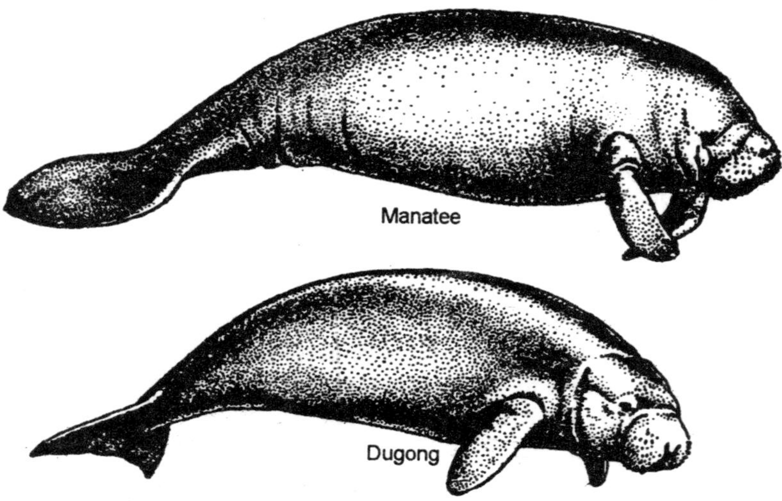

Fig. 6.2. Two Sirenian representatives.

external ears are reduced or absent, External genitalia and mammary tests are drawn into the body, Limbs are enclosed within the body and extremities are flattened the tail is short and the head is flattened, and the eyes are situated well forward. The neck, in particular, is thick and muscular and considerably more flexible than in most large carnivore. The skin also shows adaptation to a water environment, and the hair is flattened. Pinnipeds never groom the pelage with the mouth or tongue. The terrestrial carnivore body is a picture of contrast. The canid body, especially the large one, is designed for long distances running.

Subcutaneous Fat

Subcutaneous fat forms a substantial part of all pinniped bodies. Skin and *blubber* make up more than 25% of the weight of the Weddell seal, *Leptonychotes weddelli* and almost 50% of body weight of the southern elephant seal. Blubber the notified adipore layer provides reserve energy during fasts and lactation, thermal insulation buoyancy, and the padding necessary for a stream lined profile. The proportion of body fat to total weight inland carnivores is apparently much smaller, even in dormant or hibernating bears, skunks, and raccoons.

Brain

Although the pinniped brain has not been studied in great detail, it is evidently large, like that of terrestrial carnivores, but more spherical and more highly convoluted. The sea lion brain resembles that of bears, while the seal brain looks more like that of cats ad dogs. Compared to a dog's brain, the cerebellum is large, probably because of the increased coordination demanded in swimming. The auditory nerve is large and the olfactory lobes are reduced. Pinniped eyes (the walrus excepted) are larger than those of land carnivores, and they function well at low levels of illumination. Seal vision is excellent both in water and on land.

Teeth

Pinnipeds have fewer and more uniform teeth than most land carnivores (except for the walrus, a special case once again). All species have the pronounced upper and lower canines but lack the carnassial cusps characteristic of land carnivores. The postcanines in most species are rudimentary pegs functioning to hold prey that is swallowed whole. Variation in pinniped teeth reflects the type of flesh the animals feed on and as in many mammals, it is a useful taxonomic indicator. Generally, the teeth, mouth, jaw, and associated structures of pinnipeds are designed for grasping tearing and swallowing prey whole or in chunks, rather than for chewing shearing or crushing large bones.

In contrast with land carnivores, the deciduous teeth or milk teeth of pinnipeds disappear before or soon after birth. Several physiological adjustments to aquatic life have bee made on pinnipeds that set them apart from land carnivores. These adaptations in respiration circulation renal physiology, and the like are important and interesting, but it would be too much of digression to cover these topics here.

PREY, FOOD HABITS, AND FEEDING BEHAVIOUR

Pinnipeds, deviated from land carnivores in form and function because of adaptations that took place as a result of seeking food in the seal. It is therefore most important to compare these two animals groups from the point of view of hunting and feeding habits, and in relation to their respective prey.

Prey

All carnivores including pinnipeds and land carnivores are meat eaters, but only the marine mammal eat flesh exclusively. Seals and sea lions feed mainly on crustacea, molluscs, fish penguins, and other sea birds. Although diet varies with the species, the majority of pinnipeds feed on a variety of prey. The Alaska fur seal, *Callorhinus ursinus*, is the best studied pinniped fro the point view of food habits best studied pinniped from the point of view of food habits. It catholic diet is representative of the wide range of fishes and squid eaten by the Otariids (the earred seals or sea lions and fur seals, as opposed to the Phocids, the earless, true seals). Alaska fur seals feed on more than 30 kinds of marine organisms: at leas 27 species of fishes. 1 species of octopus, and 5 species of squid.

The most common prey of fur seals are northern anchovy, *Engraulis mordax*, squid and Pacific herring. (*Clupea harengus*). Various rockfish, Pacific hake, *Merluccious productus*, Pacific saury, *Cololabis saira*, salmon, *Oncorhynchus* spp., and American shad, *Alosa sapidissimia*, are eaten in lesser quantities. Hermit crabs, amphipods, and various dividing sea birds (e.g., Rhinocerous Auklet, *Cerorhinca monocerata*, Redthroated Loon, *Gavia stellata*, and Best Petrel, *Oceanodroma leucorhoa beali*) are eaten occasionally, but they seem to play only a minor role in the seals' diet. Fur seals, like most other pinnipeds, are *opportunistic feeders*. What is eaten depends on the seasonal distribution and the abundance of prey. Since fur seals commonly feed on schooling fishes, stomach contents of several animals collected in the same area often contain only one type of fish. The Steller sea lion, *Eumetopias jubata*, and the California sea lion, *Zalophus californianus*, eat similar fishes and squid, especially where the feeding range of the three species overlaps. Steller males in the Bering Sea and off Alaskan islands supplement their diet with Alaska fur seal pups. Most true seals, such as the harbor seal, *Phoca vitulina*, and the grey seal, *Halichoerus grypus*, are mainly fish eaters, but they occasionally fed on an assortment of crustacea, octopus eels and molluscs.

The northern elephant seal, *Mirouna angustirostirs*, is a deep dividing seal that feeds seal primarily on squid skates, rays, small sharks and ratfish *Hydrolages collier*. The small seal, *Pusa hispida*, which inhabits the circumpolar Arctic coasts, feeds on up to 72 different species of small pelagic amphipods, euphausians, and other crustacea, as well as on small fishes. Only two of the 32 different pinniped species are rather specialized feeders that exploit one type of prey species almost exclusively. The Crabeater, *Lobodon carcinophagus*, feeds on krill, small shrimp like animals that it catches in quantity and strains from the water through its multicusped cheek teeth, much as a mysticete whale sieves krill through its baleen. The walrus, *Odobenus rosmarus*, feeds primarily on three genera of bivalve molluscs, *Mya*, *Saxicava*, and *Cardium*, for which it forages in shallow coastal waters less than 40 fathoms deep. Bivalves on the sea bottom are examined and sorted by the lips and whiskers; feet and fleshy parts are torn off and swallowed whole or sucked out. But when molluscs are scarce, the walrus may eat young ringed seals, bearded, seals, *Erignathus* apparently feeds on Cetacea.

Occasionally a narwhal, *Monodon monoceros*, or a beluga, *Delphinapterus lcucas* is eaten. However, it is not known whether walruses actually kill them or feed on corpses. According to *Mech* "probably every kind of backboned animals that lives in the range of the wolf has been eaten by the wolf". Wolves, *Canis lupus*, eat mice, mink muskrats, squirrels rabbits, various birds fishes lizards snakes , grasshoppers, earthworms, and berries. But predation on small animals plays only a minor role in the wolf's diet. The main prey of the wolf, like those of the African wild dog. *Lycaon pictus*, are large animals. The wolf's primary prey in the United States and Canada are : white tailed deer, mule deer moose caribou elk Dall sheep, bighorn sheep and beaver. Wild dog kill mostly Thomson's gazelles juvenile wildebeests and Grant's gazelles and occasionally, warthogs and zebras.

African foxes feed mostly on rodent and small reptiles birds's eggs, insects and vegetables, matter, jackals eat similar foods in addition to small mammals and carrion. Additional information on other canids can be found in *Fox*. Of all the land carnivores, bears have the most diverse tastes. The grizzly, before man virtually annihilated it, was an omnivorous opportunist *part excellence*. The grizzly ate almost anything and everything that was available. This long list included: meat, fresh or putrid from whales, water birds, fishes, elk, deer, antelope gophers lizards, frogs and domestic livestock: a wider variety of plant materials than herbivorous animals ate, some of which were various berries, clovers, nuts, wheat corn, potatoes tree bark and various bulbs and assorted foods like honey, ants and their larvae, yellow jacket nets, and mushrooms. Polar bears, *Thalarctos maritimus*, have a much more restricted diet. Although they consume some vegetation and carrion during the summer, they feed primarily on the ringed seal during most of the year.

Size of Prey

Among large carnivores there is a remarkable thing about their feeding habits is that the prey is often larger than the predator. Wild dogs, whose average weight does not exceed 18 kg, bring down impala and reedbucks that are double or triple their size and zebras that are even larger. Wolves weighing 36 to 45 kg kill moose and bison that may weigh over 500 kg. These canids overcome large prey by hunting in packs that rangé from a few animals up to a score or more of them. One dog usually selects a quarry from a retreating herd of gazelles, and the other dogs follow it . The quarry is coursed for 1 to 3 km until it becomes exhausted. The lead dog catches up to it and grabs it or bowls it over. Once overtaken, they prey is fallen on by all the following dogs, and it is dismembered an eaten with great dispatch. Wolves bring down deer or moose in a similar way. The first wolf to catch up to the fleeing prey attempts to grab hold of the rump, flanks, necks or nose. Once the prey is down, other attack from every side.

Although the technique my vary with prey species, the strategy of the individual quarry and the composition of the hunting pack, it is clear that these pack hunters get help from each other in getting their food. Lone wolves or even small packs are at a disadvantage in bringing down large prey. In black backed jackals, *Canis mesomelas* two hunters are more than four times as successful as one in bringing down gazelle fawns as observed by Wyman. In pinnipeds the size of prey eaten varies greatly like land carnivore, but except for the dobious possibility that walruses kill small whales, pinnipeds always kill and eat animals that are much smaller than themselves. Prey size determined how alaska fur seals, Steller sea lions, and harbor seals eat their food. Small fishes less than 30 cm long, such as anchovy, herring saury, or squid are consumed whole under water. Except for very small food items like lanternfish, the prey is swallowed head first. Larger prey measuring more than 30 cm long, such as hake, rockfish, or salmon, is brought to the surface, grasped by the head shaken violently and reduced to chunks that are swallowed piecemeal.

As fur seal get older and larger, large fishes up to 92 cm make up an increasingly high proportion of their diet. Thus the size of the prey increases with the growing age of fish seas. In most of pinnipeds the hunting is done in groups, but, unlike the case with canids, grouping does not enable them to exploit larger animals. Rather, grouping seems to make it easier for them to exploit large aggregations of small prey. *Fiscus* and *Baines* saw Steller sea lions leaving their hauling grounds in compact groups of several hundred to several thousand animals. They swam out to feeding areas, where they dispersed into smaller groups of less than 50 animals containing both sexes and mixed sizes of animals. Massings of this type fed on large of sea lions feed or squid.

Casual observations suggest that groups of sea lions feed more efficiently on schools of fish because cooperation among the hunters enables them to herd and control the movements of the school. When large fish schools are absent, sea lions feed singly or in small groups of two to five animals. In some cases the hunting is performed in a very special way before Steller sea lion females leave the rookery and their pups to go to sea and feed, they may engage in activities that resemble and that may have an analogous function to, the prehunt greeting ceremony of wild dogs.

Several females may gather at water's edge, where they mill about restlessly in close contact with each other. They may vocalize repeatedly and engage in what appears to be low intensity aggressive behaviour before swimming off as a group to feed. In both species, these activities may reinforce group cohesion and unity as well as synchronize the time of departure. Similar observation have also been made in canids and pinnipeds. They may travel several kilometers to get a meal. Van *Lawick-Goodall* and *van Lawick-Goodall* report that wild dogs did not start their firs chase until they were 8 km from where they started. The chase itself covered a distance of $5^{1/2}$ km. Wolves may pursue moose or caribou for up to 5 to 8 km but usually give up sooner. They may range up to 32 km from their den. Steller sea lions and California sea lions travel much further from their rookeries or hauling grounds to feed. The former have been seen feeding as far as 112 to 136 kms from land and the latter as far as 62 kms from land.

Usually, only small groups of less than 30 individuals are seen this far from land. Large groups of sea lions (100 or more) are seldom seen feeding more than 16 to 24 km from a hauling ground or rookery. It would be it interesting to compare land carnivores and pinnipeds on the success of the respective hunting and fishing expeditions, but the difficulties inherent in observing seals feeding in the water makes this impossible at present. However, some indirect evidence is pertinent. Although both pinnipeds and the larger land carnivores are nomadic, the former travel greater distances in their annual feeding migrations. The record holder is the Alaska fur seal, which migrates from its breeding grounds in the Bering Sea to its winter feeding grounds along the shores of the eastern and western Pacific as far south as San Diego, California, and Hokkaido, Japan, a distance of some 5000 km. Wolves may follow caribou on their annual migrations.

Times of Hunting

Wild dogs and Wolves, *Ceron alpinus*, prefer to hunt in early morning or early afternoon or evening, although hunts occur on moonlight nights. Wolves an bears may hunt by day or night. Most of the smaller foxes are nocturnal feeders. Fur seals and sea lions are primarily night and early morning feeders. Stomach contents of animals collected at various hours after sunrise show decreasing amounts of food. However, in some areas where large schools of fish are present, daytime feeding also occurs. Typically the midday hours are spent resting, and feeding activities gradually resume in late afternoon. The harbor seal is a daytime feeder.

Amount of Food

The amount of food varies for animal to animal. Canids eat great amounts of food in a single feeding bout. A wolf may gorge 9 kg of meat in single meal; this may represent 205 of its body weight. However, estimates of little more than half that amount per day per wolf may be more representative. *Estes* and *Goddard* estimate that wild dogs average 2.72 kg of meat per day, or approximately 6.7% of their body weight. But these canids may have to go without eating for

Fig. 6.3. (a) Phoca (Seal); (b) Odobenus (Walrus).

several days at a time. Five to 7 days fasts have been recorded in wolves, and one wolf went for 17 days without food. It is common knowledge that domestic dogs can go for several days without eating, particularly when sexual activity is probable. Grizzlies and black bears in the more northerly parts of North America may go dormant for varying periods of time during the winter go dormant for varying periods of time during the winter when food is scarce. In general all of the land carnivores will feed regularly if possible. Even bears will actively feed through out the year when food is available.

Fasting

Among aquatic a carnival mammals the pinnipeds feed daily except at certain times of the year, usually the breeding season, when they undergo long fasts. Thus, feasts and famine are even more extreme in their annual cycle. when they are feeding fur seals and sea lions ingest less food relative to body weight than the canids mentioned above.

Spaulding estimates that fur seal, seal lion, and harbor, seal food requirements range from 2 to 11% of body weight per day with 6% being the average. During the breeding season, fur seal and sea lion males fast for approximately 6 weeks; harbor seals do not appear to fast during this time. The record for long fasts in pinnipeds goes to male northern elephant seals. They may go without food for 3 months. Throughout this time they are actively fighting and attempting to copulate, and they may lose over 450 kgs. Females of this species fast for an average of 34 days during the period in which they give birth and nurse their pups daily for 28 days. A female's total body weight maybe reduced by almost 50% during this fast. During the nonbreeding season, males and females fast again for approximately 30 days while undergoing the annual molt.

Specific Feeing

Most canids and ursids scavenge for food at some time, Perhaps bears exploit carrion more than any other land carivore. Carrion was a major part of the California grizzly's diet more than 100 years age, before kept the California coast clean of pinniped and whale carcasses that repeatedly washed ashore. There are numerous reports of 12 to 15 bears feeding at the same time on a single whale carcass. Many of the larger land carnivores typically desert remains of a kill and cache it away and return to et from it later. Polar bears are an exception to this rule. They do not cache ringed seal carcasses that they kill, despite the fact that they may only feed on the blubber and leave the meat behind. In marked contrast to these feeding habits of land carnivores, pinnipeds do not cache food, nor do they scavenge although the walrus may be an exception to this rule. Pinnipeds consume most of their prey entire and discard only the heads of large fish, particularly rockfish.

Predation

One drawback that pinnipeds experience in their marine habitat to which the larger land carnivores are relatively immune is predation. In carving their niche in the sea, pinnipeds became exposed to large predators who were also adapted to aquatic life and more formidable than themselves. At sea, the main predators on pinnipeds are large sharks, particularly the great white shark, *Carcharodon carcharias*, and the killer whale, *Orcinus orca*.

All pinniped species living in the Pacific are preyed on by these animals. The degree of predation is unknown. When pinnipeds return to land or ice to give birth and nurse their young, they are vulnerable to predation by large land carnivores such as polar bears and grizzly bears, perhaps wolves and of course, man. Small carnivores such as foxes and skuas, gulls, and hawks, may prey on their young. Being adapted primarily for locomotion in water, they are no match defending themselves against large land predator. Thus, it would appear that predator pressure was important in causing pinnipeds to breed in remote sanctuaries free from these predators; that is, offshore, islands or rocks, sandbars and ice floes, and offshore breeding areas are limited.

SOCIAL BEHAVIOUR AND REPRODUCTION

It is likely that pinnipeds were forced to share limited breeding areas, and this led to the high degree of sociality and the large social gatherings that we find in pinnipeds today. In terms of the sheer number of animals that congregate together, the pinnipeds are much more social than any land carnivore. As many as 2 million Alaska fur seals can be found every summer on two tiny Pribilof Islands in the Bering Sea. Three thousand northern elephant seals maybe found packed together tightly on one beach during the peak of the breeding season. Two thousand California sea lions may be found sleeping together in close contact during the nonbreeding season. The largest wolf or wild dog pecks are miniscule in comparison. Pinnipeds are among the most *polygynous* mammals.

Apparently, this mode of life developed very early in their history, and it was closely tied to their amphibious habits. When females began clustering in time and space, the conditions became ideal for promoting male-male competition, *polygyny*, and sexual dimorphism. Males must have competed to mate with as many females as possible. Those who kept other males away from areas containing females sited more pups.

In their short breeding seasons on traditional rookeries, fur seals and sea lions developed a social structure characterized by territoriality among males. Males without territories do not copulate, or they copulate only rarely; those who secure the most well placed territories—territories containing the most estrous females—do most of the breeding. Another system that developed was for males to dominate other males and thus gain access to more estrous females wherever, they were situated. This was the strategy adopted by northern elephant seal males. They fight for social status in a dominance hierarchy. The highest ranking male or males depending on the number of females present—locate themselves nearest the females and keep all other away. One or a few of the highest ranking males nor breed at all. One male may dominate a haem for several years and inseminate more than 200 females. Grey seal society is similar to that of elephant seals. Thus, in most of the well studied pinnipeds, males are *polygynous* and are either territorial or exhibit a dominance hierarchy during the breeding season. This is a reflection of the manner in which the largest and strongest males go about monopolizing a large number of females.

Since size and physical aggression convey such a great reproductive advantage to males, these traits have become increasingly elaborated in time. Sexual dimorphism is greater in some seals and sea lions than in other mammals. For example, the Alaska fur seal male is five time larger than the female. Among terrestrial carnivores the Canids are "*social*" in quite a different way from pinnipeds. Social organization is characterized by seasonal or permanent pair bonds in foxes and jackals, and by cooperative hunting and group life in the wolf, dhole, and wild dogs. Black bears are essentially solitary, but female cub associations are long lasting, and strong and but female cubs inherit their mother's feeding territories. Wolf packs are organized into separate male and female social hierarchies.

Although the female hierarchy is more ambiguous, it is clearly more pronounced than the size-related dominance observed in female elephant seals. Hunting dog packs have similar hierarchies, particularly during the breeding season, but it is in the female hierarchy that is most evident. In wolves and wild dogs, the dominant breeding female may prevent subordinate females from breeding or may drive them out of the pack and kill their pups. Dominant females may also enlist the help of males, and they may prevent other adults from feeding the subordinate female or her young.

Territorial Behaviour

Although many canids and ursids are territorial in the sense that they defend a feeding or home range against other individuals or groups of individuals, this behaviours is quite different from the defense of rigid individual territories by fur seal males and Steller sea lion males against conspecific males during the breeding season.

7

SOCIAL AQUATIC ANIMALS

HUMPBACK WHALES

At the turn of the century, humpback whales *Megaptera novaeangliae* became one of the main targets of the voracious international whaling industry. These playful whales were easy pickings for whaling ships. Humpbacks grow to 50 ft. long and weigh 40 tons or more. Fat content is lower than in other large whales but they are easier to find and catch. Their regular and easily located congregations for breeding in winter, and feeding in summer, meant that they were more exposed to hunting than the solitary blues and fins. In Antarctic waters alone an estimated population of 22,000 southern humpbacks was reduced to barely 3,000 individuals by the early 1960s when, thankfully, a ban on humpback whaling was introduced. The humpback whale, though, has a global distribution with migrations to the food rich Arctic or Antarctic waters in summer, and to the breeding grounds around the tropics in the winter.

In summer the northern race of humpbacks spend its time in the Arctic, filtering out small fish and shrimp-like crustaceans called krill. Apart from a few social interactions when meeting others, they are usually silent. It is after their migration south, to breed and calve in warmer waters, that their beautiful sounds are heard. In the Atlantic ocean, groups of whales can be spotted off the islands of the West Indies, around Bermuda and off the west coast of Africa. In the Pacific, southern California and Hawaii are known gathering places. This seasonal predictability and easy discovery nearly led to their extinction.

Since the turn of the century the worldwide population is estimated to have been reduced by 95% by whaling. Humpbacks are currently protected throughout their ranges by International Whaling Commission regulations. Except for limited aboriginal kills in the Caribbean and Green-land and inadvertent deaths when being caught in fishing nets off the east coast of North America, they are now safe. Ironically it was the whalers who first noticed that their prey was sensitive to sound. The disturbance of a noisy oar in its rowlock would frighten a whale away. There is even record of an underwater camera shutter release scaring a humpback whale 20 metres away. Their protection came not a moment too soon, for extinction would have robed us of a remarkable animal now known worldwide for its beautiful underwater songs. The songs were first recorded by O.W. Schreiber in 1952 from a US Navy underwater listening post on the submarine slope of Kauai, Hawaii, but he didn't recognise which marine creature was responsible. William Scheville at Woods Hole Oceanographic Institution later identified the sounds as coming from humpback whales.

The humpback subsequently became a world bestseller on an LP record called *Songs of Humpback Whale*. The scientist responsible for drawing scientific attention to their songs is Dr. Roger Payne of the New York Zoological Society who, together with his wife Katy, a musicologist turned zoologist, has been involved in a intensive study of humpbacks, their songs and behaviour. Roger Payne's interest in bioacoustics began with the directional sensitivity of bats' ears; continued with the owl's ability to locate prey by hearing; and then followed with studies of the way some moths can avoid bat sonar. His first encounter with whales was on a rain washed beach near Tufts University where a small porpoise had been stranded. Payne was appalled by what he saw. 'It had been mutilated. Someone had hacked off its flukes for a souvenir. Two other people had carved their initials deeply into its side, and someone else had stuck a cigar but in its blowhole. I removed the cigar and stood there for a long time with feelings I cannot describe.' He resolved at that moment to learn about whales, so that some day he might be able to have at least some effect on their future.

The opportunity came in 1967 when Roger and Katy Payne visited Bermuda to look at migrating whales and met Frank Watlington, an acoustics engineer at Columbia University Geophysical Field Station. Watlington was responsible for recording anything whether natural or man-made that produced sounds in the sea, in particular under-sea explosions. He played them underwater recordings of humpbacks and lent his tapes. While out in a row-boat, attempting to get close to humpback whales without frightening them, Roger Payne heard the sounds of whales amplified through the bottom of the boat. He noticed that one of the whales was singing phrases similar to those on Watlington's tape, and it suddenly became apparent to Payne that he was listening not to random sound but to regular repeating patterns. The humpbacks make a variety of grunts and squeaks too—indeed, they appear to have a whole repertoire of vocalisations—but these repeating patterns are by far the most interesting. They could be considered as true songs for they consist of long complicated repetitive sequences, much like bird song except that unlike most bird songs, each one can last from five to more than 30 minutes. (It even resembles bird song if speeded up.) A whale might stop singing or resume singing at any point in the song so it is difficult to determine a beginning, middle or end. Each song may be sung over and over again without breaks for many hours. A whale recorded in the Caribbean by Howard and Lois Winn, of the University of Rhode island, sang non-stop for 22 hours, breathing in the intervals between phrases.

It was still going strong when the Winns pulled up hydrophones and went home. Humpback whale songs are the longest and most complicated of animal songs known to man. Day and night the Paynes collected their sequences of whale songs by dangling hydrophones or underwater microphones from outriggers to each side of their small sailboat. The sounds were then recorded on tape-recorders and taken back to the laboratory and analysed with the help of Scott McVay at Princeton University. Together with Frank Watlington's recordings they have an almost continuous record of the Bermuda whales for a period of over 20 years. In an attempt to analyse the sounds, the Paynes divided each song into identifiable parts.

The smallest part is a unit (equivalent to a musical note). Units are grouped into small repeating sequences called phrases. Groups of similar phrases are known as themes. Off Bermuda, eight to ten themes make a song (later off Hawaii, as few as four to five themes were fond to make a song), and the song is repeated without pause as a song pattern. By breaking down the whale sounds in this way the Paynes were able to make a series of interesting observations. It turned out that all humpback whales in a particular area sing the same song.

Whales from the Pacific Ocean, however, sing a different song from those in the Atlantic, although the laws on which they base their songs appear to be the same in both populations. A song, for example, might consist of five themes (A-B-C-D-E) which always follow in the same order. If a theme, say C, is dropped, the remaining themes are always sung in the same sequence (A-B-D-E). In much the same way, the laws that govern the structure of sonnets, or the composition of western music are the same whether the piece is composed in New York or London. The laws governing humpback whale songs are, however, complicated and little understood and it is not known whether they are inherited culturally or genetically. The migration routes of the Atlantic and Pacific populations are unlikely to cross, hence the dialect- like differences. But there are separate centres of whale breeding activity in the Pacific Ocean.

Baja California and Hawaii, 3,000 miles away, for example, are two separate breeding areas. The question is, do whales from each area meet up and share similar songs or are there dialect differences even between Pacific whales? A photographic technique identifying distinctive markings on the underside of humpback flukes, used by James Darling of the University of California at Santa Cruz, together with song analysis, helped to give an answer. A study in 1979 revealed that Baja and Hawaii humpback whales were singing the same song. Two years later a follow-up showed them to have made identical modifications to the song.

The whales from separate breeding areas must have been in contact, for songs could not physically travel 3,000 miles from one breeding ground to the other. Analysis of fluke-pattern photographs confirmed that whales feeding of south-east Alaska would appear in the breeding areas in both Hawaii and Baja California. The degree of interchange between these areas is still unknown. The discovery had other implications. It meant that whales might visit anyone of several breeding areas, a factor important for conservation. If, one year, a breeding area was affected by some man-made or natural catastrophe and all the whales killed, whales which that year had visited an alternative area might turn up after the site had recovered and repopulate the area. Also, when counting whales to establish stocks and whaling quotas, it would be quite possible to count the same whale twice—once in each breeding area—and get an overestimate of numbers. By studying whale sounds the researchers had identified a conservation loophole. Another surprising discovery was that the songs are continually changing.

All the whales in an open population change their songs in the same way so that each individual is up to date with the current vogue—a kind of top-of-the-pops. The changes are progressive and rapid, each component perhaps changing every two months, although many elements of the song are changing at any one time. Taking a folk-song analogy, in the first month an individual might sing: 'London's burning. London's burning. Fire-fire, fire-fire...' and so on, while two months later one phrase would be modified so the song becomes: 'London's burning, London's burning, London's burning, Fire-fire, fire-fire.....' etc. Humpback whales are composers, much as humans are, except that the whales do not create new songs, they evolve something different from what they already have by making minor modifications.

During a season a song may change components, add extra parts and drop in pitch. A component may undergo rapid changes for several months and then be left alone while other parts of the song are modified. Curiously, newly created phrases are sung more rapidly than older ones. Sometimes a new phrase is created by taking the first and last parts of older phrase and dropping the bit in the middle, much as we shorten 'I would' to 'I'd'. What selective advantage a whale may gain from changing its song is unclear. Whether it is dominant trend setter who introduces the changes will be difficult to find out, but what is clear is that an entire song is renewed totally after about eight years. Humpback whales sing only on the breeding grounds and occasionally during the migration. Only one isolated and misguided individual has been heard singing in the feeding areas in the Arctic or the Antarctic. What happens to the song, therefore, between breeding seasons? Whales, like elephants it seems, never forget.

In the new season the song is picked up usually as it was left the previous season, remembered by each male in the population. Most of the changes that take place occur during the singing season and not in between. (There is only one case of a substantial change in a theme, apparently between seasons). With such sophistication in song development, the question often asked is whether whales are highly intelligent creatures capable of indulging in intellectual pursuits and communicating in complex languages, even to the point, as some have suggested, of being wiser than man. True, whales have large brains, in fact the largest brains of any animals that have ever lived. How they use this large brain is unknown. It is tempting to suggest that whales reason about the world just as man does, but they are constrained by the limits of the environment in which they live. Whales lack hand and therefore brainpower cannot be channelled into the creation and use of tools, for example. They do, however, need to develop senses and skills to maintain their efficiency under the sea. Perhaps a part of their large brains is used in running this sophisticated sound system capable of discriminating songs above the backyard jungle of the noisy ocean, interpreting the content, learning the structure, remembering the detailed phrasing, processing the information, and may be modifying for future use.

The various frequencies of which a song is composed will attenuate differently in the water; higher frequencies will travel less far than low frequencies. Signals will sound quite different depending on whether the sender is near or far away. A whale would have immensely complicated computations to make in order to decipher the message as it was when sent. Clearly acoustic requirements in the sea could tie up a substantial part of the whale's brain capacity. (Bats, on the other hand, are involved in extremely complicated sound processing, using a brain of not more than one gram). At the very least, study of the humpback whale song is going to give some clues as to the way whale minds are working. Roger Payne's latest work is concerned with trying to relate songs to behaviour.

Already the whale researchers have noticed that, during certain phrases of a song, the humpback will make particular movements of a flipper or the body. Underwater photographer Al Giddings recorded seeing a singing humpback 60 fts. down moving its flippers back and forth in time with its song phrases. They seem to be dancing as they sing. Unfortunately, there are usually no other whales around to watch so perhaps they are merely flexing and stretching their muscles as they make the sounds. Whale song was originally thought to be associated with courtship behaviour at the breeding grounds. Writing in *Marine Bio-acoustics* in 1964 William Schevill noted, 'The sonorous moans and screams associated with migrations of (humpback whales) past Bermuda and Hawaii may be audible manifestations of more fundamental urges...' Whale renditions have been compared with those of birds, crickets and gibbons in that they probably convey such information as species, sex, age, location, identity, readiness to mate, and readiness to engage in aggressive behaviour with rivals. Aggression is not a word often associated with whales but it is becoming clearer that whales are more aggressive than was first thought. Researchers from the University of Hawaii have described 'raw freshly bruised dorsal fins' and 'head nodules that appeared red bruised'. Some have suggested that an assembly of singing humpbacks is reminiscent of a lek, where males gather on a communal courtship display ground to which females come in order to breed. It could be that females choose their mates on the basis of quality of singing, although there is no evidence of this as-yet.

Behaviour associated with singing whales has been studied by Peter Tyack of Rockefeller University. In 1977, while recording humpbacks around Hawaii with Roger Payne, Tyack would go out in a small boat, looking at random for whales, put the hydrophone over the side and record the sound when he could heard one whale dominating. 'When you get into the water near a singing humpback whale it can be a scary experience. Your lungs resonate with the sound; it is very loud and you feel it throughout your body—you feel it more than hear—it is very eerie to be sitting 30 ft below the water just having your body reverberate with the sound of a whale. Occasionally when one of us is alone in a boat and has been out there for a few hours, he'll notice the whale next to the boat, ten or 15 ft away. The whale will often surface, come right up within inches of the boat itself, seeming to look the boat over, often staying for up to half an hour at a time, spy-hopping, lifting its head up to look into the look into the boat, lifting its flippers up; it is very strange to have your wild animal come up to you when you don't expect it'.

Often Tyack was able to identify the singing whale because, when it surfaced, the sound level would be reduced. Occasionally, he noticed, the whale would stop singing and then surface with another whale, the two whales moving off together. Following the whales in a small boat revealed little of their behaviour, so at the end of the season Tyack was left with little more than a tantalising hunch that something interesting was going on. In the intervening months before the next season, Tyack and his co-workers devised an observational technique which was going to reveal fascinating information about humpback whale behaviour, in particular some insights into the function of song.

The technique involved the combined operations of observers on a hill overlooking the whales, with others following individual whales in small boats (Boston whalers with outboards). A bay was chosen on the sheltered west coast of Maui, Hawaii, where large concentrations of whales appear each winter. In that way the overall patterns of movements could be seen from the hill while individual identities and vocalisations could be recorded in the boats. Each team was in radio contact with the other. In the spring of 1979 they tried their experiment. The first singer they followed stopped singing, joined with another whale and the two went off. During the season, 13 out of 28 whales that were followed after they had stopped singing joined with other whales. Indeed, the recording team learned to expect interesting things when a singing whale they were pursuing stopped singing. By observing whale behaviour in this way they were able to put together a picture of leviathan life below the sea. A singing whale is almost always alone, separated by several hundred metres from any other whales. It moves slowly, turning this way and that while singing.

It occasionally moves towards other whales but avoids any other whale that is singing. Singers do not appear to hold duets or to interact vocally in the way. They also do not seem to have strict territories. Singers are not found singing at the same 'song posts' on different days. In a few cases, Tyack observed singing whales approached by single non-singing whales. The original singer would stop singing, move along for a while with the intruder and then silently, but rapidly, swim away, seemingly displaced from his singing post. The new arrival would then start singing in the other's place. If a singer approaches a small group of whales such as cow and calf, he will often turn deliberately towards them. The cow and calf sometimes steer away, other wise they continue on the same course, but they rarely move towards the approaching singer. With a two year breeding cycle, where a cow may conceive one year, have a calf the next, and then wait for a year until the calf is weaned before ovulating again, a cow may not be sexually receptive, and so will actively avoid a singing male. The singer often pursues the group and will attempt to catch up. If he is able to join them he immediately stops singing and becomes, an escort to the cow.

The cow, the calf and the ex-singer then swim along slowly and silently together, the calf keeping close to the cow. Escort whales were once known as 'aunties' and were thought to be female helpers. Sometimes they will interact with gentle flippering or rolling, behaviour previously observed in association with sexual activity in gray and right whales. Normally any signs of aggressive behaviour are absent in cow-calf-and-escort groups. On one occasion, however, when Tyack was in the water with the whales, he somehow got between a cow and her calf, the calf having swum over to investigate Tyack. The mother, clearly upset, turned sharply towards Tyack and in so doing bumped into the surfacing escort whale which could not get out of the way in time. The escort produced a sound and the cow responded by emitting a series of grunt s for about half a minute which might have been translated as 'watch out where you're going!" Occasionally the cow and escort will dive below out of sight leaving the calf at the surface for a period of ten to 15 minutes, returning to the calf for five minutes and disappearing again. Whether mating takes place deep down in the ocean is not known, although researchers believe that this must be the case. Females ovulate at this time and males show an increase in testes weight (compared with the summer feeding condition) and increased sperm production.

In addition females calve in winter and the gestation period is about one year. Songs sung towards the end of the season tend to be longer than those at the beginning. A similar observation in songbirds has been accounted for by an increasing concentration of testosterone in the blood. It is thought that similar events could be happening with humpbacks. If the cow-calf-and-escort trio swim into the vicinity of another singing whale, the singer may begin to approach the group. Invariably the cow-calf-and-escort speed up and alter course away from the singer, but the singer seems highly motivated to join them. He will speed up dramatically, even while singing, and pursue the group, turning this way or that in an attempt to catch up. As soon as he has reached the cow-calf-and-escort he will also stop singing and a sudden change in behaviour takes place in the group. They begin to swim faster. The two escort whales engage in some kind of competition in which the secondary escort attempts to displace the principal to escort from his position next to the cow. With the cow and calf up front and the two escorts behind the group may be swimming at ten to 15 km

Fig. 7.1. Humpback whale cow-calf escort group.

per hour. Until now the original trio were silent but as soon as aggressive activity starts, a whole barrage of sounds can be heard. The whales will thrust at each other with their flukes or ram into each other as they jockey for position. The impact of fluke on blubber can be heard as a very loud slapping sound.

Occasionally they blow bubbles underwater, an activity associated with aggression in many other species of whales. They also produce vocal sounds which the researchers recognise as social calls. There are many types of sounds—sometimes segments of songs out of context, high pure trumpeting calls and low grunts but as yet it is difficult to sort out their meanings. The social sounds are loud and can be heard up to nine km away, thus advertising the location of the group. Other lone singing whales, on hearing the disturbance, will stream into join the group. As more and more whales join, the activity increases. The rapidly enlarging group becomes very rowdy and engages in a great deal of aggressive behaviour. At the surface, flippers, flukes and heads can be seen being thrown out of the water.

Below, the group appears to keep a particular structure. The cow and calf are the centre of activity with the cow as the nuclear animal and the principal escort alongside. The rest of the group take up stations around the pair and continually challenge the position of the principal escort by attempting to place themselves between him and the cow, the 'nuclear animal'. By now there might be up to 15 animals in the group. Sometimes a secondary escort will displace the principal escort which will swim off. He may try to regain his position and is sometimes successful. However, the position of principal escort in a group is not long lasting.

According to whale researcher Hal Whitehead of the University of Cambridge, who first described the 'nuclear animal-principal escort' structure, the principal escort retains his position for only seven or eight hours. If one of the large groups has been swimming along together for a while, the activity seems to flag. Animals gradually lose interest and drop away. They take up well-spaced positions in the ocean and recommence singing. The cow-calf-and-escort group swims on. In order to confirm some of these observations. Tyack and his colleagues set up play-back experiments. One immediate problem was to get sufficient volume of sound, and the help of the US Navy in the form of an enormous loudspeaker was the answer.

In one test Tyack played back the social calls of an active group. The result was surprising and at times a little frightening. Singing

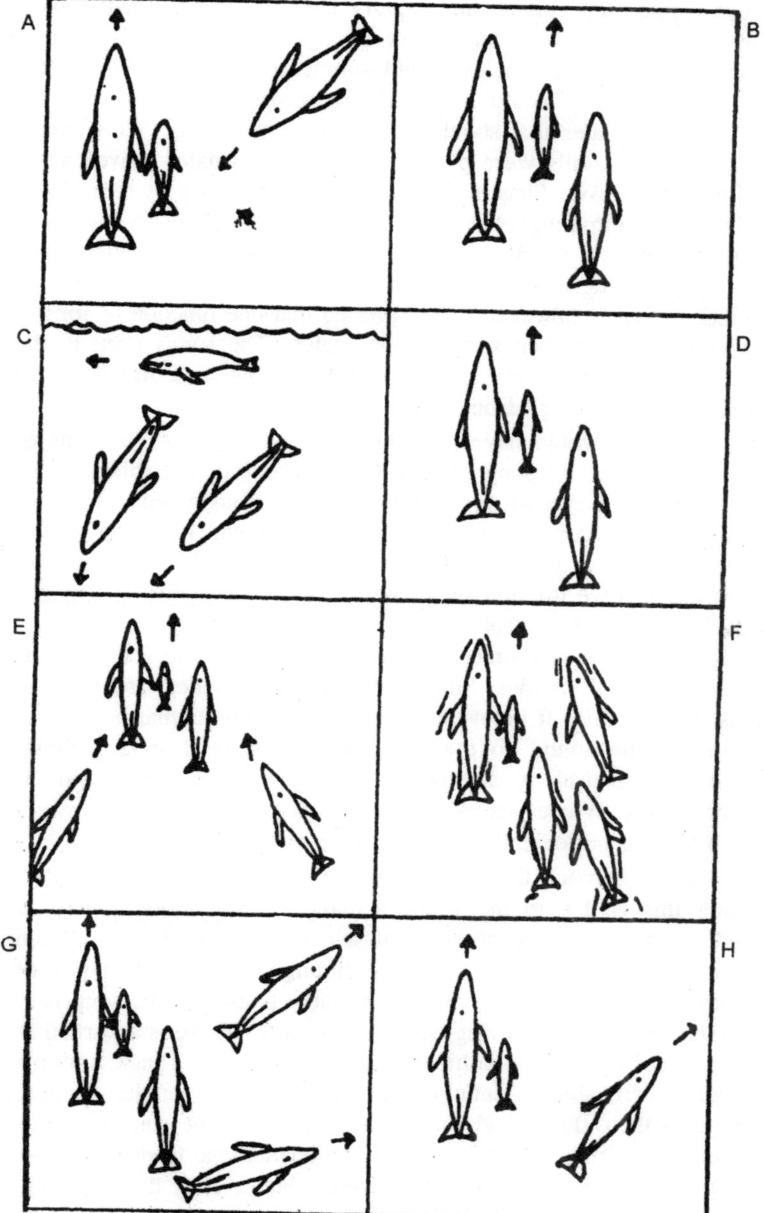

Fig. 7.2. (opposite) Humpback whale: (cow-calf-escort groups.

whales would stop singing and rapidly home-in on the loudspeaker, charging the boat at high speed. Reaching underwater speeds of up to 12 km per hour, the whales would submerge and dive five metres from the boat. They would then circle for some time as if looking for the group of whales that should be making the sounds. When songs instead of social calls were played, virtually all whales moved away from the loudspeaker. Singers react to the play back song by terminating singing and swimming elsewhere. It was difficult to follow these individuals without the song as a guide so it is unknown whether they are simply spacing themselves from the playback and then starting to sing again. Tyack, however, has suggested that one function of song is as a spacing mechanism for courting males. The songs used in the playback experiments were always recorded in the same month as the experiments were carried out.

Although it would have been interesting to see the reactions of the whales to an old song. Tyack and Payne felt that this kind of interference with a wild population might be considered a form of vandalism. Still a complete mystery is the way in which singing whales produce the sound. It is presumably a vocal sound like our own but this would involve air and bubbles are not obvious during singing. Inside the whale's head there is a great deal of complicated plumbing. It could be that air is shunted back and forth across the vocal chords in these tubes and cavities. Singers perform their underwater arias at depths of 80 to 100 ft below the surface. A deep submarine canyon will reflect the sound so that it reverberates as if in a gigantic underwater cathedral. A shallow sea bottom produces almost studio-like recordings. Whales stay singing below for up to half an hour at a time, returning periodically to the surface to breathe.

As they approach the surface the sound gets quite a bit fainter. Whether this is due to the whale actually singing more quietly or caused by some acoustic property of sea water is not known. When breaking the surface they do not stop singing or break the rhythm of the song, but during four or five identifiable pauses in the song they will take breaths before diving below once more. There is a particular song theme in which breathing tends to occur so Tyack and his colleagues were able to anticipate a singer's arrival at the surface. As they tip their flukes and sound, the volume of the song becomes much louder, literally vibrating through the boat. In the feeding grounds in the polar region the whales are relatively quite. They do, however, occasionally indulge in social conversations using sounds similar to those emitted by the large rowdy groups.

During a preliminary study in the Glacier Bay area of Alaska, Tyack observed that the sounds are usually made when the whales are in groups, particularly when individuals meet and join or split up. Bill Dolphin, at Boston University, has made a more extensive study but is not sure whether the sounds are greetings or keep away calls. Whales frequently feed in organised groups, sometimes in line or chevron formations, so competition for resources is unlikely to be the reason for any spacing calls. Social calls could be used to keep a feeding group together but there are no observations of this as yet. Humpbacks have one usual form of feeding in which sounds are often heard, called bubble-net feeding. Starting low in the water, the whale will lay a circular column or net of bubbles in the water. As the circle is closed the whale will come up right in the middle of the bubble tube engulf the food in the centre. Tests on the density of food in the bubble net have shown it to be up to 50 times higher inside the circle than in the surrounding water. Pleats in the lower jaw swell out with each mouthful. The whale's jaw is then raised and the pleats flatten forcing the water out between the baleen plates and leaving a concentrate of krill and small fish which can be swallowed. According to Charles and Virginia Jurasz from Glacier Bay, Alaska, the diameter of the bubble is selected and nets with a variety of mesh sizes can be made.

Southern Right Whales

While Peter Tyack and his colleagues studied humpbacks off Hawaii, another group of Rockefeller University whale-watchers, working with Roger Payne, directed their attentions to another great gathering of whales, this time southern right whales *Eubalaena australis*. Reaching lengths of 50-60 ft these enormous blue-blacked whales are characterized by the 'bonnet', a collection of white, crusty patches on the top of the head and on the snout which often provide anchorage for whale lice, barnacles and sea anemones. These callosities, as they are known, probably serve similar functions to human eye-brows and other facial hair, in the case of the whale deflecting water from entering the blow-hole when at the surface. Those on the snout seem to function as 'antlers' during aggressive bouts. The centre of activity was the Peninsula Valdes in Patagonia— an enormous cape enclosing two shallow, almost land-locked and very desirable bays; desirable for whales that is, for Peninsula Valdes is renowned for its high winds and rough seas. The windier it is better southern right whales like it.

As the wind increases and white-horses appear on the sea's surface, the whales burst into life and literally play in the storm. An individual

Fig. 7.3. Physeter (sperm whale).

will leap from the water, crashing back in an explosion of spray. Sometimes a huge tail is seen sticking vertically out of the water. Observations have revealed that southern right whales actually sail in the wind. High winds are associated with much activity including lob-tailing (slapping the water with the tail fluke), flipper-slapping and breaching. During a storm, underwater noise increases, particularly as a result of waves crashing on the shore, thus making the lower frequencies with which southern right whales communicate. By slapping the surface, individuals can keep in touch. Investigating sound communication and its importance in the social behaviour of southern right whales are Rockefeller University's Christopher and Jane Clark.

Originally a biochemical engineer, Chris Clark got involved by accident—he let Roger Payne borrow his pick-up truck, and after a few visits to the Payne household became hooked on whales. Roger Payne and his family had already spent five seasons recording southern right whales at Peninsula Valdes, and their enthusiasm was clearly infectious. Chris Clark's expertise in electrical engineering was to help solve the problem of following whales under water. Using an underwater hydrophone array linked to a portable mini-computer, Clark and his colleagues were able to observe the whale's movements and interactions from the cliff-top while at the same time plotting the positions of the whales making sounds. The most prolific sound to be heard in the Bay is a simple low frequency, tonal call which is thought to be used for contact over long distances. As individuals approach one another they exchange an increasingly rapid series of calls, and having met stop calling altogether. These calls can be heard at anytime of the day during the five months of the year in which the whales congregate in the bays.

Contact calls have a frequency between 100 and 200 Hz, which happens to coincide with the quietest range of frequencies in the usually noisy bay. Southern right whales in the bay appear to be using a low-noise sound window through which to communicate and so make contact with others over great distances with the minimum of effort. Whales calling in the bay, a shallow cathedral-like dish 25 kilometers long by 15 kilometers wide, can be heard plainly from one side to the other.

In deeper water the sounds should travel even farther. Often, several whales will come together to form a tight active group. Five 40-ton males and three females, for example might be swimming in an area the size of a gymnasium. In these active groups the sounds tend to become more excited, rising in pitch, with mixtures of high melodic notes and low, pulsed growls—the higher the pitch, the higher the excitement.

It is difficult underwater to work out which whale is making which noise, but it is already clear that when a large number of males get together aggressive growling sounds dominate proceedings. It has been suggested that they are competing for the female, prior to mating. If two males are left with a female, bouts of growling are heard. Eventually one male departs and the sounds change to more high-pitched melodic notes. Chris Clark describes the vocabulary as a continuum, rather than a set of discrete sounds except, that is, for the stereotyped contact call. Curiously southern right whales don't have an alarm calls, although frequently harassed by killer whales. Indeed, the calls seem not to contain any complicated messages. Right whales are grazers, and therefore do not require a sophisticated communication system to coordinate activities such as hunting. They are promiscuous rather than monogamous, so few social sounds are required. Their only need for calls is to bring the 'herd' together. Once he had identified the southern right whale's repertoire and suggested possible functions for the sounds.

Chris Clark used play back experiments to determine whether the interpretations put on the calls were indeed correct. His first task was to see if whales responded to contact calls. A variety of sounds was collected for the experiment. In addition to the right whale contact calls themselves, Clark tried background noise, pure tones at frequencies contained in whale calls (i.e. 200 Hz), humpback whale songs, and even Handel's Water Music! An underwater loudspeaker was placed on the bottom of the bay, not far away from the hydrophone array, in front of the observation hut. To make identification easier, the observers waited until there were only two or three whales in the area. A whale would be spotted by the pattern of callosities on its head and its calls recorded for a quarter-of-an-hour before the playback experiment. Playbacks were started when the whale had passed the loudspeaker and hydrophone array, and was heading out of the sea. The observers noted any sounds the whale made in response to the playback, and its direction and speed. The whales responded only the playbacks of southern right whale calls. They would stop, turn around

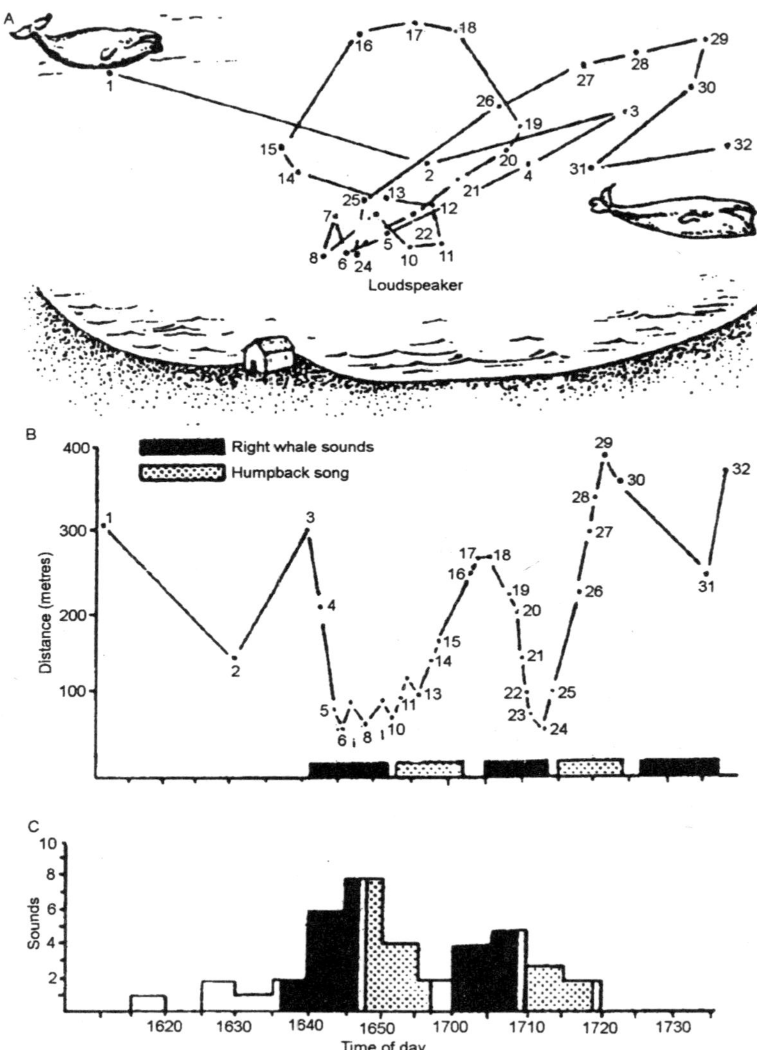

Fig. 7.4. Response of Southern right whale to the playback of right and humpback whale sounds: A–The path of whale. B–The distance of the whale from the loudspeaker. C–The number of sounds made by the whale per 5-minute interval.

dramatically in a flurry of foam, and swim at full speed back to the loudspeaker, at the same time producing more sounds. When other playbacks were used, whales would ignore them and continue to swim away.

On one occasion, at the end of the season, when few whales were in the bay, a lone right whale was exposed to playback of a group of whales. It responded by returning to the speaker. No more sounds were played and the animal began to swim off. After a little while in turned back and swam to the area of the speaker. It continued to do this until dark when Clark left it alone in the bay silently searching for its fellows. Having shown that southern right whales can differentiate their own calls from other similar sounds, the Clarks hope to continue playback experiments in order to determine the biological functions of the sounds in a whale's acoustic repertoire.

Scattered Herds

For many years a mysterious low frequency sound has been heard in the oceans of the world. The sound is pulsed in 'blips' with a frequency close to 20 Hz. With a bandwidth of only 3 Hz, the '20 Hz signal', as it is known, is almost pure tone. It is also of high intensity, so loud that scientists at first believed it could only come from a non-biological source such as a surf on a distant beach. The sounds, though, increase in late afternoon, reaching a peak around midnight, and waves on beaches don't usually have such a pronounced diurnal rhythm. Listening to underwater sounds off Bermuda, B, Patterson and G.R. Hamilton noted the spacing of the '20 Hz'. Trains of pulses were heard for about 15 minutes, separated by two-and-a-half minutes of silence, a pattern reminiscent of the breathing cycle of a large slow swimming whale. Using an array of hydrophones, William Schevill and his colleagues from Woods Hole Oceanographic Institution, Massachusetts, were able to home in on the sources of the '20 Hz signal' and each time it turned out to be coming from a fin whale *Balaenoptera physalus*, the second largest animal ever known to have lived. But why should fin whales want to make these very loud low frequency sounds? Most whales are thought to be social animals.

The smaller toothed whales, like dolphins and killer whales, group together in schools, pods or herds, sometimes in large numbers. The baleen whales, on the other hand, are only found travelling alone or in small bands of no more than 20 animals. Could it be, though, that the great whales are indeed in herds, but in herds many hundreds of miles across? Roger Payne and Douglas Webb at Woods Hole believe there could be something in this hypothesis. They believe that fin whales might be in contact with each other using these low frequency sounds. They do not suggest that meaningful and complex messages are winging their way across the ocean, but simply that acoustic signalling might

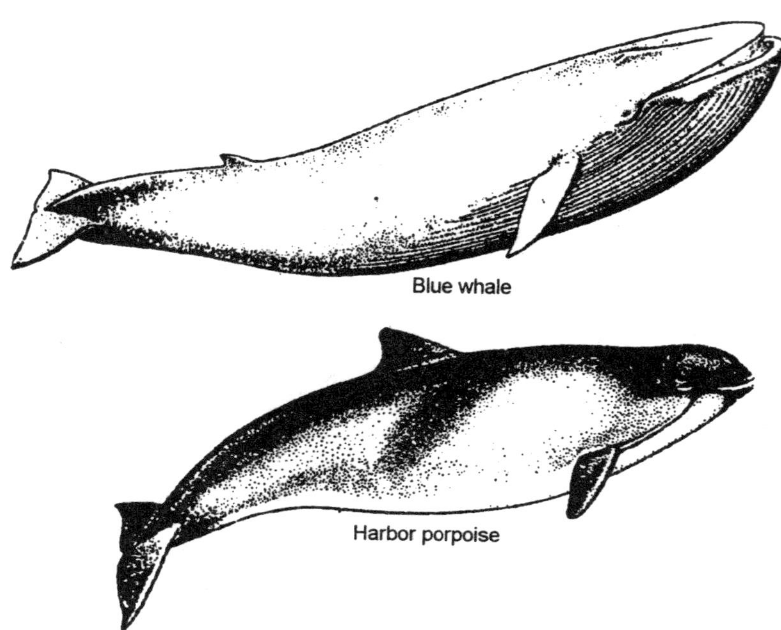

Fig. 7.5. Blue whale and Harbor porpoise.

be used by whales to locate one another, to aid future rendezvous, and to keep them together in their widely dispersed groups. A lone whale, therefore, just might have company, albeit a hundred miles away over the horizon. Two fin whales, several kilometers apart, emitting individually pulsed 20 Hz signals were recorded by Patterson and Hamilton off Bermuda.

The researchers were able to follow the direction the whales were heading with the aid of a multiple hydrophone array. The first whale called for about three hours while going south. The second whale, about five kilometers to the east, then began to call and the first whale changed direction towards it. Was the first whale being guided to the second by sound? Follow-up work in this area is prohibitively expensive so as yet there is no answer, but the simple and repeated patterns of pulsed sounds are ideal for long-range communication.

Fin whales, unlike humpbacks and grays, do not appear to breed always in the same areas, so there must be a mechanism to bring individuals together. The 20 Hz signal seems to fit the bill. It is lower in frequency than the noise generated during turbulent storms; loses little energy when bounced off the sea bottom; and is apparently the best frequency for propagating through a surface-frozen polar sea. But the ocean is a noisy place. The whales would have to make themselves heard over considerable background chatter, particularly in today's ocean where man's supertankers and other technological advances pollute the ocean not only with oil but also with noise. How do they do it? There is one published case of a small underwater explosion from just four pounds of dynamite, detonated off Australia and detected at Bermuda at about 12,000 miles away. That is unusual, but calculations made by Payne and Webb indicate that fin whales could communicate over considerable distances by making use of deep water sound channels. These acoustic channels are the result of physical characteristics of the ocean. Differences in water densities, salinity, temperature and ocean currents, produce a channel at a particular depth which tends to trap sound. It works as a kind of underwater voice-tubed by concentrating sound energy in a narrow beam rather than diffusing it over a large area. Without the help of a deep ocean sound channel, individuals could speak with one another over a distance of about 50 miles. Making use of a cylindrical propagation in a deep water sound channel, the distances could reach a maximum of about 500 miles.

In pre-steamship days, these distances would have been further increased to 140 miles and 3,500 miles respectively, and these are conservative estimates. Clearly the fin whale acoustic system must have evolved in quieter ocean and it is conceivable that man's maritime developments could have seriously impaired communication between whales and upset their lifestyle—a further set back following their wholesale slaughter at the beginning of this century. So far, researchers have not demonstrated that whales actively seek out the sound channel, but inevitably their calls will spill into the channel and the sounds propagate over long distances, audible to any whales listening at channel depths. The sensitivity of hearing in baleen whales is thought to be excellent albeit in the lower frequencies. Fin and humpback whales have many more fibres in the nerves from the ear to the brain either man or the bottlenose dolphin. They appear to lack the high-frequency or ultrasonic hearing capabilities of dolphins but they may hear very low sounds in the infrasonic frequencies.

The blue whale *Balaenoptera musculus*, for example, the largest creature ever known to lived on earth, is thought to emit very low frequency grunts, but surprisingly there has been suggestion that it is also capable of vocalising in ultrasonic frequencies. The minke whale

B. acutorostratus one of the smallest of the baleen whales, calls with a very low note. The Californian gray whale *Eschrictius robustus* makes moans, bubble sounds, knocks and 'a metallic - sounding pulsed signal'. They also produce clicks which are probably used for navigating and locating food. The sperm whale *Physeter catodon*, a large odontocete or toothed whale, produces clicks which are thought to travel long distances under the sea. William Watkins and William Schevill at Woods Hole have been analysing the sounds made by sperm whales. They have a multiple array of hydrophones and can locate accurately the positions of individuals. Groups of these large, toothed whales are heard to make clicking sounds and each whale produces its own distinctive pattern of clicks like a morse code.

Using this unique identification and location technique, Watkins and Schevill have been able to track individual sperm whales and observe their behaviour. They have found, for example, that sperm whales surfacing within ten meters or so of each other spread out like an inverted funnel when they dive again, and so are separated by much greater distances as they reach the bottom. Returning to the surface they emit more clicks and gather together once again in a close-knit group. Sperm whales are more obviously gregarious than some of the other large whales, and their click signatures would be a important in a complex social structure where individuals might want to contact each other during mating, feeding and so on. Groups of sperm whales are thought to be led by a dominant animal which would use sound to guide its family group.

Whalers were aware of messages passing between sperm whales. They were sure that distress or alarm calls were given by harpooned animals which seemed to elicit responses from others many kilometres away. Although we still know surprisingly little about the behaviour of humpback, southern right, sperm, fin, blue and minke whales, we do know that, despite international agreements, they are still threatened by the activities of our own species. By the time we understand sufficiently about them to begin the urgent work of conservation, it may be too late. The damage may be irreparable. In the future their beautiful calls might exist only on long-playing records and the research tapes of scientists.

Dolphin Language

Killer whales, pilot and sperm whales, dolphins and porpoises are all classified together as the odontocetes—the toothed whales. Characteristically they are all noisy animals, but the voice of the dolphin is perhaps the most familiar. All over the world, 'singing dolphins' have thrilled and amused audiences in dolphinaria and marine circuses. As far back as the 4th Century BC Aristotle wrote of dolphin squeaks and moans, but in AD 1983 we still don't know exactly what they mean, or indeed even how they make their sounds. It is probably the mystery and apparent charm of these animals that has most contributed to the stories about their abilities.

The almost complete lack of understanding of their behaviour, coupled with their perpetually smiling faces, makes them irresistible as objects of affection. It is perhaps because of this close affinity between man and dolphin that dolphin behaviour is often interpreted in the same way as human behaviour. According to some researchers, the mythology obscures the science. The early Doolittle obsession with wanting to talk to dolphins stems from the fact that dolphins have large brains, as large as those of humans. Surely with a brain that size, so the theory goes, a dolphin has something worth while to say? This is further supported by the revelation that the neo-cortex—the part of the brain with which we create, innovate and reason—covers 98 % of the surface of the dolphin's cortex. In man the figure appears to be 96%.

Unfortunately, anatomy also reveals that the dolphin's neo-cortex is much thinner than in humans, so this line of reasoning is rather inconclusive. It was thought at one time that, contained within the dolphin's wide range of vocalisations, were complicated messages that could rival the complexity of human speech. Dolphins were thought to have mystical, paranormal powers and were attributed with super intelligence. They were thought to be among the cleverest animals on this planet. Doyen of the 'smart dolphin movement' is John Villy, a doctor who has studied such topics as neurophysiology and hallucinogenic drugs. His object has been to discover a means of communication between dolphin and man in order to understand dolphin language, culture, philosophy, and even their system of ethics. In an early experiment, Lily had dolphins mimic English. They could follow a count up to ten, and almost say some simple English words.

The recordings are amusing, but mynah birds and budgerigars can do better. Lilly's early mimicry experiments, however, did turn up some interesting information. Dolphins, it seems, showed a remarkable ability for rapid mimicry on some occasions an animal would begin mimicking before the signal it was copying was finished. The time domain, therefore, was quite different from that in budgies. In addition, in another experiment a researcher would read out a series of syllables, consisting of vowels and consonant sounds, and the dolphin would mimic the same number of sounds with the same durations with 70% accuracy.

Dolphins obviously appreciate quantitative aspects of human speech much better than do mynahs and parrots. English, though, was clearly not the language with which to investigate the linguistic abilities of these remarkable creatures. Today, teams from California, Florida, Hawaii, and The Netherlands are carrying out numerous costly experiments using signs, symbols and computer-assisted sound equipment, all witness to man's need to communicate with another animal. There is, of course, a serious side of this kind of research. By 'conversing' with the study animal, information can be sought about its cognitive characteristics and an understanding gained, perhaps, about its intellectual abilities and limitations. Lilly himself is involved with the JANUS (Joint Analog Numeric Understanding System) project in California.

A sophisticated computer-assisted sound system receives and transmits bleeps and whistles to and fro between captive dolphins and their human researchers. Dolphin sounds are matched to computer generated sounds, and to visual letters and other symbols which the dolphins can see on an underwater screen at side of their tank. The dolphins react to the symbols on the screen, cause the sounds or symbols to change or simply match the sounds with dolphin sounds. The aim of the experiment is to create a new form of 'language' that is mutually accessible to both man and dolphin. Tests have started with 48 or so sound and symbol combinations, known as morphemes. Each sound symbol combination is associated with an object, a place or an action. Lilly is using the computer interface to link the dolphin's high frequency (1,400-4,000 Hz) vocalizations with the human's relatively low frequency (200-2,200 Hz) speech and to change the time domain so that the dolphin's utterances can be directly translated into the relatively slow delivery of human speech. Lilly estimates it will take about five years to work out a human-dolphin dictionary which could be used to communicate across the species boundary. Flipper Sea School in Florida., once host to the famous television dolphins of the same name, has adopted a similar approach to training in order to explore a dolphin-human communication system. A whistle language, similar to that used to train sheep dogs, is being used to instruct the dolphins.

In addition, one dolphin is responding to human words as well. At the Dolphinarium, Hardervijk, The Netherlands, Wilhelm dodoc Van

Heel has been working with a Killer Whale called Gurden. Frequency modulated tones associated with objects in the tank were—played to the killer whale. When the signal was given Gurden was expected to touch the relevant object. This she did. If the wrong object was presented with a particular sound signal the whale became very upset. Eventually Gurden was able to imitate the object-linked signals and touch the correct objects. Van Heel them introduced sound signals, to represent action words, verbs, and the whale went on to recover objects on sound cues.

The remarkable thing about this particular whale, though, according to the television programme *Talking Whales*, was that Gurden began to talk back. During a training session she was heard to give the sound signal to fetch the object in the tank, a copy of the signal the researchers had given her. This was the first time the conversation had been two-way. In Hawaii, Dr. Louis Herman, director of the Marine Mammal Research Laboratory of the University of Hawaii, together with Dr. Douglas Richards and Dr. James Wolz, has been working on another major project with two female bottlenose dolphins *Tursiops truncatus*. The dolphins have learned to respond to visual and verbal signals in much the same way as a sheep dog; but there the similarity ends, for the dolphins talk back. The back ground to Louis Herman's current research interests includes many years of observation on the dolphin's sensory systems, particularly its capacity to see and to hear, and studies of its learning and memory abilities.

The dolphin, for example, has a good auditory memory, being able to remember long lists of sounds presented to it. Herman therefore wanted to know if language ('the most complex method known of information transfer from one creature to another') could be added to the growing list of dolphins' abilities. Instead of looking at the way a language might be produced by a dolphin, the research team concentrated on how an animal might understand language. To do this, Herman copied the technique used by second-language teachers who instruct pupils by getting them to carry out tasks. The level of understanding can be measured by how well the task is carried out. This involved entirely a new method of training. At a dolphinarium a normal way to get an animal to learn a new tricks to entice it gradually with food. To teach a dolphin to go through a hoop, for example, the animal is presented with a hoop, rewarded initially for poking its head through, and then rewarded progressively as more and more of its body passes through until it swims through completely.

A gesture or sound signal would then be provided to initiate the entire behaviour of swimming through the hoop. Herman's technique differs in that the dolphin is first taught the word for 'hoop' and then the word for 'through'. The words can be combined into 'through hoop' and without special training the dolphin will know what to do. If a new word such as 'gate' is introduced and linked to 'through' the dolphin knows to go through the gate as it did the hoop. An object word and an action word are taught to the dolphin and combined with other words in new contexts; this gives the researchers flexibility in the training system. Unlike sign-language experiments with primates, where chimps and gorillas are invited to converse with the trainer, thereby producing 'words' which might be open to misinterpretation.

Herman is simply examining the cognitive characteristics of dolphins with what he feels are more objective interpretations of the results. Herman is simply asking questions. In what do dolphins specialise? What are their limitations? What are they good at? How do their success and failures compare with those of other animal? What role does intellect play in a dolphin's world? In short, Herman's main object appears not to be language but to create a tool for discovering a whole range of behavioural abilities in dolphins.

Dolphin Sounds

What of dolphins themselves; what are they saying to each other? Probably the most famous experiment in dolphins communication research was carried out in 1965 by Jarvis Bastian of the University of California at Davis. He placed a pair of bottlenose dolphins in adjacent tanks so that they were isolated visually, yet could still hear one another. The female was taught to push paddles in order to receive a reward. The male was offered another set of identical paddles but received no training. He learned, however, to push the right paddle.

The only way the male could have obtained the information would have been from the female next door. Throughout the tests both animals were heard to emit many whistles, squeaks and clicks. Although the information must have come from the female, unfortunately the results were not conclusive proof that deliberate communication had been taken place. The male may have picked up sounds inadvertently made by the female and quite independently trained himself to use these to his own advantage a remarkable feat in itself. There have been many stories of dolphins in the wild using sophisticated communication system. It is reported that dolphin schools will avoid boats out to capture them. It has been recorded even that dolphins will steer clear of a particular type of boat simply because the shape is similar to those that are out to them harm. It was thought that somehow or other dolphins were able to tell each other 'to avoid that boat over there; one of the school was killed by a similar-shaped boat last week'.

On the other hand, dolphin schools are hauled out and killed in their hundreds by tuna fisherman around the shores of Japan. No complicated communication system saves them. If messages carrying sophisticated information are passing to and fro, why don't they avoid the trap? Indeed, why do so many small whales get themselves beached to die of exposure in the sun? The paradox remains. What is clear is that dolphins have a varied vocal repertoire. There are squawks, whistles, squeaks, burps, groans, clicks, barks, rattles, chirps, and moans. The bewildering array of sounds can roughly by divided into two types—pulsed and unpulsed sounds. The pulsed sounds include clicks and burst pulses. The bursts may be included into chuckles, chirps and click-trains. Trains of click sound like rusty hinges being opened, or the whine of machinery, while some burst pulses have been variously described as 'raspberries' or 'Bronx cheers'. The more continuous sounds are the whistles and squeaks of frequency-modulated pure tones which may last for several seconds.

It was thought by the early researchers that the whistles are the main communication sounds. This may have been because the whistles are easier for humans to hear and to study, the clicks being mostly in the ultrasonic frequencies. But this was perhaps yet another piece of scientific mythology associated with dolphins and was brought into doubt when it was revealed that many odontocetes, including river dolphins, sperm whales and many other openwater dolphins, have not been heard to use whistles; they only emit pulsed sounds. Many of the pulsed-click sounds are used for echolocation and navigation; some, though, are thought to have social functions. Male bottlenose dolphins are harbour porpoises *Phocoena Phocoena* have known to be give pulsed 'yelps' during courtship.

Similarly Atlantic spotted dolphins *Stenella plagiodon* 'squeak' during training sessions in captivity. Frightened or distressed dolphins emit pulsed 'squeaks', which could be a type of alarm call. Aggressive 'buzzing' trains of clicks are heard when two males confront one another. What seem like exchanges of burst-pulses have been recorded between individuals in a school of Hawaiian spinner dolphins *Stenella longirostris*; so, too, between pilot whales *Globicephala melaena*, and between narwhals *Monodon monoceros*. All three species, are known to whistle. As yet, the social function of click-sounds has been title studied. There are, however, a few observations which allow some generalisations to be made about those animals that use whistles and clicks and those that use just clicks.

Heaviside's dolphin *Cephalorhynchus heavisidii*, the harbour porpoise, the finless porpoise *Neophocaena phocaenoides*, and the pigmy sperm whale *Kogia breviceps*, have pulsed sounds only and are all odontocetes that aggregate into relatively small groups of three to 20 individuals. Hawaiian spinner dolphins and bottlenose dolphins, on the other hand, often form huge schools of hundreds of animals that have a tendency to forage together. They are the whistlers. The very high frequency clicks travel less far in the water than in the relatively lower frequency whistles, and would be more suitable for sending messages between individuals of a small group. Whistles would carry better between dolphins in a large and spread-out group. There are suggestions that emotional meanings can be read into these sounds. Abrupt, loud sounds, for example, are given in aggressive situations. Sometimes these vocalisations are accompanied by non-vocal 'jaw-clapping' or 'tail-slapping'.

Intimate chuckling sounds are heard during bouts of caressing and touching. Dolphins touching, for example, make a sound much like the noise you get when rubbing a finger against a wet balloon. There are signature whistles. A dolphin will produce a whistle that seems to be its own whistle, which appears to be used as a long-range individual identifier-signal across the width of the school, to let every dolphin known where all the others are placed. In an experiment in 1974, in the USSR, two neighbouring bottlenose dolphins, linked by sound only, both produced whistles and rhythmically related clicks.

It was suggested that dolphins, like songbirds, have an opening identification portion of the call, the whistle, followed by a more complex message portion, the clicks. The birdsong parallel was continued further in the interpretation of the results of an experiment with tropical spotted dolphins *Stenella attenuata*. A male was captured and recorded. It produced almost continuous bouts of whistling which were probably alarm or distress calls. The calls were played back to the school from which it came and they fled, instantly. The same calls played to another school elicited curiosity and not flight. The captured dolphin's own school detected danger from the familiar, but frantic, call of the subject, whereas the 'strangers' were unable to appreciate the significance of the calls. Does each school, then, have its own vocabulary of calls, may be as a result of mimicry within the school? Dolphins are good mimics, as witnessed in early communication experiments, and related species, such as killer whales, are known to have distinct local dialects.

Killer Whales

The evidence of dialects among killer whales *Orcinus orca* was demonstrated by Dean Fisher and John Ford at the University of British Columbia in Vancouver. One of their first discoveries was that the sounds of any given killer whale pod are very stable. Unlike humpback whales, killer whales do not change their songs through time. In a variety of situations the animals make the same associated sounds over and over again. In several different pods, John Ford has been able to identify, on average, 12 distinct stereotyped or discrete calls which are exchanged between whales when spread out and foraging may be over an area of a couple of miles. It is thought that these calls are emitted in order that individuals in the pod can keep in touch with each other, although out of direct visual contact. It is now known yet whether each of the 12 sounds has a different meaning because, below the sea, it is difficult to link sounds with any particular pattern of behaviour. The Vancouver research team feel that the most likely information being communicated includes their position, individual identity, emotional or active state, and pod identity through the dialect. It looks unlikely that a more highly structured and sophisticated message, like human speech, is being exchanged. One call is given, the other in the group respond, and then they all switch to another type of call. Dialects have arisen as a result of the isolation of one pod from another. Killer whale pods are essentially extended family groups.

Once an animal is born into a pod it rarely leaves the group. In this way a pod may grow up to 50 strong, although the average family groups consists of between six and 15 individuals. If an ancestral pod grows to a certain size, it is likely (although not proved) to split into

Fig. 7.6. Phocaena (Porpoise)

two or more smaller groups, which spend progressively less time together. Initially the calls given by the groups would be similar but as time went by, probably over several decades, the dialect of each group would drift away into its own distinct sound and shape. Around Vancouver, for example, resident pods hunting in the same bays and having their own clear patrol areas tend to have very similar calls, whereas visiting migratory or transient groups can be heard to have distinct calls indicating a quite separate ancestry.

Hawaiian Spinner Dolphins

One group of dolphins whose social interactions have been studied fairly extensively are the Hawaiian spinner dolphins of the Pacific. Much of the early work was with dolphins in captivity, but increasingly today researchers prefer to work with natural, relatively undisturbed groups of animals in the wild. Dolphin watchers, Professor Ken Norris and Sharron Brownley, from the University of California at Santa Cruz, have been observing dolphins in the wild. in particular the Hawaiian spinner dolphin—easily identified by its twisting leap. They have been looking at the structure of dolphin schools, at where individuals move, how they interact socially, and listening for the sounds they make during different behaviour periods throughout the day and night. Spinner dolphins, according to Sharron Brownley, are very convenient animals to study for they do everything together.

They sleep, play, feed and travel together, and over the course of the day act in a predictable manner. At dawn a large group might arrive from night-time feeding and split into smaller groups as it gets close to the shore. In bays around the Hawaiian islands they spend the day resting, playing and generally socialising. In the late afternoon activity increases with what is known as zig-zag swimming. Individuals swim back and forth until the entire group move once more out to sea in order to feed during the night. Throughout these activity periods the spinner dolphins are vocal, with different sounds and sound patterns at different times of the day. When they are resting they are quiet. They swim slowly, close together, and make few sounds, usually a few clicks. Gradually during the afternoon excitement rises and they begin to increase the number of whistles and burst pulses. The whistles are thought to represent individual identities, while the burst pulse signals indicate to each other their emotional states, whether angry or playful. Each activity period blends with the next. The sounds then change from predominantly burst-pulses and few whistles to more whistles and fewer burst-pulses.

On the surface, activity hots up considerably as dolphins are seen leaping and twisting out of the water. Underwater they roll over each other and play, all the while emitting sounds. The more physical activity, the more noise. Sharron Brownley feels that when the animals go into zig-zag swimming it is almost as if they are trying to decide whether they are ready to go to sea to hunt. They swim slowly to the entrance of the bay and then rapidly swim back. Cape hunting dogs in Africa go through a similar ritual before hunting. Here cooperation is the key to successful hunting. Before leaving for the plains they go around twittering to each other, getting progressively more excite about the hunt and therefore more ready to hunt as a unit. The pitch of excitement rises and the level of noised rises until of a sudden they all take off and track down their evening meal. Similar kinds of prehunting calls are heard from other animals that hunt in groups, such as wolves and hyaenas. Here a dominant animal assembles the group.

In wild Hawaiian spinner dolphins schools it is difficult to identify whether an animal is male or female, let alone dominant or submissive, although researchers believe there may be group elders which shape the direction and guide the school. In other dolphin species, such as bottlenose dolphins, dominance hierarchies have been observed where a large female is dominant over the other females and sub-adults and a large male is dominant over the entire school. In Hawaiian spinner dolphins, however, when one dolphin starts calling, the rest tend to chip in too so there are periods of hectic vocalisations followed by periods of silence.

Sharron Brownley suggests that the bursts of noise may indicate how unified the group feels. When all the dolphins chime in at the proper time they know they are all ready to hunt. If some do not join the chorus then it might indicate they are not ready to go. The chorus then starts again until everybody is alert and ready. The peak period of vocal activity, then, is late afternoon. The noise level rises noticeably. The wild cacophony of sound is so incomprehensible and so full of all kinds of sounds that Ken Norris has nicknamed it the 'Yugoslavian-news-report'. Non-vocal sounds also accompany the activity. Animals leap from the water, slapping their heads and tails, and generating loud percussive noises which may help whip up excitement in the school. The leaping and spinning seems to be infectious, spreading rapidly through the group. Both visually and acoustically, leaping, slapping and whistling serve to tell where each is located and whether ready to go. Once at sea they use click for navigation and hunting.

Dolphin Whistles

Whistling associated with feeding has been observed with many different odontocete schools, both wild and captive. In dolphinaria bouts of whistling coincide with the main feeding periods. Douglas Richards at the University of Hawaii made recording of a pair of newly arrived bottlenose dolphins and found that they whistled least at night (the period of low activity for this species) and most early in the day. A year later whistling became synchronised with routine feeding and training sessions. Dolphins riding the bow-wave of large boats or humpback whales are heard to whistle, as are individuals in unfamiliar situations, such as those stranded, captive, or otherwise isolated from the school. The level of excitement is accompanied by a comparable level of whistling.

Encounters with unusual objects or potential predators, on the other hand, will elicit silence or reduced vocal activity. When two school meet, whistling activity may increase or decrease dramatically. It is not clear why. Animals harpooned whistle continuously, as will mothers separated from babies. Captive dolphins whistle when introduced to their new tank, although they quickly settle down. Many attempts have been made to catalogue dolphin whistles. Often, several identifiable whistles are recorded together with a multitude of minor variations. It has been suggested that these animals have a graded system of vocalisations, where each sound type overlaps with another to form a series of sounds rather than discrete signals. The same kind of grade series has been described for certain monkey's and apes.

Dolphin Clicks and Echolocation

The other main category of dolphin vocalisation, pulsed sound, if often, and in most cases mainly, used for echolocation and navigation. Echolocation is sometimes considered as autocommunication or communication with self, and broadly meets the definition, used earlier, that an animal has communicated when it has transmitted information that influences a listener's behaviour. Most of the toothed whales are thought to be able to interrogate their environment with sound. Using echolocation clicks a dolphin can see with sound. By bouncing sounds off an underwater target and analysing the signal it gets back, a dolphin is able accurately to locate the object, determine whether a target is dead or alive, and if alive and potential food may be able to stun it, and sometimes kill it, with a high intensity beam of sound. That dolphins are able to echolocate was revealed in 1942, and although since then experiments have mushroomed, we still know very little of the nature and function of a dolphin's echolocation system. One of the workers who has been at the centre of dolphin research, right from the early days, Ken Norris of the University of California at Santa Cruz. His pioneering work was with captive animals in dolphinaria or marine circuses; indeed, the only way to fund serious research was to teach dolphins new tricks. One of his first tests was for a TV show.

Ken Norris was allowed the use of a dolphin on the understanding that it could be televised, and therefore also it had to be entertaining. The researchers noticed that when a dolphin swam between a pair of hydrophones placed in the tank the sounds would increase in volume if it swam directly towards one hydrophone. Then as it turned and swam towards the other, the sounds would appear to come up in the second hydrophone. Clearly sound was being emitted in front of the animal, but proof was needed that it was being used for echolocation. It was necessary, therefore, to blindfold a dolphin in order to determine whether it could use sound alone to get about in it tank. The researchers had many failures for its is very difficult to tie anything on to make anything sticks to a dolphin. The solution to the problem was a pair of rubber suction cups, one placed over each of the dolphin's eyes. The test animal, a bottle nose dolphin named Kathy, swam of across the tank blindfolded as if nothing was strong. With ease she picked her away through a maze of poles without every touching one, and was

Fig. 7.7. Bottle-nose dolphin.

able to locate small objects on the far side of her ten meter tank. Small pieces of fish dropped right next to a barrier in the maze would be scooped up without touching the obstacle.

Another test with Kathy was a discrimination test. She was invited to distinguish between a horse capsule filled with water and a piece off fish about the same size. Kathy took the fish every time. Dolphins clearly can use a sense other than sight to navigate and locate food. As dolphins have all but lost their sense of smell, and extra-sensory perception has zero scientific credibility, sound was considered to be the likely candidate. One of the problems, though, is understanding how dolphins make these sounds. They don't have vocal chords and are rarely seen to blow bubbles when vocalising. This is a controversial area of research. One school of thought has held that the sounds are produced in the larynx just as in the other mammals, but echolocation clicks seem to emanate from the forehead rather than the throat. Indeed, if microphones are placed around the head of a 'clicking' dolphin the sounds can be triangulated to deep in the forehead, at the back of the nostrils. If probes are placed in the muscles of the larynx and nostrils has been found that during sound production the larynx is quiescent while valves in the nostrils show muscle activity. Ultrasound scans have given similar results.

At Boston University, R. Stuart Mackay and H.M. Liaw projected narrow beams of low-intensity ultrasound at a frequency too high for the dolphins to hear and were able to identify the structures that moved during sound production. The apparatus was a modified foetal heart monitor of a type found in most maternity hospitals. The observers were able to see the nasal plug and the vestibular, nasofrontal and premaxillary air sacs vibrate with clicking or buzz sounds. Nasal diverticula on the right side vibrated all the time when clicks are produced while the left nasal diverticula vibrated only some of the time. The vestibular sac inflated as the clicking sound was made, probably as a resonator. The nasal plugs were thought to be the site of the original sound production as air moved upward, presumably from the lungs. When the blowhole is closed, air recycles to the vestibular sac.

Clicks seem to originate in the right diverticula. Whistles are thought to be generated in much the same way as human whistles except that they are 'blown' internally. Dolphins have a pair of nostrils inside the head which come together under a single blow-hole. The and ancestors of these animals probably possessed paired external nostrils, just as most mammals do today, but as dolphins evolved into divers they needed a way of storing air if they were to make and use sound underwater. They developed a covering over the nostrils, blow hole, with a complicated series of nasal sacs and valves below. Air is breathed in at the surface, and on diving, the blowhole is closed.

The air trapped in the dolphin's respiratory system can then be passed across the valves in the nasal sacs to produce sounds and then recycled through a complicated series piping to be used over and over again as the dolphin echolocates underwater. The click sounds are thought to leave the dolphin's body through the forehead. At the front of the forehead is a large fatty body known as the melon which focuses the sound, much as an optical lens focuses light, about a metre in front of the animals's head. Donald Malins and Usha Varanasi, of Seattle University have found a concentration of unusual lipids, made of isovaleric acid, at the centre of the dolphins forehead. The three dimensional arrangement of these small molecules, rarely found in the lipids of other animals, suggest to the researchers that the area is, indeed, a 'sound lens'. When the sound returns having bounced off a target, it does not enter an ear canal but is picked up by the thin bone of the lower jaw as a vibration, travels along fatty tissue in the hollow lower jaw, and thereby is transferred directly to the middle ear. Sound production, transmission, and reception is optimised for life underwater. The rapidity with which the clicks are emitted means that humans cannot hear individual packets of sound, rather we hear trains of pulses much like a creaking old door hinge closing.

One of the remarkable things about dolphin echolocation is the rapidity with which the ear and brain process these clicks. A click-train may be made from up to 700 units of sound per second. A dolphin is capable of mentally separating these units, listening to the individual echoes, and decoding information while interrogating a target. In the human ear the sounds would fuse together in our minds at 20-30 clicks per second. The sound is also known to enter solid structures. In experiments, dolphins have been able to tell the difference between a copper and aluminium plate painted the same colour. They also have the ability to distinguish a hollow aluminium tube from a solid one, although both tubes looked identical from the outside.

Perhaps one of the most interesting aspects of dolphin research centres on the sound beam itself. Modern dolphins have narrow beams of sound; the bottlenose dolphin, for example, has a beam width of about 9°, a thin pencil-beam of sound. The Indus river dolphin *Platanista indi*, a more primitive species, emits a 65° beam. Part of a evolution of these creatures might have been a narrowing of the beam width, giving a greater distance penetration for the same amount of energy. The concentration of energy in narrow-beam dolphins has become so intense that the prey being detected has become affected by it, leading to an entirely novel way of catching lunch. Together with Bertel Mohl, of Aarhus University, Denmark, Ken Norris has been pursuing the idea that some of the toothed whales may have the capacity to debilitate their prey with sound. The sound put out may be so intense that the prey may be killed, or at least immobilised to prevent escape. The idea is not new. Drs. V.M. Bel'Kovitch and Yablokov, in the USSR, calculated that dolphin sounds should have enough energy to-incapacitate prey. Further support came from the work of Dr. A.A. Berzin with sperm whales. Berzin looked at a number of sperm whales caught by the whale fishing industry.

Strangely, some of them came up with what looked like congenital deformities of the lower jaw. The lower jaw was curved so the animal was unable to close its jaws together to catch prey. In the stomachs, however, were plenty of squid and the whales looked otherwise perfectly healthy. Squid swim much faster than sperm whales, anyway, so it was doubly surprising that these enormous animals, the size of an omnibus, were able to take the ton or so of squid per day needed to survive. Berzin concluded that the jaws were not essential for feeding and that sound was being used to stun the squid. In Hawaii, Whitlow W.Au and A.E. Murchison measured the intensities of sounds emitted by a bottle nose dolphin that was asked to carry out extreme-distance discrimination tests. A sphere, about the size of a tangerine, was placed about 143 metres from the dolphin. The animal was able to locate the sphere, but interestingly, when the echolocation sounds being produced were measured, they proved to be many orders of magnitude higher than any sounds recorded from a dolphin before. Indeed, they were close to the finite limit sound, that is, the limit at which any more energy put into the water would simply turn to heat. With that kind of sound level Ken Norris asked the question—will those sounds kill prey? Tests with man made sound beams showed that fish could be killed with a high intensity sound beam.

At the Marine Biological Laboratories at Plymouth, large squid could be killed quite rapidly by sound. In theory, then, dolphins could kill with sound, but the next question was whether they used this weapon. Norris considered that it would be odd for a creature to have evolved such a weapon if it didn't use it. Further tests were carried out with dolphins in capacity. Live fish were placed in a tank with three Hawaiian spinner dolphins, and sure enough after the short while the dolphins began to spray the fish with sound. The fish used were much larger than the dolphin's normal prey so it was not expected that the dolphin's would kill them. Norris and his colleagues, though, watched for any signs of debilitation.

The dolphins continued to direct echolocation skills at the first school for over an hour. A dolphin would make a run at the school attempting to put the fish right on the tip of its beak. The fish school would split, head for the tail of the dolphin and reform. Slowly, though, the fish school became depolarised. The researcher noticed after a while the fish were not all facing in the same direction. Then individuals began to wander away from the school, seeming totally disorientated. Similar observations have been made in the wild. Striped dolphins *stenella coerulealba*, for example, have been seen to circle an anchovy school, spray them with sound, and then cut through the school shovelling fish into their jaws at will. The anchovies do not attempt to escape. It is possible that the fish eventually suffer from a build-up of waste products in the blood due to their strenuous efforts to escape but an interesting observation made by a biologist on a fishing boat off Vancouver does not appear to support that. A large salmon was clearly visible in the water below the boat.

As the researcher watched, the salmon suddenly stopped swimming and directly behind a group of killer whales approached. One of the whales scooped up the salmon and swam on. The salmon made no attempt to escape. Another piece of anecdotal evidence is that many scuba drivers in dolphin circuses have felt a light touching sensation on the backs of the necks which they have attributed to the echolocation activity of their charges. If the interpretation is correct, dolphins are clearly formidable killers, but one problem individuals in a large dolphin school might have is the danger of zapping another dolphin. In a school of actively feeding dolphins it would be very easy for one animal seriously to upset another, and dolphins are known to get angry with each other. To look for signs of 'echolocation manners', Ken Norris watched and listened carefully to three individuals in a tank. In more than a hundred crossing of one dolphin with its fellows, it never once echolocated them.

As a dolphin crossed in front, the actively echolocating animal switched its gear off, turned its head away, and then switched back on again. With such a devastating weapon in the dolphin's head, it is important that 'echolocation manners' are a part of the social organisation of a dolphin school. Large schools of dolphins may consist of hundreds or even thousands of animals which hunt in a 'line-abreast' formation. In this way a broad expanse of ocean can be echolocated in the search for food. Sound is thought to be used also in locating good food-gathering areas. Dolphins, it seems, can listen to the underwater background noise associated with the increased number of organisms at submarine escarpment and sea-mounts. There is also evidence that small schools of bottlenose dolphins sometimes detect and follow the sounds being made by pilot whale schools.

The pilot whales, with their longer diving times, appear to be better at finding food and the dolphins take advantage of this superior food-finding ability. Having located a school of fish, a large school of dolphins will change its shape during the attack although they will always dive in synchrony. Sound signals must keep the school in step. Killer whales have been seen to work together in a similar way to capture prey. Observers of harpooned baleen whales off Australia have described pod members hanging on to lips, flukes, and lying on the blowhole in attempts to immobilise the whales. On other occasions killers will encircle prey animals like seals or walruses, coordinating their moves by sound. When approached by killer whales, other cetaceans often become silent, flee quickly or 'spy-hop', that is they rise out of water and search the surface.

8

GROUP MOVEMENT IN BIRDS

Any movement two areas is called migration. As a rule, it is a response of an animal population to changes in environmental conditions. Birds are more uniformly than any other group of animals. Nearly all orders of birds include species which perform migrations; in other vertebrates and in the lower group of animals the migratory habit occurs in species scattered through a smaller proportion of orders. Among birds there are two common kinds of migration, *daily* and *seasonal*. A daily migration is a movement to and from a familiar place such as a roosting area. A seasonal migration, on the other hand, involves a passage at one season from a place of hatching and a return at another season to the same general area. This section of the book is concerned entirely with seasonal migration.

For centuries the phenomenon of migration has been primarily associated with birds; indeed, judging by the references to the subject in the earliest literature, migration was first observed in birds. Since civilization developed in a temperature region of the world where migratory movements are especially pronounced, it is hardly surprising that bird migration has long received attention. Man could not help noticing the flocks of birds in the spring and fall, the seasonal disappearance of some species and reappearance of others; nor could he help being interested in what he saw and eager to investigate what he could not understand. Today there is an enormous literature on bird migration, based on extensive studies in Europe and North America. And yet the causes and processes of migration are not fully known.

The root of the problem is that bird migration whether it occurs by day or night, is "an unseen movement." One must investigate its mechanisms indirectly through laboratory research on physiological and environmental influences; the analysis of migrant birds mist-netted or otherwise captured, of birds killed during migration by television towers and other man-made hazards, and of returns from banded birds; mathematical calculations based on kinds and numbers of birds observed through a telescope as they fly across the face of the moon; correlation of meteorological data with known migratory activity; deductions derived from direct field observations, radar surveillance, and tracking by radiotelemetry; and experiments on homing and direction-finding. The following pages present briefly some of the important facts and concepts of migrations and suggest a few studies that students may undertake in conjunction with class work.

Causes of Migration

All modern birds—even those incapable of fight—are descended from volant stock. Early in their history birds had the power of fight and presumably could migrate with facility. Today many species in the temperate regions of the world are strongly migratory, exhibiting mass movements away from both poles as day-length, temperature, and food supply diminish, then reversing the movements at the season when there is a general augmentation of these environmental factors. Such migrations are more evident in the temperate region of the Northern, or Continental, Hemisphere since more species are involved. Because the north-south migrations of the Northern Hemisphere include many well-known and conspicuous birds, which seem to move in the same general way, there has been a tendency to conclude that migration, like bird flight, has developed along the same line in all species, with deviations for adaptive purposes. A study of the migration phenomenon in all groups of birds soon shown the fallacy of this reasoning one finds that:

1. Some species migrate when the temperature is mild and the food supply ample, and others when the opposite conditions are true.
2. In certain species, some populations migrate while others do not.
3. Some species migrate in directions other than north-south.
4. In some populations, some individuals migrate while others do not.
5. Some species migrate irrespective of day-length, as in tropical lands.
6. Some individuals migrate in some years but not in others.
7. Some species migrate as a result of seasonal alternation in rainfall and drought.

The only conclusion one can safely reach after considering the above peculiarities is that there is no one line along which migration in all birds developed and that there must be different causes of migration in different groups of birds. A number of authorities have sought causes of migration in historical factors. There of the several causes suggested bear mention. (1) Bird migration, at least in the

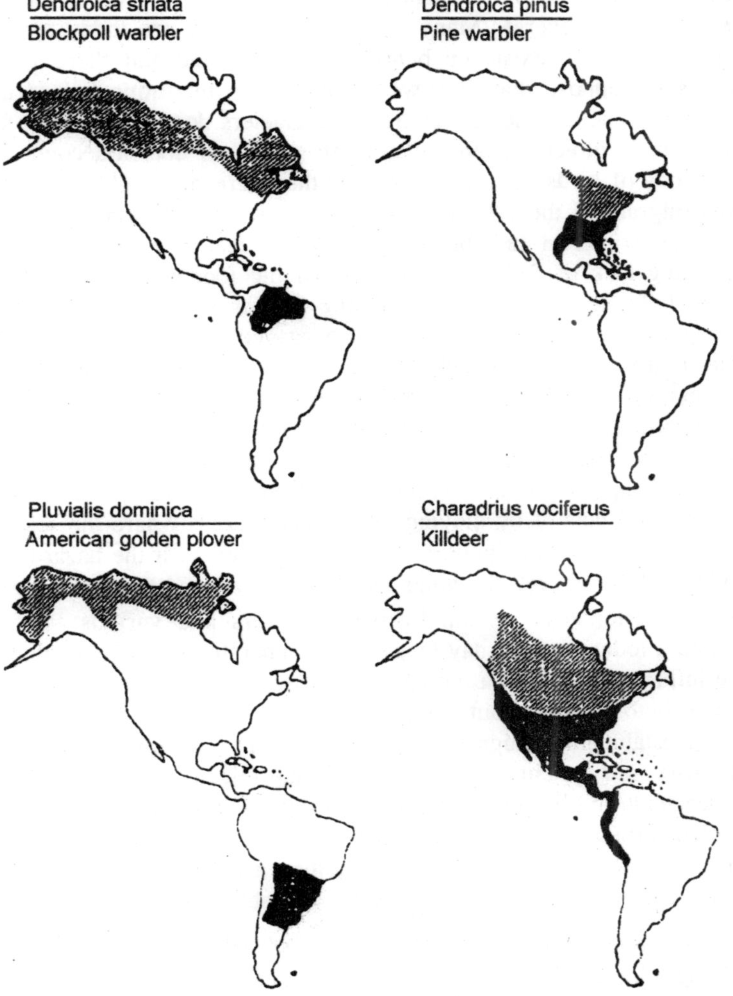

Fig. 8.1. A comparison of long-distance (left) and short-distance migrants within two groups, the warblers (Parulinae, top) and plovers (Charadriidae).

Northern Hemisphere, was initiated by the effects of the Ice Age (Pleistocene). Prior to the coming of the great glaciers to the polar regions, birds lived the year round in the Northern Hemisphere where they originated. Though forced to retreat with the advance of the glaciers, they nevertheless continued to return to nest in the summer because of an innate attachment to their homeland. Objections to this suggestion are several. Many of the birds which are today typical migrants were in existence before the Pleistocene and there is no reasons to suppose that they were not already migrating before the glaciers advanced. The retreat from the glaciers does not account for migrations in directions other than north-south, nor does it account for migrations of birds in tropical regions that were never glaciated. (2) Birds migrate in the fall because of the oncoming shortage of food during the winter in their breeding areas. Not that they "know" of the impending lack of food; it is the fact of the food shortage that has caused fall migration to evolve among species which, owing to the shortage, would fail to survive if they stayed for the winter. Birds return in the spring because the northern areas provide more favourable and less crowded conditions for nesting and rearing families. (3) Intraspecific and inter-specific competition for food, territory, nest, sites, and so on may have been important factors initiating migration. When, as suggested by *Cox*, individuals among normally sedentary species could benefit by moving into adjacent areas where the season was favourable and competition reduced, they did so if the hazards of moving did not exceed the gains in their survival and reproduction.

Unquestionably, certain historical factors and various factors prevailing today such as day-length, air temperature, and food supply have influenced migration indirectly by affecting the environment, but no one factor can account for migration or has played the principal role in establishing migration. In view of the current diversity of migratory movements, it is evident that migration has evolved independently in different bird populations experienced an unfavourable situation during a seasonal period in the area it occupied and could gain an advantage by shifting to another area for that period, it gradually developed a migratory patter, probably with a genetic basis.

Preparation and Stimulus for Migration

Migration is synchronized with the annual seasonal changes, but it will not take place until the bird is internally prepared and outwardly responsive to a stimulus. Before migrating, a bird must be ready to meet the energy requirements for prolonged flight. It accomplishes this by eating amounts of food in excess of its daily needs and thereby storing energy in the form of subcutaneous fat. At the same time, the bird must become predisposed to migrate by developing a condition commonly called *migratory restlessness*. In the spring, the physiological process leading toward this state is influenced by the pituitary gland whose activity at this time is stimulated by the total effects of day-length; in the fall, the bird reaches a similar metabolic state during a period when the pituitary is "refractory." When the bird has attained the necessary physiological and behavioural conditions, an outside stimulus is required to trigger migratory behaviour.

The stimulus is probably some meteorological factor, or combination of factors, such as a change in the temperature of the air, the direction and velocity of the wind, or the onset or passage of a cold front. Normally an adult migratory bird goes through the special metabolic cycle twice a year, once before the journey to the nesting area and once before the journey away from it. If the bird does not reach the necessary physiological and behavioural condition, it cannot migrate. Furthermore, if the external stimulus for migration is absent, the bird tend not to migrate even though physiologically and behaviourally capable. For a summary of knowledge concerning the nature and mechanisms of periodic preparation and stimulus for migration.

Diurnal and Nocturnal Migrations

Migrations proceed by day or night. Many birds—e.g., loons, geese, ducks, gulls, and shore birds–travel by day or night, apparently indifferent to daylight or darkness. But this is not the case with many other birds. Herons, hawks, eagles, falcons, crows, hummingbirds, swifts, and swallows migrate only during the day, while nearly all passerine birds (excepting crows, swallows, and a few others) migrate primarily during the night from after sunset until dawn. At least two explanations have been advanced for the development of nocturnal migration. (1) Movement by night affords birds, which normally live in thick vegetational cover and rarely take long flights away from it, the protection of darkness against their diurnal predators. (2) Movement by night affords birds the opportunity of using all the day light hours of feeding, thereby enabling them to build up sufficient energy resources for sustained long-distance flights.

Effects of Weather on Migration

Normal weather alternates between fair and inclement conditions. The movements of the vast majority of migrants show a close correlation

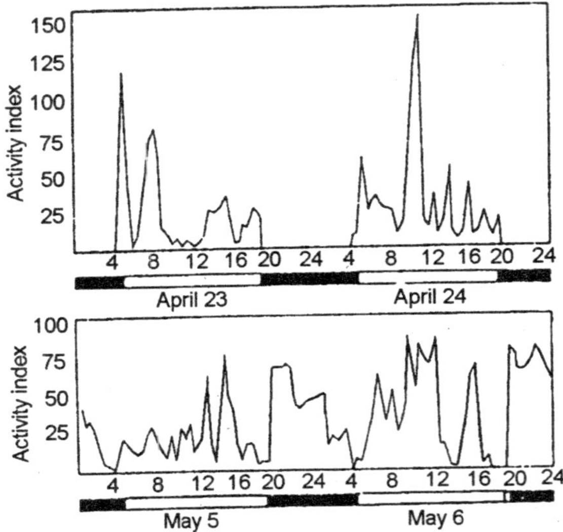

Fig. 8.2. The measurement of Zugunruhe in a White-crowned Sparrow (Zonotrichia leucophrys) before and during its normal migratory period. The black bar represents darkness and the open bar daylight.

with these conditions which are largely governed by barometric pressure patterns, temperature, and wind directions. To understand how migration takes place with relation to weather, the student must first be acquainted with a few basic facts about weather elements and their sequence over a ground area. Perpetually sweeping across the continent in an average easterly direction are air masses that vary in velocity, depending on the season and numerous other factors, from 500 to 700 miles a day.

In these masses, which may be visualized as roughly circular are centers of low barometric pressure ("Lows"), with generally warmer, more moist air, and centers of high pressure ("highs") with cooler, drier air. Within the lows the air circulates counterclockwise; within the highs, air circulates clockwise. Where the lows and highs adjoin there are boundaries called "fronts." The front of an oncoming high is the cold front; of an oncoming low, the warm front. The area of generally low pressure between the cold and warm fronts is called the warm sector. Any weather map appearing in daily newspapers will show the lines of equal barometric pressure (isobars) as roughly concentric circles around lows and highs, and cold and warm fronts as heavy lines with marks indicating the direction they face. Areas where there has been precipitation are shaded. The table, "Relation of Weather

Elements to Migration," demonstrates in a general way the sequence of spring and fall weather elements in a general way the sequence of spring and fall weather elements when a warm and then a cold front pass over a given area in conterminous United States.

The accompanying sketch, "Pattern of fronts and Wind Directions," shows how the sequence indicated in the table might appear on a wether map. The movements of the elements in both the table and the sketch is to the right (east). Thus the set of weather elements characterizing "A head of Warm Front" will be followed by the elements of "*Warm Front*" and so on. The rapidity with which one set of elements–e.g., the elements of a warm front–arrive, prevail, and finally disappear in an area is dependent on the width of the front and the speed with which the front travels.

The width and speed of the front are in turn dependent on the season of the year,. the temperature discrepancy between fronts, the

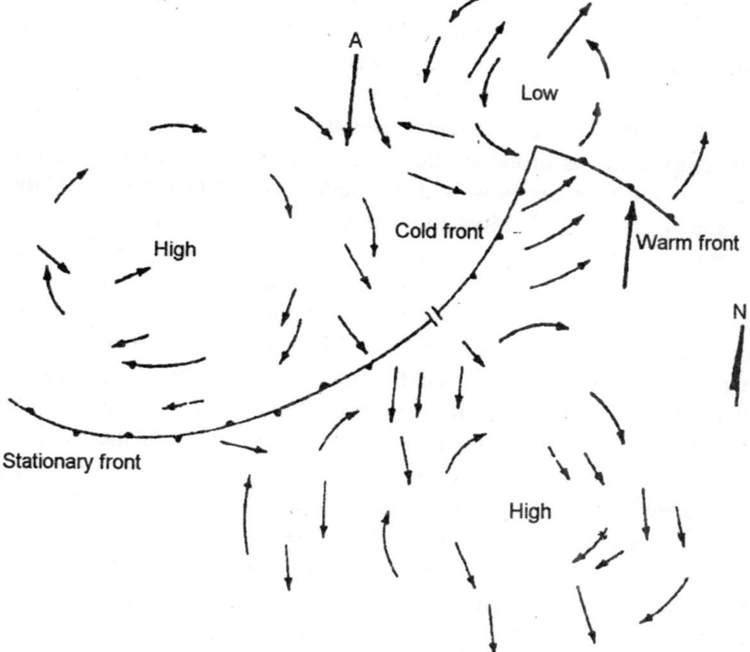

Fig. 8.3. An example of how the distribution of fronts and pressure cells can provide favourable winds for a migrating bird. Large arrow A notes a situation with favourable winds for a bird moving south, while large arrow B shows a situation favourable northward movement.

wind direction, and continent- wide weather conditions at the time. Migration in the spring usually takes place with warm weather. Studies of spring migratory movements in eastern United States and Canada show that movements begin at the onset of warm fronts, when barometric pressure is dropping and warm moist air from the Gulf of Mexico and Caribbean Sea is flowing in from a southerly direction. As each low (with mild temperature and southerly winds) passes, movements proceed in full force. On the approach of cold frints, the movements usually slow down–though they may occasionally be heavy, as when a cold front advances against a western flank of warm air. When cold fronts arrive, movements stop. Not until the highs have passed will movements begin again. Migration in the fall usually takes place with cold weather. In the Chicago area migratory movements in September and October start immediately after the passage of cold fronts when there is a flow of continental polar air from a northerly direction. Movements then proceed in full force until the first part of each low passes; thereafter the movements decrease somewhat in intensity and may even cease together as the next cold fronts approach. The table, "Relation of Weather Elements to Migration," relates spring and fall migration movements to the sequence of weather elements.

Many species are reluctant to initiate migration under an overcast. Air temperature is probably the principal factor in starting migration; wind direction or air flow is a decidedly critical factor in regulating migratory movement. Most movements take place when the wind is favourable–i.e., blowing in the direction of flight –and after migrants have been held back for long intervals by such weather conditions as fog, rain, and headwinds. In the spring the weather that ordinarily accompanies cold fronts is especially obstructive to movements, forcing migrants to take to the ground without delay. The arrival of any migration movement in an area is more often controlled by the weather at the print of departure than during its course. Thus at the height of the migration season, when the weather appears suitable in his area, the student will not always observe migration movement because inclement weather at the point of departure or some intervening point has arrested the flight. The study of weather maps in conjunction with observations on movements is very useful. By means of weather maps one can frequently anticipate a migration movement in a given area.

For instance, if during the height of spring migration the map indicates a cold front moving in from the northwest, one may expect its arrival to stop and hold numerous transients in the area. Weather maps also assist in explaining the failure of a migration movement to appear. If there is no pronounced migration movement in an area for a long period, even though the season and weather are favourable, maps may reveal that a pressure area along the migration route has become quasi-stationary, either "damming up" migrants (in case it is favourable) allowing migrants to move along from day to day without massing in conspicuous waves. Exceptional weather conditions with unusually high winds often deflect migrants from their usual routes and at the same time carry many non-migrating birds from their regular ranges. The hurricanes or heavy northeast storms that occasionally move along the eastern Atlantic seaboard, by the counterclockwise direction of their winds, force many southbound migrants and sea birds far inland.

As a result, ducks, geese, and gulls are reported in great abundance in places where they do not ordinarily occur, and such sea birds as petrels show up far inland. Two very severe northeast storms in December, 1927, and January 1966, bore spectacular numbers of Eurasian Lapwings (*Vanellus vanellus*) over the North Atlantic from their migration route in western Europe to the vicinity of Newfoundland. The student will find instructive the paper by *Bagg* showing in detail,' with a series of weather maps, how the great storms brought the Lapwings to North America.

Regularity of Migratory Travel

Despite the effects of weather on migration, migratory travel over a period of years is regular on the average. This is apparent to a student observing bird population from year to year in any given area of North America north of Mexico. If one keeps records for several years of the days when common summer-resident species first arrive in the spring, and computes average dates of arrival of each species will appear. Similarly, by keeping records of departure of transient spring species–i.e., dates when species going through the area are last seen—he will known approximately when these species depart each year. There is much less regularity in the arrival and departure of transient species in the fall, thus the securing of records is difficult. Fall migration is more prolonged. Slight variations in weather conditions have stronger effects.

Cool late-summer weather, for instance, may induce species to arrive surprisingly early and warm weather may cause them to linger very long keeping track of early and late individuals is complicated by the fact that many species are customarily silent in the fall and have

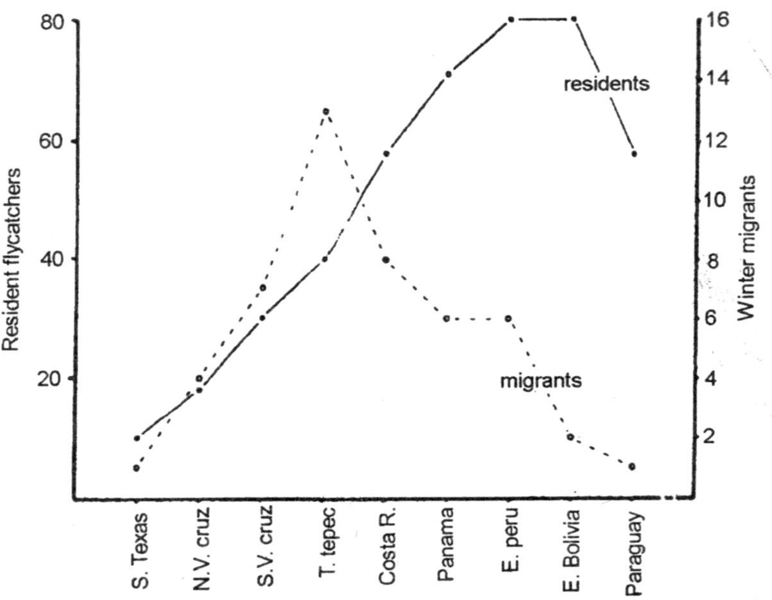

Fig. 8.4. Distribution of resident and migrant species of flycatchers (Tyrannidae) on New World wintering areas.

inconspicuous plumage. While it is possible to obtain average dates of arrival and departure, fall dates are apt to be much less useful and meaningful on account of the great discrepancy in annual dates and the problems of finding early and late individuals. In conterminous United States, north of the southern tier of states and in Canada, there are two so-called *migration waves*. The first, in the early spring, is made up of "hardy" birds—many fringillids and a species or two from various other bird groups; in the second, a month later, there is a preponderance of insect-eating birds—various species of flycatchers, vireos, warblers, and other groups.

The migratory movement has a wave effect in that the total population of birds rises and recedes. Thus soon after the first few individuals, called *stragglers*, make their appearance, the population steadily approaches maximum density, which may prevail for a day or more. Thereafter the population dwindles until only those birds which stay for the summer are left. Migration waves also occur in the fall but the migration movements are more prolonged and the crests much less apparent. In the spring, the migration population of any one species, provided it is large it is large in the area studied, will show the wave effect. Almost invariably the species makes its initial appearance with a straggler or two.

A few days to a week or more later the described for groups of species. Such a population wave, as it moves northward in the spring, may be many miles in width. The author of this book took a trip southward from Minnesota through Iowa and Missouri in late March, before the Robins (*Turdus migratorius*) had appeared commonly in Minnesota. He recorded on occasional Robin in northern Iowa; great numbers through central and southern Iowa; and a steady decrease in numbers through northern Missouri until in central Missouri there were only scattered individuals, presumably the birds that were to become summer residents there. He estimated the width of the migration wave to be roughly 225 miles. In making the first spring and fall spring and fall studies of birds, the student should pay special attention to the local movements of transient species. He should note the dates each species is seen, make counts or estimates of the number of individuals of each species observed on each date, and keep a record of weather conditions. This information will give him a proper conception of the length of time each species remains in the area, the way a species population rises and recedes giving the wave effect, and some of the relationships of weather conditions to population trends.

IRREGULAR MIGRATIONS

The many cases of movements among bird populations, which either do not conform to the usual seasonal migration pattern or are not sufficiently well understood to seem a part of the pattern, are loosely classified as *irregular migrations*. In certain permanent-resident species there may be mass movements of particular populations with some periodicity. The Blue Jay (*Cyanocitta crystals*), especially in the northern part of its range, shows some migratory movement. Each fall, small numbers usually move south past Hawk Mountain in Pennsylvania; in 1939 there was an exceptionally heavy migration (over 7,000 individuals counted), which may have been due to a shortage of beechnuts and acorns is northern forests. In northwestern Oklahoma, the Bobwhite (*Colinus virginianus*) shows a distinct seasonal population shift by moving from summer habitats in the uplands to pass the winter in bottom-lands and dunes where there is better cover. Their movements, involving distances up to 26 miles, are apparently heavier during severe winters. Populations of species which are permanent residents may on occasion show a mass movement–*invasion* or *irruption* without periodicity.

Now and then there is a winter when considerable numbers of Snowy Owls (*Nyctea scandiaca*) leave the Arctic and invade southern Canada and northern conterminous United States. The cause of this behaviour is sometimes attributed to a sharp reduction in the lemming on which the Snowy Owl preys to a large çxtent. Crossbills (*Loxia* spp.) and Evening Grosbeaks (*Hesperiphona vespertina*), normally residents of the Coniferous forest Biotic Community, occasionally appear in flocks as for south as Florida, presumably due to a failure of their food supply. Sometimes invasions involve mainly young birds. At Cedar Grove, Wisconsin, on the west shore of lake Michigan,

Fig. 8.5. Distribution and migration of Arctic terns from North America. This species is distinctive for its migratory pathway, which takes it across the Atlantic and southward as far as Antarctica.

Mueller and *Berger* reported a southern invasion of Goshawks (*Accipiter gentilis*) in the years 1961 through 1963.

Most of the birds that they trapped and examined were apparently hatched in 1961 shortly after a decrease in snowshoe hares and grouse, their principal prey. *Mueller* and *Berger* hypothesized that these young birds had come south after being displaced by adults already well established in a range that could not support large wintering populations of the species. Young of numerous species, after attaining full growth, often wander in the late summer and fall for great distances. The movement, called *Juvenile wandering*, is explosive in that the birds move in all directions from the hatching area. Among the species particularly noted for this behaviour are egrets, herons and gulls. Some young egrets and herons actually travel several hundred miles north of their place of hatching in the south.

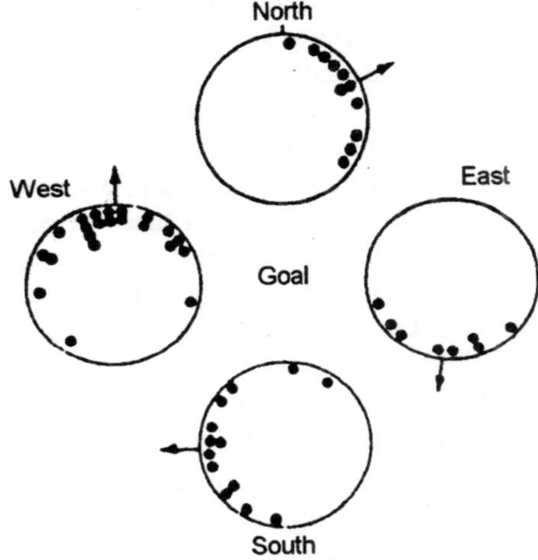

Fig. 8.6. The observed departure bearings of pegions whose clocks had been advanced by six hours and then were released 30 km to 80 km north, east, south and west of their home lofts. Solid arrows note mean directions.

At the conclusion of the breeding season in the big colony of Herring Gulls (*Larus argentatus*) at Kent island, New brunswick, an impressive number of immature birds go northward along the coast, although the majority seem to take a southerly direction. Many first-year Herring Gulls, reared in colonies on islands in Great Lakes, proceed in late December to the coast of Georgia, Florida, and the Gulf of Mexico, and quite a few reach the coast of Mexico; second year and older Herring Gulls tend to remain on the shores of the

Great Lakes within 300 miles of their colonies. Juvenile Sooty Terms (*Sterna fuscata*) from the large nesting colony on the Dry Tortugas, islands lying directly west of Key West, Florida, move across the Atlantic to the Gulf of Guninea, West Africa, and do not straggle back to the western Atlantic until they approach breeding age; the adults, after nesting, tend to confine their dispersal to the Gulf Mexico and the Caribbean. For the extensive wandering of by the young birds, the most plausible explanation is that they cannot compete sufficiently well with older birds for food and must therefore keep moving until they find an adequate supply for themselves.

Reverse Migration

Migratory movements may be reversed, proceeding in a direction opposite the one expected for the season. A good example of ***reverse migration*** occurs in the fall on Nantucket Island, Massachusetts, and Block Island, Rhode Island, where many nocturnal migrants (mostly passerines representing well over 50 species) sometimes pass rapidly through during the daytime and leave in a north or northwestward direction, "for the mainland, into the wind. *Baird and Nisbet* interpret the movement to be the result of south-bound migrants, carried toward the Atlantic Coast by strong northwest winds, attempting to fly back and redetermine or regain their preferred lanes of passage overland. Another example with a different interpretation comes in the spring from Point Pelee, a peninsula projecting nine miles southward into western Lake Erie from Ontario, and from Pelee Island that lies about eight and one-half miles southwest of Point Pelee and nearer the Ohio mainland. Time and again many small land birds have been seen returning southward over Lake Erie from the tip of Point Pelee.

Levis reported such a movement in mid-May at Pelee Island. Here for several hours he watched large number of birds representing 35 species (mostly passerine) streaming southward into a headwind. The movements are not based solely on visible evidence; they have actually been proven. Birds, banded at Pelee prior to starting south, were later recovered at Pelee Island and on the Ohio mainland. The particulars in these reverse flights may be birds which, during the preceding night, overshot their destination or were swept past it in high winds and, consequently, are attempting to return to it.

Rate of Migratory Travel

Most passerine birds fly at ground-speeds averaging 18 to 25 miles per hour (mph). Stronger fliers such as ducks, hawks, falcons, shore birds, and swifts attain much greater speeds. Any bird can accelerate its speed in special circumstances as when frightened or driving earthward. In general, the normal unhurried cruising speed of a bird is much slower than suggested by published records, most of which, until recently, were estimated by observers in automobiles or airplanes moving parallel to the bird's line of flight. *Schnell,* using Doppler radar equipment similar to that operated by law enforcement agencies in determining speed of automobiles, measured the ground flights-speeds of 17 species of birds in northern Michigan. He recorded on windless days two speeds of the spotted Sandpiper (*Actitis macularia*) at 25 mph; four speeds of the Eastern King-bird (*Tyrannus tyrannus*) at 21 mph and one at 13 mph; three speeds of the Cedar Waxwing (*Bombycilla cedrorum*) at 21,23 and 29 mph; and three speeds of the Red-winged Blackbird (*Agelaius phonenceus*), one at 17 mph, and two at 23 mph. Had the wind been blowing, the speeds might well have been slower for birds flying into it and faster for birds flying with it.

Strong winds can significantly affect flight-speed as Schnell proved with the 267 speeds of the Herring Gull (*Larus argentatus*) that he recorded in different wind velocities. Speeds, he found, averaged 25 mph in winds less than 6 mph, but averaged 18 mph in winds less than 6 mph, but averaged 18 mph (extremes 7 and 39 mph) into winds of 6 to 15 mph and 34 mph (extremes of 21 and 49 mph) with the same winds. During migration, according to radar surveillance by *Bellrose*, birds appear to reduce their flight-speed somewhat proportionately to the increase in favourable wind-speed.

Apparently they adjust their flight efforts in relation to the degree of wind-speed whereas the groundspeeds of birds in the daily activity flights, as shown by Schnell, are definitely influenced by winds. Many of the stronger flying birds show great ability for fast migratory travel. *Mc-Cabe,* in an airplane going at an air speed of 90 miles per hour, was overtaken by two flocks of sandpipers flying at an estimated air-speed of 110 mph. *Speirs,* once estimated the average ground-speed of the Oldsquaw (*Clangula hyemalis*) at 61.5 mph and the air-speed at 50.5 mph. Birds homing to their breeding sites, after being displaced at great distances away, demonstrate impressive ability for sustained speed for many hours. A female Purple Martin taken from her colony at the University of Michigan Biological Station in northern Lower Michigan and released in Ann Arbor, Michigan, 234 miles to the south, at 10:40 pm, was back feeding her young at 7:45 am, having de the return flight in not more than 8.6 hours at an average speed of 27.2 miles per hour (Southern, 1959).

A Manx Shearwater (*Puffinus puffinus*), removed from its nesting burrow Skokholm off the west coast of Wales and released in Boston, Massachusetts, reached its burrow, after at least 3,2000 miles in 12 days and some 13 hours, or an average of 250 miles a day (*Mazzeo*, 1953). Another sea bird, a Leach's Petrel (*Oceanodroma leucorhoa*), average about 300 miles a day for nine days form its point of release at Prestwick, Scotland, back 0 its nesting burrow on New Brunswick's Kent Island in the Bay of Fundy. Of the 18 Laysan Albatrosses (*Diomedea immutabilis*), taken from their nests on Midway Island—one of the Hawaiian Leewards in the north-central Pacific–and released at widely scattered points in the northern Pacific, 14 returned, one from the Philippines, a distance of 4,120 miles in approximately 32 days, and one from Whidby Island off the coast of Washington, a distance of 3,200 miles in 10.1 days at an average speed of 317 miles a day.

Presumably, not one of these birds homing to its breeding site had the fat reverses for energy that migrants acquire prior to their long journeys. There is much additional evidence, obtained by other means of the bird's ability for sustained speed during long distances of migration. *Cochran, Montgomery*, and *Graber*, using radiotelemetry, tracked migrating *Hylocichla* thrushes nearly all night from Illinois northward into Michigan, Wisconsin, and Minnesota. Although they found considerable variation, most flights were at air-speeds–i.e., speeds with relation to the wind aloft—between 25 and 35 mph. These were usually less than ground-speeds—speeds with relation to the earth and thus suggested that the birds were aided by favourable winds.

One of the most remarkable records of a long, sustained flight is that of a banded Ruddy Turnstone (*Arenaria interpres*) released by Max C. Thompson at St. George Island, one of the Pribilofs in the Bering Sea, on August 27, 1965, and shot four days later, on August 31, at French Frigate Shoals in the Hawaiian Leeward Islands. Assuming that this Individual covered 2,300 miles between St. George Island and French Frigate Shoals is in a steady Bee-Line flight, average speed was 575 miles a day. Recently, radar studies have provided many reliable estimates of the rate at which migratory birds travel. W.R.P.

Bourne assessed the air-speed of Lapwings (*Vanellus vanellus*) In their flights during June over the southern North Sea to England at 35 knots (40 mph). Lee at the Isle of Lewis in the Hebrides, Scotland, showed that the air-speed of Wheatears (*Oenanthe oenanthe*) from Iceland approximated 20 knots (23 mph); Redwings (*Turdus iliacus*), from 30 to 55 knots (61 to 63 mph). *Bergman* and *Donner* demonstrated that the still-air-speed for the Oldsquaw (*Clangula hyemalis*) at low attitudes over the Gulf of Finland was 40 knots (46 mph) and for the Common Scoter (Oidemia nigra), 45 knots (52 mph). Once the birds reached a higher attitude inland, their speed increased by about 10 percent.

After recording air-speeds of passerine birds off the coast of Norfolk, England, for a whole year, *Tedd* and *Lack,* on analyzing the results, found evidence of a seasonal difference: in the spring, the speed averaged 27 knots (31 mph), 4 knots faster than in the fall. At Cape Cod, (*Massachusetts, Nisbet* and *Drury* found another seasonal) difference in that the directions of migration in the spring were much less diverse than in the fall, thereby suggesting much less time lost in passage. In the late fall, long-distance migrating ducks commonly pass from breeding area to winter quarters in a short series of mass movements, each of which carries them many hundreds of miles in one continuous flight. They start each flight immediately after the passage of a cold front when temperature has dropped and the sky is clear, but they may overtake bad weather as they proceed.

Sometimes, owing to the triggering effect of extremely low temperatures resulting from a strong flow of polar air, the mass movements are spectacular both in numbers of birds involved and distances covered. *Bellrose* documented one such migration in 1955 that moved with unusual rapidity from the Great Plains of Canada to the marshes of southern Louisiana. The exodus began from Canada or October 31; early on November 1 the flight was in full force through the Dakotas; and on November 2 the vanguards has reached northern Tennessee and Arkansas shortly after sunrise and Louisiana later in the day. Many thousands of ducks made the flight from Canada to southern Louisiana, a distance of 1,200 to 2,000 miles, in two days, or roughly 35 to 50 hours, at an average speed of 40 miles per hour. No doubt some of the birds covered the distance without stopping, accomplishing their migration in one flight. In undertaking long-distance migrations, many small land birds tend to begin with short flights and complete them with longer flights.

Indirect evidence of this procedure was reported by *Caldwell*, *Odum*, and *Marshall*, after comparing the fat reserves of six pieces of tropical-wintering North American passerines killed during fall migration by television towers, one near the Florida Gulf Coast and the other in Central Migrants. All the birds killed by the Florida tower showed significantly greater amounts of fat, strongly suggesting that these migrants began with low to moderate fat reserves that allowed only sort flights and then increased their reserves until they had acquired a maximum amount for the long, non-stop flights such as across the Gulf of Mexico.

European birds migrating south across Africa build up fat reserves of 30 to 40 per cent of their body weight by the time they set out across the Sahara. By making longer flights as they near their destinations, birds gradually speed up their migrations. *Cooke* provided evidence for this acceleration when he analyzed migration dates of North American species, mostly passerines, approaching their northern nesting areas. "Sixteen species, " he wrote, "maintain a daily average of 40 miles from southern Minnesota to southern Manitoba, and from this point 12 species travel to Lake Athabasca at an average of 72 miles a day, 5 others to Great Slave Lake at 116 miles a day, and 5 more to Alaska at 150 miles a day." From all these studies and reports, several generalizations on the rate of migratory travel emerge. Strong winds can affect ground-speed of birds in their daily activity fights but not in migration. Birds make long, sustained flights, usually at increasingly higher speeds at higher altitudes. Spring migration proceeds at a greater rate with less time loss than fall migration. Larger birds such as ducks accomplish their migrations in a short series of a few mass movements, occasionally in one non-stop mass movement. Small land birds, however, tend to begin their migrations in many short flights, gradually building fat reserves for long, non-stop flights, thereby accelerating their migrations.

Mortality in Migration

Migrating is dangerous for all birds. In their long flights over land or water they are likely to meet disaster through vagaries of the weather. When forced to land by cold fronts, frequently they must accept environments where, because of inadequate cover, they are easy victims of predators. While migrations are adjusted to the normal alternation of fair and inclement weather during spring and fall, sudden and unseasonable changes in the weather occasionally have serious effects. Once during fall migration in the vicinity of Lake Huron, untold numbers of birds crossing this huge lake were forced into the water and drowned because of a very quick drop in temperature and an exceptionally heavy snowfall. After the storm, one observer reported an estimated 5,000 dead birds washed up on a one-mile stretch of shore. In their flights north in the spring, birds are sometimes caught in severe storms and killed by becoming first exhausted and then being rain or wet snow.

After a blinding March snowstorm in Minnesota as many as 750,000 Lapland Longspurs (*Calcarius lapponicus*) were found dead on the ice of two lakes, each of which covered only a square mile. Man has created awesome hazards for migrating birds by erecting lighthouse with strong light beans and by illuminating various tall structures such as the Washington Monument in the District of Columbia and the Empire State Building in New York City. During nights in the spring and fall when migration is proceeding at a low elevation because of an unsurmountable cloud layer, passing birds are attracted by the brightness and, approaching it, soon become blinded and fly into its source, killing themselves. Under certain circumstances, even street lights can be a hazard. Vast numbers of parulid warblers, driven ashore on the Texas coast by a northeast storm while migrating northward across the Gulf of Mexico during a night in early May, met their death by flying into street lights.

In a part on Padre Island, *James* counted more than 900 dead birds under just one light pole plus an estimated 100 on the adjacent pavement. There were nine other light poles in the vicinity with similar tolls. Airport ceilometers indirectly cause mortality among small, nocturnal migrants. These instruments, which are used to determine the cloud ceiling, direct a narrow, extremely brilliant beam of light straight upward. At night when the clouds are low, they produce a bright spot on the cloud ceiling than can be seen for a considerable distance. On mornings following a heavy, nocturnal migration, numbers of birds varying from three to over a thousand have been found dead near spots where ceilometers are used.

Three ornithologists, *Howell, Laskey* and *Tanner,* who have investigated many of these accidents, explain the cause as follows: when there is a pronounced migration and a low ceiling, the beam attracts the migrants. After circling through the bright light, many circle back to fly in and about it. While in the beam their bodies reflect light attracting still other migrants. Blinded by the light, the birds die by collision, either with each other, with the ground, or (rarely) with a building. A direct cause of mortality among small, nocturnal migrants are television towers erected to heights of 900 to 1,000 feet or more and, as is often the case, situated on hills or bluffs where they reach even greater heights above the local terrain. Each tower is supported by guy wires and has a system of steady, flashing red lights, mandatory on all tall structures that are potential hazards to airplanes. When migrants re flying under low ceiling, they are attracted to a television tower because of its lighted area and become reluctant to leave. Just as birds, released at night in a lighted room with doors an windows open, continue to fly about in the room rather than escape into the darkness, the migrants fly through the tower framework and circle out to the edge of the lighted area, then return toward the light. Mortality results when the birds strike the dark guy wires while circling. The above observations and explanations come from *Graber*. For *Tordoff* and *Mengel* and *Stoddard* and *Norris*.

ALTITUDES OF MIGRATORY FLIGHT

The recent analysis of migratory flight by means of radar shows not only that birds move at attitudes averaging higher than formerly believed, but also that their attitude varies widely depending on the circumstances. *Nisbet*, studying radar heights of nocturnal fall migrants above Cape Cod, Massachusetts, and the outlying ocean, found that the most frequent height was usually between 1,500 and 2,500 feet. About 90 per cent of the birds, probably small passerines, were below 5,000 feet. On some nights they were lower than 2,500 feet, while on others they were up to 6,000 or 8,000 feet. The presence or absence of cloud cover may determine the attitude chosen by birds.

Bellrose and *Graber* discovered during their radar studies of nocturnal migrants in central Illinois, that birds are prone to migrate at higher altitudes when the skies are overcast than when they are clear. If the clouds are not too high, the birds apparently attempt to surmount them; but if the clouds are too high, they usually continue to fly, sometimes in the clouds, although usually immediately under them. When the birds are flying under overcast and, consequently, at much lower attitude, they can usually be heard from the ground. Birds fly higher by night than by day. Lack first noted this tendency, by radar, among migrants crossing the southern North sea between England and the Continent. Later, *Eastwood* and Rider at the Bushy Hill station in England proved that the tendency is significant. From their considerable data they were able to show that 80 percent of the birds fly below 5,000 feet at night and 80 percent below 3,500 during the day. They further demonstrated that migrating birds, in a 24-hour day, have a tendency to fly at the lowest altitudes in the afternoon and the highest just before mid-night.

From radar studies and tape recordings to call notes in Illinois, Graber has concluded that migrants reduce their altitude after mid-night to 1,500 feet or less, although they continue their flight until day light. After their descent to lower altitude, they increase their calling, their descent to lower altitude, they increase their calling. This helps to explain why nocturnal migrants can be heard more frequently as dawn approaches than earlier in the night. There are seasonal variations in altitudes. *Bellrose* and *Graber* found that migrating birds in Illinois fly higher during the fall than during the spring, possibly because the winds during the fall are more favourable for southward migration at higher altitudes.

Eastwood and *Rider*, on the other hand, found the reverse to be the case in England. They suggest as one reason for this seasonal difference that flocks of fall migrants include many young birds whose flight capabilities are inferior to those of adults and which are, consequently, unable to achieve the higher altitudes of the more mature spring migrants. Birds migrate higher over land than sea. Common Scoters (*Oidemia nigra*) and Oldsquaws (*Clangula hyemalis*), in their spring passage over southern Finland and the Gulf of Finland, were noted by *Bergman* and *Donner* to fly at altitudes that averages 3,400 feet over land ranged from 300 to 1,00 feet above water. Passerine migrants, in passing from sea to land in England, were shown by *Eastwood* and *Rider* to make a similarly significant though not as great change in altitude, climbing from a median height of 1,700 feet to a median height of 2,200 feet.

Birds have long been known to reach very high altitudes during flight. Direct observations from aircraft proved that large birds can fly over the highest mountain ranges – for example, the Himalayas between central Russian and India. The Yellow-billed Chough (*Pyrrhocorax graculus*) was actually found on Mt. Everest at an elevation of 27,000 feet. The use of radar in determining heights of migration shows that while most birds rarely exceed 8,000 to 10,000 feet, a small proportion of migrants, particularly the stronger fliers, nonetheless attain great heights, in some cases astonishing. In his radar studies of migrants at Cape Cod Massachusetts, *Nisbet* recorded a number of birds on several dates in September and October, usually before mid-night, or before and after sunrise, at altitudes between 8,000 and 15,000 feet and a few birds as high as 20,000 feet. These may have been sandpipers and plovers, flying over the ocean, toward the Lesser Antilles and eastern south America.

Using an especially powerful and accurate height finder at Norfolk in southeast England, *Lack* observed at sunrise on 16 dates in September a thin scattering of birds extending fairly uniformly up to at least 15,000 feet. Probably the highest migrants were small shore birds such as the Dunlin (288) which had left Scandinavia the night before. Bearing in mind that the oxygen content of the air at 18,000 feet in 50 percent less than the air at sea level, one wonders whether high-flying birds suffer "altitude sickness." When a man sets out to climb a lofty mountain the acclimates himself gradually over a period of days, but birds take off and reach a comparable elevation in a matter of a few hours. Do birds have special adaptations that enable them to avoid altitude sickness? The answer may come someday from experimental studies of bird flight in pressure chambers.

Course of Migration and Migration Routes

Many species breeding in North America north of Mexico have their winter ranges far south and southeast; in southern Mexico, Central America, and South America. To reach their winter ranges, most of these species proceed at night form their breeding ranges to southern United States in broad fronts without notable regard to topographical features, Radar and–other observations confirm that their movements trend southeast–toward their winter ranges. This direction, coming as it does in the wake of a cold front, is with the wind since the movements take place with the onset of a warm front. Many species, however do not exactly retrace their course in the spring, but fly instead somewhat to the west of their fall passage. Evidence of this elliptical course–going to southern United States one way and coming back another–is borne out by kills at television towers.

Certain specie are well represented in the spring migration but seldom or not at all in the fall, and *vice versa*. Owing to the narrowing of the North American continent southward and the intervention of the Gulf of Mexico and Caribbean Sea, all species moving southeastward through the United States toward their wintering ranges in southern Mexico, Central America, and South America converge on special routes. There are five altogether. Certain species use mainly one route, others two or three; no species is known to use more than three. The routes are as follows:

Route 1: From the coasts of Newfoundland, Nova Scotia, New England, and New Jersey southward over the Atlantic Ocean to the Lesser Antilles and the northeastern coast of South America. A few shore birds use this route.

Route 2: From Florida southward over the Bahamas, Hispaniola, Puerto Rico, and the Lesser Antilles to South America. The few birds which frequent this route are seldom far from land as there are many small islands along the way.

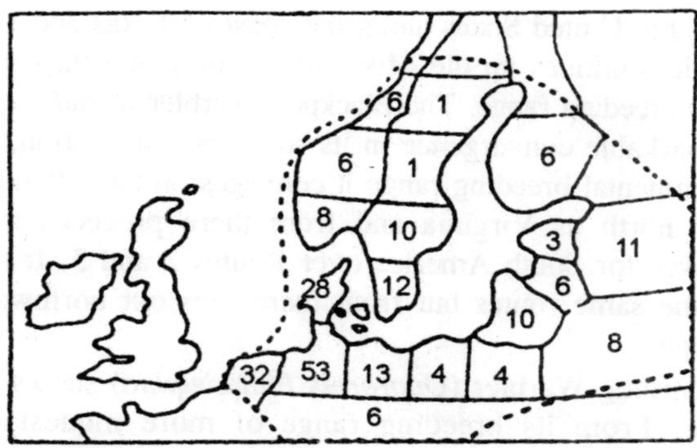

Fig. 8.7. ***Breeding range of starlings that spend the winter in the British Isles. Numerals indicate the breeding season recoveries (May-Aug) of starlings previously ringed during winter (Dec.-Feb.) in the British Isles.***

Route 3: From Florida southward over Cuba and Jamaica across 400 miles of the Caribbean Sea to South America.

Route 4: From the shores of the Gulf States across the Gulf of Mexico to the Yucatan Peninsula and southern Mexico. This is the route most frequently used by the many species of birds from eastern United and Canada.

Route 5: From the Texas, New Mexico, Arizona, and California through northern Mexico. The majority of birds from western conterminous United States, western Canada, and Alaska use the western side of this route.

Species very greatly in their course of migration and use of routes. A few species move south over one route and return by another; A few species move eastward or westward before going south; and a few species have spectacularly long routes that take them as far south as far south as southern South America. Among a few species which breed in western, United States–e.g., the Western Kingbird (*Tyrannus verticalis*) and the Scissor tailed Flycatcher (*Muscivora forficata*)–some individuals in the fall move eastward across the Gulf states to winter in Floria whereas most of the population moves directly south into Mexico. To illustrate the wide variation in migratory movements and the choice of routes, the migrations of six species are described below.

The American Golden Plover (*Pluvialis dominica*) passes eastward from its breeding range to the Atlantic Coast, where it turns southward over Route1. Once in South America, it flies directly across the continent to its winter range. It returns, however, by another route, coming up across northwestern South America and the Gulf of Mexico and reaching the United States along the coast of Texas and Louisiana. From there it continues up the Mississippi valley and through central Canada to its breeding range. The Blackpoll Warbler (*Dendroica Striata*) shows a remarkable convergence in its southward migration. From its vast transcontinental breeding range it conveges on the Atlantic coastal plain as far north as Virginia and from there proceeds to Florida where it leaves for South America over Routes 2 and 3. It returns to Florida by the same routes but from there fans out northward to its breeding range.

The Mourning Warbler (*Oporornis Philadelphia*) shows a similar convergence. From its breeding range of more modest extent it converges on southern Texas, then goes southward along the eastern portion of Route 5 through eastern Mexico and Central America to its winter range. It returns via the same part of Route 5 and fans out from Texas northward. The American Redstart (*Setophaga ruticilla*) shows little convergence. Instead, it passes southward more or less directly over a broad front of nearly 2,500 miles, eventually using Routes 2, 3 and 4. It returns the same way.

The Connecticut Warbler (*Oporornis agilis*) migrates in an eccentric manner. From its breeding range it flies directly eastward to New England, then to South America along the Atlantic coastal plain and eventually Route 3. From its winter range is South America it returns over Route 3 to Florida, but from there it passes diagonally to the Mississippi Valley and northward to its breeding range.

Altitudinal Migration

Some bird populations of high mountains in both temperate and tropical regions move down to the slopes and valleys when winter sets in at higher elevations. In the descent of a few hundred feet they accomplish what many other populations do in their latitudinal migrations of many hundreds of miles. Several observers have noticed that mountain birds tend to move to higher slopes after the nesting season. Probably many are young birds which, after getting their growth, always show a great tendency to move about. The reason for this may be the agonistic behaviour of their elders, but more than likely it is another example of juvenile wandering.

Distance in Migratory Travel

Extremes in distances traveled by migrating birds are represented on the one hand by high-mountain species, which merely pass up or down slopes of several hundred to a few thousand feet, and on the other hand by the Arctic Term (*Sterna paradisaea*). It makes the longest flight of any bird, migrating thousands of miles from the Arctic, where a part of the population breeds, to its wintering area adjacent to the pack ice around Antarctica. This species nests as far north as the northern tip of Greenland at 83 degrees South Latitude and has been recorded as far south as 74 degrees South Latitude. Most of the individuals from Greenland, Canada, and northeastern United States undertake the passage by flying across the North Atlantic to the continental shelf of western Europe, and then south over the coastal waters of West Africa and finally across the Antarctic Ocean. They return by the same route. While a few species breeding in conterminous United States, Canada, and Alaska go as far south as central and southern South America, the majority migrate no farther than northern South America.

In fact, many species in northern conterminous United States, Canada, and Alaska move only to the distances traveled by birds and their size, flying abilities, or habits. A number of small species journey to Mexico and Central America, outdistancing a great many species which to all appearances have much greater capacities for travel. Still to be explained is why certain species, sometimes closely related, have routes that are half the length or less. It has been postulated that the long migration of certain species to northern latitudes for nesting is to take advantage of increased daylight and the consequent shortening of the period in which the young are confined to the nest.

No satisfactory theory has yet been advanced to account for the arduous passage of such birds as the Bobolink (*Dolichonyx oryzivorus*) across the tropics from the northern to the southern temperate region, or the Arctic Term from one polar region to the other. When a migratory species has an extensive breeding range that includes parts of northern conterminous United States, Canada, and Alaska, the more northern populations of that species move farther south for the winter than the other populations do. For example, among the several subspecies of the Fox Sparrow (*Passerella iliaca*) breeding along the Pacific Coast from Alaska to Puget Sound, the subspecies nesting farthest north have been found wintering in southern California, passing by other subspecies, which either do not migrate at all or move only to central or northern California. Apparently the more northern breeding populations of a widespread species acquire a stronger migratory habit than the populations breeding in more southern areas that have milder, year-round climate. In some species there may be sexual differences in the extent of the migration. *Howell* found that females in the eastern race of the Yellow-bellied Sapsucker (*Sphyapicus varius*) outnumber the males in the southern part of the winter range by about three and one-half to one.

Flight Lanes and Concentrations

In North America migratory movement is continent-wide. There are probably no areas over which birds do not pass in their latitudinal migrations. Prairies, forests, mountains, lakes, and inland extensions of the oceans fail to stop or divert migration altogether. Even so, as diurnal migrants move northward or southward across Canada and the conterminous United States, many species tend to favour or be influenced by certain topographic features which trend in a north-south direction. Some species fly along ridges; others follow the coasts, large rivers, move along peninsulas, or pass from island to island across large bodies of water.

In both Canada and the United States there are many places where the topography is such as to cause narrow flight lanes in which migratory movement is especially conspicuous. Particularly in the fall, hawks and a few other large birds follow the crests of north-south ridges, riding on the updrafts as they proceed southward. The traffic in these lanes is unusually heavy on clear, windy days following the passage of a cold front, because there is considerable wind deflected upward, thus making these lanes advantageous to travel. One of the best known places is the Kittatinny Ridge is eastern Pennsylvania.

At one point, Hawk Mountain, where the ridge becomes suddenly high and slender, the birds are brought together and closer to the ground. As a result of this narrowing of the flight lane, observers at hawk Mountain have been able to count over 22,000 hawks moving by in a singe season. Large bodies of water constitute barriers to day-migrating birds and thus cause flight lanes to curve around them. The Great Lakes are a good example. In the fall, south-bound hawks approaching their north shores from western Quebec and southern Ontario take the shortest courses around them that geography and air movements will allow. If the birds are migrating in great numbers, as on the second days after cold fronts when there are steady westerly winds and ample sunlight producing thermals, large numbers can be seen in continuous passage at such points as Port Credit on the northwest shore of Lake Ontario, Port Stanley on the north side of Lake Erie, Cedar Grove on the west side of Lake Michigan and Duluth at the westernmost extension of Lake Superior. Peninsulas projecting into large bodies of water that lie athwart the direction of migration many become funnels for land birds in diurnal passage. Hawks moving northward in the spring through Michigan at the Straits of Mackinac.

Here, if the weather is rainy and windless, they settle on trees and other perches. With the advent of the next clear day and a favouring wind, the hawks begin to spiral higher and higher until a bird peels out a heads northward over the Straits with the others following. Small numbers of north-bound Blue Jays (*Cyanocitta cristata*), in order to get across Lake Mendota at Madison, Wisconsin, converge first at Picnic Point and then spiral upward until "barely visible to the naked eye" before crossing the 1.7 miles of open water to Fox Bluff on the north shore. Cape May, the southern tip of New Jersey between the Atlantic Ocean and Delaware Bay, is noted for its hordes of land migrants, large and small, which gather from August through November when northerly winds are strong. Some of the migrants held up here are those which regularly follow the Atlantic Coast southward, but many are birds which have been drifted by wind southeastward from their usual flight lines.

Frequently, migrants at Cape May may be seen in the day flying north and north westward into the wind as they skirt Delaware bay before continuing their journey. Similar concentrations may be observed at Cape Charles, a south-pointing peninsula separating the waters of Chesapeake Bay from the Atlantic. Night migrants do not follow topographically determined flight lanes to any significant degree. Instead, as radar surveillance shows, birds migrate at night without regard to what lines below. In the spring, land migrants returning to the United States form across the Gulf of Mexico vary in their manner of arrival in accordance with weather conditions. If the weather is mild and the wind favourable, birds bound for more northern destinations continue Inland from the Gulf for considerable distances before coming to land.

The coastal area thus appears to be an ornithological "hiatus." But if a cold front with strong northerly winds moves in over the area while migration is in progress, migrants are forced to come down to the first land reached, with the result that the coastal area is flooded with birds that linger here until he weather again becomes favourable. A somewhat similar situation occurs during spring migration at Point Pelee. Here, since it is the nearest land, north-bound small land birds, on meeting a cold front from the north while they are over Lake Erie are forced to descend and remain until the cold front abates. At such times Point Pelee swarms with birds.

Flocking During Migration

During migration bird species show wide differences in flocking habit. A number of diurnal migrants, notably many hawks and other predators, are little inclined to move in groups, preferring to travel solitarily. But the majority of migrants, diurnal or nocturnal, exhibit the flocking habit. Certain species migrate in flocks strictly of their own kind. These are usually birds whose flight-speed, feeding habits, or roosting preferences are so individual as to make them incompatible traveling companions. The Common Nighthawk (*Chordeiles minor*) and Chimney Swift (*Chaetura pelagica*) are good examples of birds which migrate in their own company. Neither waxwings nor crossbills wind migrate with other birds, but Cedar and Bohemian Waxwings (*Bombycilla cedrorum* and *B. garrulous*) have been seen in the same flocks and so have Red and White-winged Crossbills (*Loxia curvirostra* and *L. leucoptera*).

Some of the largest birds –e.g., pelicans, cormorants storks, swans, geese, and cranes–noted for V-shaped or linear flock formations likewise tend to travel in unmixed groups. The majority of species traveling in flocks, weather unmixed or mixed, give call notes. This is especially true of nocturnal migrants. Some species such as *Hylocichla* thrushes utter calls heard only in night migration; other species such as the Bobolink and Dickcissel (*Spiza americana*) give calls that are the same as ones heard in the day-time on their breeding grounds. If the calls are distinctive as in the case of the Bobolink and Dickcissel, it is possible to identify the species, but the calls of most species are link and Dickcissel, it is possible to identify the species, but the calls of most species are faint chips and Lips that sound more or less the same to the human ear.

Unmixed or mixed flocks of some species of smaller birds–e.g., certain sandpipers and plovers–fly in compact formations, all individuals in the flock simultaneously performing almost identical maneuvers. Many more species of smaller birds, including the majority passerines, travel in flocks that are loosely formed, though still cohesive, the individuals proceeding in the same direction. At night, the cohesion of flocks is probably maintained by call notes. At the same time, the call notes serve to space out individuals in each flock so that they will not collide with one another. No flock of migrants appears to have a persisting leader. The direction taken by a flock, represents a compromise by each individual to the directional preference of the other individuals of the flock.

At night, call notes convey directional information from one individual to the other. When a nocturnal flock shifts its direction or lowers its attitude. Or when it is disoriented for one reason or another, all its members greatly increase their rate of calling. Whether or not flocks remain intact for the duration of migration has not been determined. Nor is it known for certain whether flocks remain together during the winter. In all probability, most flocks, if they are comprised of common, widely distributed species, change form day to day during migration and in the winter split up into smaller groups, or combine with other flocks to form large groups. Flocking during migration is undoubtedly advantageous to the individuals concerned.

Just as flocking among resident birds provides group protection against predation or increases the success in finding and exploiting food sources, flocking in migration greatly facilities the attainment of destination. Younger birds traveling with more seasoned adults benefit from their experience. As another advantage, *Hamilton* suggests that groups of birds of birds on an average determine their direction with greater accuracy than single individuals. Thus flocking assists any migrant that goes a long distance, or any migrant required to pinpoint its destination on a small land area in mid-ocean. Hamilton cites as examples: (1) The Broad-winged and Swainson's Hawks (*Buteo platypterus* and *B. swainsoni*). Both of these North America. By contrast, the Red-tailed Hawk (*B. jamaicensis*) and certain other buteos, which rarely leave the North American continent, seldom move in appreciable groups. (2) The Long-tailed Cuckoo (*Urodynamis taitensis*). After breeding in New Zealand, this species gathers in large flocks for the exceedingly long, non-stop passage to winter quarters on tiny islands in the west-central Pacific. By contrast, the Yellow-billed and Black-billed Cuckoos (*Coccyzus americanus* and *C. erythropthalmus*) of North America, which have only to reach a broad tropical region of South America for the winter, migrate solitarily.

V-shaped formations help to conserve energy by creating favourable air currents for all individuals in the flock except the leader. When fatigued, the leader drops back and is replaced by another bird in the flock. An alternate advantage hypothesized by *Hamilton* is that the V-shaped structure serves as a means of communication, enabling the individuals to profit fully form the collective direction-finding of the group. By flying in parallel alignment, each individual moves in the same direction as the leader. If, as in flocks of geese traveling, in poor visibility, difficulties in establishing direction arise, the leadership changes frequently in order that collective "judgment" may prevail in maintaining the proper course.

Sometimes the V-structure gives way temporarily to a crescent form; forward flight remains in the same direction but is slowed until a new leader takes over and the V-shape is resumed. Unmixed or mixed flocks may contain only immature individuals, only adults of one sex; or they may contain individuals of all ages and both sexes. Flocks of geese and cranes may be comprised on one family or several families. During the fall migration the adults may precede the immature birds, the birds-of-the-year. *Hagar* reported that adult Hudsonian Godwits (*Limosa haemastica*) withdraw from their nesting grounds in central and northwestern subarctic Canada in late July; the immatures follow a month later.

Among passerine species there is convincing evidence assembled by *Murray* and *Jehl* from several thousand migrants, mist-netted in the fall at Island Beach, and analyzed as to age, that adults and immatures travel at approximately the same time. However, in the case of the least Flycatcher (*Empidonax minimus*), *Hussell*, *Davis* and *Montgomerie* concluded from an analysis of 182 individuals trapped in the late summer at long point, Ontario, that most of the adults migrate in advance of the immatures: the most of the adults during the second half of July and first half of August; the majority of immatures form the second half of August to the end of September. In the spring migration many of the first flocks of certain species coming north have a preponderance of adult makes which reach the breeding grounds and establish territories before the rest of the population arrives. *A.A. Allen* once carefully studied the spring migration of Red-winged Blackbirds (*Agelaius phoeniceus*) at Ithaca, New York. Since the males and females have distinct plumages and birds hatched the previous year (i.e., the immature birds) differ sufficiently in colour from the adults, he was able to age and sex. While his findings may not hold for all Red-wing populations, It is presented below as a useful guide and basis for comparison.

Migrant adult males	March 13-April 21
Resident adult males	March 25-April 10
Migrant females and immature rules	March 29-April 24
Resident adult females	April 10-May 1
Resident immature males	May 6-June 1
Resident immature females	May 10-june 11

In watching spring migration in his own area, the student will find that not all species follow such a schedule. In fact, males and

Females of quite a few species migrate together. Only rarely, however, do females precede males.

Direction-Feeding

How migrating birds determine their direction when migrating or night over areas unfamiliar to them is one of the most fascinating aspects of migration. Before considering the subject of direction-finding in migration. It is worthwhile to review some of the problems associated with homing in birds. The basic means by which birds orient themselves, whether migrating or homing, are much the same. A great many experiments have demonstrated the remarkable ability of wild birds to return to their eggs or young after being removed great distances and released. Years ago *Watson* tested the homing ability of Sooty and Noddy. Terms (*Sterna fuscata* and *Anous stolidus*), which nest on the Dry Tortugas, the islands in the Gulf of Mexico west of Florida. These species come to the Tortugas form tropical seas and are seldom seen farther north. Two nesting Sooty Terms and three Noddy Terms were captured, marked, and transported northward in a ship to point off Cape Hatteras, about 1,000 miles by sea from the Tortugas. Here they were released. Just five days later the Solly Terms were back on their nests; one of the Moddy Terms shoved up after a few two more days.

Some of the more recent experiments demonstrating the sustained speed of homing flight by the Purple Martin, Manx Shearwater, Leach's Petrel, and Laysan Albatross (See above) attest at the same time to their precision in direction-finding. The inducement of birds to home is not necessarily provided by their eggs or young; it an be the breeding are or home range. Breeding Brown-headed Cowbirds (*Molothrus ater*), which are brood parasites, will home from maximum distances of 250 to 380 miles.

An adult female Bobolink (*Dolichonyx oryzivorus*), escaping from captivity in September at Berkeley, California, was recaptured the following June at Kenmere, North Dakote, where it was originally tapped as a breeding bird. Six hundred and sixty Golden-crowned and White-crowned Sparrows (*Zonotrichia atricapilla* and *Z. leucophrys*), captured while wintering in the San Jose area at California, were immediately carried by plane to Laurel, Mary-land and released. Fifteen were known to have come back the following winter. In the interim they had presumably found their way in the spring to the nesting grounds in northwestern Canada and Alaska, then returned to Califormia in normal migration. Homing flights are more or less routine with homing pigeons which are common Pigeons (*Columba livia*) specially bred for racing. Their precision and speed of return to the home loft is developed by training and experience.

Sometimes the experience of only one flight to the home loft is sufficient to determine the proper direction for successive flights. In his work on homing Pigeons, *Matthews* found that certain individuals, trained to maintain a giver direction, can adhere to that direction over unfamiliar terrain and that certain individuals, trained to maintain a given direction, can adhere to that direction over unfamiliar terrain and that certain other individuals can fly straight toward the home loft from unfamiliar territory regardless of the direction of the home loft. In the course of their training, pigeons must be made familiar with the area around the loft so that they will have a broad or reasonable target. Although homing has been amply demonstrated in both wild birds and racing pigeons, the perplexing question remains: How do homing birds find their way? In seeking an answer, Griffin and Hock attempted to determine how displaced and released resting birds find their way by following them in an airplane and watching their behaviour.

For their experiment they selected the Gannet (*Morus bassanus*), a marine bird which rarely occurs Inland; being large and white, it was easy to follow. Taking 17 individuals from their nests on an island off the Gaspe Peninsula, Quebec, they carried them to a point in northern Maine, 100 miles from salt water and 215 miles from the home island. There they released nine of the birds (the others were used as controls) and, from an airplane, traced their flights from 25 to 230 miles. The investigators were careful to keep the plane, 1,500 feet or more away from the birds so as not to frighten them.

The experimental birds flew in all directions with no significant tendency to head directly toward their nests. Their flight paths were generally gradual curves. The first birds to reach the coast within the first few hours were the first to get back to their nests. Altogether 62.5 per cent of these birds eventually reached their nests, in from one to four days. The result of Griffin and Hock's work on the Gannets suggests that the ability of birds to home - is dependent on random searching.

In a strange territory birds keep circling and exploring by trial and error until they find familiar landmarks within familiar territory. However, the concept of random searching as a means by which all birds find their way cannot be reconciled with the rapid homing exemplified by Sooty Terns and other species already mentioned (see above). Many birds, if not all, orienting themselves—determining their position with respect to their environment—and by following directional cues or navigating. In taking up the subject of the means by which birds orient themselves and navigate, it is well to consider first the question whether or not birds inherit at least part of their ability to find their way. Rowan at Edmonton, Alberta, caught young Common Crows (*Corvus brachyrhyncos*) in the late summer and kept them in captivity.

In November, when winter conditions had begun to set in and all adult Common Crows had left for their winter range is Kansas and Oklahoma, he banded the captive birds and released them. Altogether 54 individuals were set free, and in the next few days he received reports of recoveries. Apparently some of the birds had not traveled very far, but those which had gone an appreciable distance were headed toward their winter range. *Schuz* at Rossiten on the Baltic Coast of eastern Prussia tried similar banding experiments with White Storks (*Ciconia ciconia*). In the middle of September, after all the local population had departed, the released 73 young banded birds. Most of them traveled eastward toward the Black Sea, paralleling the normal flight line for the local population, though some of the birds flew in a more southerly direction and three went southwestward to Italy. *Perdeck*, over a period of four years, caught and banded over 11,000 fall-migrating Starlings (*Sturnus vulgaris*) in Holland and released them in Switzerland. Recoveries totaling 354 later showed that the juveniles took their ancestral direction southwest, paralleling the normal route, to a new winter area, whereas the adults soon separated and veered westward toward their ancestral winter range. From these experiments alone, it seems clear that at least some birds must have an innate ability to follow the normal migratory route or one parallel to it. How the birds "know" when they have flown far enough is yet to be determined. Granting that migrating birds have an innate ability to find their way does not deny that some birds acquire their ability by experience.

Young Indigo Buntings (Passerina cyanea), hand-raised in various conditions of isolation, apparently do not attain the accuracy of orientation typical of adult buntings and thus seem to depend, at least partly, one some kind of experience. Young geese and cranes, which migrate in families or groups of families, probably learn the migration routes by following their elders. They no doubt "memorize" features of the landscape when they migrate by day as they often do. But whether or not birds inherit or acquire their ability to find other than landscape features to orient themselves and navigate when traveling at night, over the sea, or above or in a heavy overcast. What are the cues? The first break-through came in the early 1950's when it was proved that the sum is a cue to orientation and even a guide in navigation.

At Wilhelmshaven, Germany, *Kramer* placed a small, circular cage high in the center of a circular pavilion that was completely enclosed except for six windows, each high enough to give only a view of the sky from the cage. Kramer put a hand-reared Starling in the cage at the time in the spring when it would normally migrate, and from below the cage he recorded the direction in which the bird, in its migratory restlessness, showed a tendency to flutter. The bird fluttered persistently in the normal direction of spring migration if the sun was shining but not if the sky was heavily overcast. When, by an ingenious use of mirrors at each of the windows, Kramer altered the sun's apparent position, the bird deflected its-direction in order to maintain the same angle as before.

Kramer and his associated soon demonstrated in subsequent experiments that birds have some sort of "Internal clock" that enables them to compensate fro the sun's daily "movement" across the sky and thus can hold the same direction of flight despite the sun's steadily changing position. *Hoffmann* experimented further on the internal clock as a basis for orientation by the sun. He trained caged Starlings to seek a food reward at a particular time of day and always in the same direction. Then he tested them without rewards at the time they had come to expect and at other times. The birds, he found, could allow for the daily movement of the sun.

Having proved this, he subjected the trained birds to a regime of light and darkness that shifted their internal clocks six hours ahead or six hours behind the natural day outside. Later he exposed the birds to the natural day. The birds responded by shifting their directions accordingly by 90 degrees. Counterclockwise if the clocks were set ahead, clockwise if they were set behind. These and later tests showed that the birds did not orient themselves by the elevation or altitude of the sun in the sky but by its azimuth or position with relation to compass direction. The birds thus determined direction by what is called "sun compass orientation." Soon after *Kramer* reported his initial experiments with Starlings, *Mathews*, found that homing pigeons, which are non-migratory, use the sun as a guide. When he released pigeons in unfamiliar territory under clear skies, they headed in the direction of home. If the sky was overcast, they failed to do so.

The concept of the sun as a compass for homing pigeons has since been supported by other experimental investigators using different techniques. For example, *Schmidtkoenig* shifted the internal clocks of pigeons by six hours in advance and six hours behind the natural day. On releasing the birds under a clear sky at varying distances and different direction from the home loft, he found that the birds deviated appropriately to the left, he found that the birds deviated appropriately to the left or right of the correct flight direction taken by the control pigeons whose clocks had not been shifted. This suggests that the sun played a role in their orientation. Experience also plays a role in the orientation. Each time a pigeon is released, the direction it takes is more accurate. Some birds may show one-directional orientation that is not necessarily related to migration or homing.

Mathews, experimenting with non-migratory Mallards (*Anas platyrhynchos*) at Slimbridge, England, discovered that when birds of any age were displaced during a clear day for any distance in any direction at any season, they demonstrated a strong tendency to fly in one direction, namely, northwest. The flights were always short and seemed to be guided by the sun, but since they had no discernible purpose other than possibly escape, Mathews called the behaviour "nonsense orientation." He believed that it might be innate or perhaps developed at an early age.

Other birds may show the same phenomenon. Migratory Mallards wintering in Illinois, on being trapped and released during a clear day or night at points far west and west of their winter-home lakes, consistently flew northward regardless of the direction of their home-lakes. Common Terms (Sterna hirundo), displaced from their breeding colony on Penikese Island Off Cape Cod, Massachusetts, on a clear day, flew southeast, irrespective of where they were released or the direction of Penikese Island. Adeili Penguins (*Pygoscelis adeliae*), when displaced from their breeding colonies at Cape Crozier or the cost of Antarctica to the featureless expanse of compacted snow in the interior, oriented themselves by the sun, departed in one direction, and ultimately reached their nest sites.

The investigators, *J.T. Emlen* and *Penney*, used breeding males, some of whose internal clocks had been artificially shifted in advance. Releasing them individually from three groups, each at widely separated points, they observed that the birds headed straight for the coast on courses which were essentially parallel without any convergence toward Cape Crosier. The birds seemed to have no information in their courses from the sun and, as shown by the breakdown in orientation among those birds whose clocks had been shifted, maintained their courses by referring to the sun's azimuth. If heavy clouds obscured the sun and eliminated all shadows, their orientation soon deteriorated.

Emlen and Penney considered such one-directional courses to be fixed and to be maintained by an inherent time sense; they likened the procedure to nonsense orientation as described in waterfowl and other birds, but believed that it might be escape orientation with survival value by steering the penguins to off-coast feeding areas whence they would be guided, perhaps in part by familiar landmarks, to their home colonies. Later experiments by the two investigators confirmed their original findings and conclusions. A few years following Kramer's initial discovery that migratory restlessness in caged Starlings was directed toward the sun. Sauer and his wife, undertook experiments in Germany to determine by what means nocturnal migrants find their way. The *Sauers* used three species of sylviid warblers, reared in captivity without ever having seen the natural sky, during the period when the species would normally migrate, the Sauers put each bird in a rotatable circular cage and exposed it to the night sky. Under a clear, starry sky, even when their cages were turned periodically, the test birds fluttered persistently in the direction that the species takes at the season of the experiments; under heavily overcast skies the birds fluttered randomly in various directions, obviously disoriented. Then the Sauers placed the cage in a small planetarium with a starry sky, adjusted to the local time and latitude.

The birds fluttered in the direction appropriate to that season depicted. There seemed to be no doubt that these birds obtained their information for direction from the starry sky. The *Sauers* carried their experiments further. Placing the birds under skies representing the non-migratory seasons, they noticed that some of the birds were completely disoriented. From these experiments and others, Sauer concluded that birds do not rely on the stars themselves for direction but on the azimuth and altitude of the starry sky. This bicoordinate system would give birds the necessary information on their location since azimuth or hour angle would denote longitude and altitude would indicate latitude. Working with mist-netted Indigo Buntings in the United States, *S.T. Emlen* repeated many of the Sauers' experiments and made still other well-confirmed some of the Sauers' studies but differed sharply from the others. Emlen found no evidence that the buntings relied on a bicoordinate celestial system for orientation but rather made use of the numerous stars, particularly "the constant, two-dimensional spatial relationships" existing between them. He believed that no single star, with the possible exception of Polaris (the North Star) could give the birds sufficient information for direction.

It is the configuration or patterning of the bright stars in the constellations such as Ursa Major (the Big Dipper) that gives the cues. While Emlen could not determine with any certainty which patterns were Emlen could not determine with any certainty which patterns were of special importance or essential to orientation, the evidence seemed to point to the northern astral sky, especially the area within 35 degrees of Polaris. What prompts birds to take the appropriate seasonal direction? Seeking an answer to this question S.T. Emlen induced physiological readiness for spring and fall migration in two groups of Indigo Buntings and then tested them simultaneously under the spring sky of a planetarium.

The group conditioned for spring migration oriented northward: the group in condition for fall migration took the opposite direction. This clearly suggested to Emlen that the internal physiological changes in Indigo Buntings, rather than differences in external stimuli (e.g., in the night sky), are responsible for their direction in spring all fall. Practically all the experiments on the responses of homing and migrating birds to the sun and stars have demonstrated that when skies are heavily overcast the birds show disorientation. The consequent implication is that birds cannot find their way when celestial cues are obscured. Very recent studies, however, indicate that this may be untrue. *Keeton* has shown that homing pigeons, while using the sun as a compass when available, can navigate accurately under total overcast and even without familiar landmarks.

Radar studies reveal that birds migrate successfully when there is a heavy overcast. If the cloud layer is too high to surmount, birds will fly under it or even in it. This being the case, how do birds orient themselves without celestial cues and if they are flying at night or in a cloud layer, without being able to see landmarks or topographical features familiar to them? The cue may be wind direction. *Nisbet* once theorized that the birds might use wind–more specifically atmospheric or wind turbulence—in orientation. In central Illinois, Bellrose, with his associates, has gathered ample experimental evidence to show that Blue-winged Teal (*Anas discors*), a long-distance migrant that commonly winters in northwestern south America, does indeed use the turbulent structure of the wind for direction.

The species migrates by day or night, and radar surveillance and visual sightings by Bellrose confirm its ability to migrate under overcast skies. In the fall and early winter. Bellrose used both hand-raised immature teal that he released at various points up to 100 miles distant from their home pens and wild immature teal, trapped during migration, that he held in pens and released from 10 to 40 days after other Blue-winged Teal had left the region. All his experimental birds were banded. The hand-reared birds had never flown over the area and none of the birds had ever viewed the landscape along the standard migration routes southward. When released under both clear and overcast skies, nearly all the birds tended to start in the same direction–with the wind. Analyzing his data from visual sightings and banding recoveries and drawing upon his knowledge of waterfowl movements from radar studies and airplane tracking, Bellrose concluded that Blue-winged Teal, and probably other species of waterfowl as well, prefer landscape features for orientation, but, if unavailable, they resort to celestial cues. Both being unavailable, they use wind direction, perhaps referring at the outset of flight to landscape cues for information on the sector of the compass from which the wind is blowing. The present knowledge of direction-finding, the highlights of which are briefly reviewed in the foregoing paragraphs, points all to clearly to the danger of generalizing on the means of orientation and navigation in migration.

Different groups of birds in their adaptations to different modes of existence may well have developed correspondingly different means of finding their way from one place to another in accordance with the prevailing ecological conditions. It is unlikely that any species orients itself entirely by one cue. A one-directional orientation may suffice for some birds while a bicoordinate system of navigation may be necessary for others. Direction-finding, though appearing to be basically inherent, may actually prove through eventual research to be acquired chiefly be experience. In all probability, direction-finding in strongly migratory birds is a highly complex procedure involving physiological and behavioural responses to not one but several environmental factors—celestial bodies, wind, and perhaps others presently considered as having no relationship to the problem.

Study of Nocturnal Migration

There are at least five field methods of studying nocturnal migration. (1) By observations of movements across the face of the moon. The principal equipment needed is a telescope with a power of 15 or greater. For methods and procedures, consult *Lowery* and *Nisbet* and for the application of accumulated data. (2) By radio-tracking. The equipment needed includes a data. Transmitter and receiver. For general information on the type of equipment required, consult *Sclater*, *Cochran* and *Lord* and Southern. For information on a transmitter adapted to small migrants, as well as an example of methods and procedures in radio-tracking small migrants. (3) By radar surveillance. This requires the use of a special facility operated by skilled technicians. (4) By tape-recording. Besides a tape recorder, this requires a parabolic reflector, microphone, and amplifier. (5) By the study of migrants killed at television towers or migrants killed as a result of airport ceilometers or other man-made interferences with migration. Of the five listed methods, the most practical for the beginning student is the study of migrants killed at television towers. It requires no special equipment.

Practically every city or large community has on its outskirts one or more television towers that almost certainly take a toll of nocturnal migrants. Much as these kills are regretted, they nonetheless provide rewarding material for the study of many aspects of migration locally. An early-morning search of the ground under a television tower and its guy wires following a night during which migration proceeded at low altitude will usually yield birds of many species. If the student chooses to make such a search as often as these accidents occur through the migration season and on successive seasons, keeping a careful record with dates of the number of individuals of each species killed, he will obtain a true sampling of the nocturnal migratory activity in the area of the tower. He may also discover the presence of species not heretofore reported, or even suspected, in the area because they have always passed through at night unseen. Given the time necessary, the student may collect specimens for making one or more special studies such as the following:

1. Succession of sexes and age groups in different species: whether males precede females and adults precede immatures, or vice versa.
2. Geographic variation in a wide-ranging species migrating through a particular are: whether the colour and measurements of individuals reveal one or more populations or sub-species.
3. Molt with relation to migration: whether certain species are in stages of molt and, if so, which feather tracts are involved. Normally birds do not molt the remiges and their primary coverts during migration, but the rest of the alar tract and all the other

tracts may be in the process of molt, depending on the species or the sex and/or stage of maturity of individuals in each species.

4. Weight with relation to the length of night-flight: whether the birds are quite heavy, having been killed soon after take-off, or quite light, having been killed after a long flight.
5. Fat condition with relation to stage of migration and duration of night-flight: whether the birds are quite obsess, indicating that their migration was well under way with long night-flights, or quite lean, indicating that their migration was just beginning with short night-flights.

Before collecting any specimens, the student must obtain both a federal and a state (or provincial) collecting permit (see Appendix A) authorizing him to take dead birds "for salvage." Collect the specimens after a kill as early in the morning as possible since the carcasses not only decompose rapidly but are soon molested by insects or eaten by house cata, crows, gulls, and other creatures hat inevitably find the ground around a television tower a promising source of food. Weight each specimen without too much delay as the body begins losing weight shortly after death. Mark the weight on a tag and attach it to a leg. Seal all specimens collected in plastic bags to avoid their dehydration, mark with the date, time of day, and place of collection, and put at once in a deep-freezer for study and analysis later.

Study of Diurnal Migration

Whenever opportunity permits, the student should watch the diurnal movements of migrants in his own area. He will soon discover how greatly birds are influenced in their daytime movements by the terrain, vegetation, and waterways. He is almost certain to note that passerine migrants, although continually searching for food, follow much the same courses and in the same direction. If the birds inhabit trees or shrubs, they choose paths in which these grow and avoid, when possible, crossing wide stretches of water and open country. For an idea of how much detailed information one can obtain in a local study and some of the procedures to follow, the student is referred to the classic work by Ball, in which he reported on six fall migrations on a small point of land of the great Gaspe Peninsula, Quebec.

9

SOCIAL NEST-BUILDING

All receptacles for eggs laid by birds are called nests. In each bird species, nests are remarkably similar in form and location: among different species they show wide diversity.

DEVELOPMENT OF NESTS

Early in their acquisition of homoiothermy ("warm- bloodedness") birds could no longer abandon their eggs, as did their reptilian forebears, leaving them to hatch in the heat of the environment. The embryos within required the steady warmth of the parental body. This imposed on birds the necessity of incubation their eggs-sitting immobile on them for long periods of time. To offset what might well have been lethal exposure to predation and other adversities of the environment, birds simultaneously developed protective measures. One was to select sites for eggs that provided adequate cover and freedom from adversities; another was to build nests that would accommodate and shelter their eggs and themselves.

From the meager information available on the ancestry and descent of birds, and from what is known about nest-building habits today, one may assume that the earliest nests of birds were on the ground in depressions which the birds scraped out with the bill and feet and molded to the shape of their bodies by repeated turning. Some of the nests may also have been on the floors of natural cavities or cavities (including burrows, holes in trees) which the birds excavated themselves. All such nests were without structure although they may have been lined with materials- e.g., plant stems, leaves, bits of shell, etc.- gathered from the immediate vicinity, and sometimes with feathers from incubating bird's body.

Among modern birds one finds both ground and cavity nests still in abundant use. Ground nests, for instance, are characteristic of loons, pelicans, gannets, swans, geese, most ducks, grouse, quail, pheasants, shore birds, gulls, terns, murres, and goatsuckers. Probably the floating nests of grebes and the marsh nests of cranes, rails, gallinules, and coots are derived from ground nests. Most birds with ground nests either have cryptic coloration to conceal them while they are on their nests, or place their nests on islands or in other areas generally inaccessible to predators. Cavity nests are typical of petrels, a few ducks such as golden eyes and mergansers, most vultures, puffins, auklets, parrots, trogons, kingfishers, and woodpeckers.

The earliest nests to be elevated in vegetation- shrubs, trees, and even marsh plants- were essentially platforms of loosely assembled plant materials, without evident structure, but with a shallow depression for the eggs. Platforms nests, perhaps representing the second stage in nest- building, differed principally from the first stage, ground and cavity nests, in being independent of a uniformly firm surface and comprised entirely of accumulated materials for holding the eggs as well as for lining the nest.

Modern birds which build platform nests include the anhingas, herons, storks, ibises, pigeons, and cuckoos. Platform nests are also built by cormorants, hawk, eagles, and the Osprey (*Pandion haliaetus*), but not always in trees. Some cormorants place their nests consistently on cliffs. A few species of hawks, both the Golden and Bald Eagles (*Aquila chrysaetos* and *Haliaeetus leucocephalus*), and the Osprey may place their huge nests either in trees (including stubs), on cliffs, or even (in the case of the Osprey) on the ground. The Marsh Hawk (Circus cyaneus) nests regularly on the ground without building a platform.

The third and final stage of nest- building was the development of cupped nests, with true structure, consisting of materials arranged and compacted for the bottom and sides and softer materials inside for lining.

In building cupped nests, birds succeeded in directly adapting structures to shrubs and trees from the crotches to the tips of branches. Presumably the first cupped nests were statant, supported mainly from below, with the rims standing firmly upright. The majority of hummingbirds and passerine species today build statant nests though a few species such as magpies have modified them by extending the sides upward and arching over the top in a dome.

Eventually, a number of species suspended their cupped nests from branches by rims and sides, without supporting them from below. These nests took two forms: (1) Pensile nests, as built by modern vireos, suspended from stiffly woven rims and sides. (2) Pendulous nests, as built by modern orioles, suspended from rims and flexibly woven sides, with the deeply cupped lower parts swinging freely.

In the rapid multiplication of passerine species and the consequent increase in competition for living space, many new species moved into heretofore unoccupied niches of the environment and re- adapted cupped nests to suite the particular situations other than the crotches and branches of trees. Such birds as phoebes and several swallows, using adhesive substances, built their nests adherent to cliff walls and similarly vertical surfaces. Larks, pipits, several thrushes, and numerous parulid warblers, icterid black birds, and fringillids constructed cupped nests on the ground in forests, in open country, or in marsh vegetation. Titmice, nutchatches, creepers, bluebirds, several swallows and wrens, and the Prothonotary Warbler (*Protonotaria citrea*) put their cupped nests in preformed cavities, or in rare cases- e.g., the Bank Swallow (*Riparia riparia*)- in cavities which they excavated themselves.

The nests of several groups of modern birds defy categorizing as to stages of development. For example, owls rarely, if ever, build nests of any sort. Unless they nest on the ground as do Snowy and Short-eared Owls (*Nyctea scandiaca* and *Asio flammeus*), they appropriate tree nests of other birds or use cavities. Chimney Swifts (*Chaetura pelagica*) construct "half-cupped" nests on vertical surfaces. To make the nesting materials stick together and the nests adhere to vertical surfaces, the birds apply their own saliva. Other species of swifts show wide diversity in nests and the use of salvia in forming them – e.g., the cave swiftlets (*Collocalia* spp.) of Asia form their nests almost entirely of coagulated saliva. In eastern Indonesia, Polynesia, New Guinea, and Australia, gallinaceous birds called megapodes (Megapodiidae) dig pits in the soil or build mounds of rotting vegetable matter in which they lay their eggs and then cover them leaving them to be incubated by the surrounding heat. The mounds of one species (*Megapodius freycinet*) may measure up to 60 feet long, 15 feet wide, and 10 feet high–probably the largest "nests" of any bird.

Classification of Nests

It is possible to set up a classification of nests indicating their probable course of development and illustrating the diversity of nest types. Such a classification for North American birds is presented below, with spaces for the names of species showing each nest type. Bear in mind that the classification is purely artificial and does not necessarily portray the phylogenetic relationships of the nest builders.

Stage I. **Ground and Cavity Nests.** Nests without structure.

A. Ground Nests. Simple depressions with or without lining. Birds with ground nests either have cryptic coloration or select nest sites inaccessible to predators.

____________________	____________________
____________________	____________________
____________________	____________________
____________________	____________________

B. Cavity Nests. Nests in caves, crevices, burrows, holes in trees, or bird boxes, with or without lining.

1. In preformed or natural cavities.

____________________	____________________
____________________	____________________
____________________	____________________
____________________	____________________

2. In cavities excavated by occupant birds.

Stage II. **Platform Nests.** Nests elevated, without structure, consisting of loosely assembled materials with a shallow depression for the eggs.

____________________	____________________
____________________	____________________
____________________	____________________
____________________	____________________

Stage III. **Cupped Nests.** Nests adapted to crotches and branches of trees, with definite structure, the materials arranged and compacted to form a cup.

A. Used in crotches and branches of trees to which they are adapted.

1. **Statant Cupped Nests.** Nests with rims standing firmly upright, supported mainly from below. The sides may or may not be extended upward and arched over the top in a dome.

____________________	____________________
____________________	____________________
____________________	____________________
____________________	____________________

2. **Suspended Cupped Nests.** Nests not supported from below but from the rims, or sides, or both.
   a. **Pensile**. Nests suspended from the rims and sides; rather stiff.

      ______________ ______________

      ______________ ______________

   b. **Pendulous.** Nests suspended from the rims and sides; rather flexible and extremely deep-cupped with lower part swinging freely.

      ______________ ______________

      ______________ ______________

B. Re-adapted to other situations.
1. **Adherent Nests.** Cupped nests whose sides are attached by adhesive substance to a vertical surface.

   ______________ ______________

   ______________ ______________

   ______________ ______________

   ______________ ______________

2. **Ground Nests.** Cupped nests on the ground. The sides may or may not be extended upward and arched over the top, making a domed structure.

   ______________ ______________

   ______________ ______________

   ______________ ______________

   ______________ ______________

3. Cavity Nests. Cupped nests in crevices, holes, bird boxes, etc.
   a. In preformed or natural cavities.

      ______________ ______________

      ______________ ______________

      ______________ ______________

      ______________ ______________

   b. In cavities excavated by occupant birds.

      ______________ ______________

      ______________ ______________

      ______________ ______________

Identification of Nests

Because nests of many closely allied species are very similar in location, materials, and structure, the student must be careful of identifying any nest, with complete certainty, without knowing the bird that constructed it. But nest of certain species, or groups of closely allied species, are sufficiently distinctive to allow identification with reasonable certainty. The student attempting to identify nests in the conterminous United States is referred to the keys to nests by Headstrom (1949, 1951). Also a good reference is the work by Campbell (1953). Although it deals with British birds, the introductory chapters suggest methods and techniques for finding and studying nests.

Nest-Building

The various phases of nest-building have been investigated the least of any subject In the breeding cycle of birds. The author's own studies and the works of many others have contributed to he outline of nest-build procedures given below.

Selection of the Nesting Site

The need of suitable support and protection governs the selection of the nesting site. Birds do not deliberately select the site to conceal the nest, although frequently concealment results from the placing of the nest in a position, protected from the destructive forces of the environment (e.g., sunlight, wind, cool night temperatures).

Almost invariably, a period of appetitive searching during which the birds move from one potential site to another - usually within the confines of their established territory - precedes the final selection of a nesting site. In species that building cupped nests in trees, the birds try fitting their bodies into crotches between branches; in species building the same type of nests on the ground, the birds scratch small depressions in the surface and attempt to mold them to suit the body contours. Often they accumulate a few nesting materials, and occasionally they partly construct nests at several sites before making the final selection. The searching for a nest site is first strongly manifest after mating when the males are singing frequently and intensively and the females are receptive to males. They defend the territory strongly during this period. Nest-searching activities frequently appear following each sexual relation. The number of copulations during the days of the searching period is unknown.

Neither has the duration of the searching period been satisfactorily determined. In some cases it lasts from three to five days; in a few cases it is more prolonged. Searching, by no means continuous during the period, occurs at intervals, especially during the early hours of the morning. These intervals of searching increase in length daily as the nest-building drive matures.

The role of each sex in searching for and selecting the nesting site varies with the species. In birds maintaining territories of Types. A, B, and D, one or both sexes participate. When both sexes participate, generally the female takes the more aggressive role. In birds with territories of Type C, the female alone searches for and selects the nesting site.

The searching period represents the phase in the breeding cycle during which the nest-building drive develops. Even though birds find suitable nesting sites during this period, the selection is not final. They select the nesting site when their arrival on a suitable site coincides with the blocking of the nest-building drive by consummatory behavior.

Beginning of Nest-building

In many birds nest-building begins with the initiation of rapid growth in the ovum and continues more or less intensively while the ovum matures. The first egg is usually laid one or several days after the completion of the nest, but in a few species nest-building continues to some extend after laying has begun. Climatic conditions may influence the beginning of nest-building. High temperatures at the start of the nesting season stimulate nest-building; low temperatures tend to inhibit it. Delayed development of vegetation may delay the building.

Process of Nest-building

The process of nest-building may be roughly divided into three steps: (1) Preparing the site or support (e.g., scratching a depression, "cleaning out" a preformed cavity, excavating a new burrow or hole). (2) Constructing the floor and sides (i.e., the "outside"). (3) Lining the nest. One or two of the stages do not occur, or occur only in part, in certain species.

Usually, the birds gather all "outside" materials for nests of Stages II and III in the vicinity. For nests of State I, the materials are frequently those within reach of he bird as it stands on the nest. But materials for lining the nests of States II and III are often sought at great distances, presumably because the special materials are not readily available close at hand. Tree Swallows (*Iridoprocne bicolor*) may sometimes fly several miles to a chicken farm to obtain the much-preferred white feathers.

Most birds carry materials in their bill, although diurnal birds of prey - eagles, hawks, etc. - use their feet more commonly. Several species of African lovebirds (*Agapornis*) carry nesting materials in their rump feathers.

Each bird builds its nest with a number of stereotyped movements. If the nest is to have structure, the movements required are necessarily more numerous and complex. In constructing a typical cupped statant nest in a tree, the passerine bird sits in the center of the cup that is taking shape and performs the following characteristic movements: Pulling a long slender piece of material (e.g., a plant fiber) with its bill inward over the rim and *tucking* it into the wall under its breast or *drawing* it alongside against the wall; *looping* a similar piece of material around a supporting branch by starting it around one side and then *reaching* around the opposite side and *drawing* it back so that it encircles the branch - and sometimes repeating the performance until the material encircles the branch several times; *inserting* short pieces of material into the rim or wall with jabs of he bill; shaping the cup by *squatting* in it while alternately *fluffing* and *compressing* the body feathers; *pressing* its head, tail, and partly opened wings down against the rim of he cup, while alternately *pushing* with one foot and then the other against the floor and sides of the cup; *turning* frequently between and sometimes during any of the above movements.

Learning plays a role in nest-building to the extent that a bird must determine, for instance, by trial and error (see the section, "Behavior.") which materials are the most suitable for the structure and the lining. The principal movements in nest-building, however, are innate, derived and then stereotyped from such sources as maintenance activities or even irrelevant behaviors. To cite one example: Harrison (1967) has suggested that the behavior called "sideways-building," in which the bird pulls materials into its nest and draws it alongside, is derived from 'sideways-throwing," an irrelevant behavior common in a number of shore birds and gulls when the birds picks up a leaf, stick, shell, or any other small object and flicks it backward on one side with a quick motion of the head.

Many more detailed studies of nest-building in different groups of birds must be made before the origin and homologies of all the movements can be established. For some of the few detailed studies available to date, the student is referred to the works by Armstrong (1955), Herrick (1911, 1935), Marler (1956), Tinbergen (1939, 1953), and others.

Participation and Behaviour of the Sexes

In a great many species the female builds the nest alone. In many others the female builds the nest with the assistance of the male. The female gathers the materials, works them into place, and molds the depression. The male's assistance usually amounts to gathering materials with or without the female and, on returning with them, passing them to the female for use in the nest. In a number of species the role of the sexes in nest-building does not conform to the above. The male may (1) build the entire nest, although this is rare; (2) build certain parts of it, such as the floor and sides, or excavate the burrow or hole; (3) share all nest-building activities with the female. Within species, the role of the sexes is subject to variation, but the variation is seldom as extreme as that between species.

There is apparently no dependable correlation among species between coloration of the male and his nest-building proclivities, nor a correlation among species between coloration of the male and his nest-building proclivities, nor a correlation between his nest-building proclivities and his part in subsequent incubation. There is, on the other hand, a rather positive correlation between a male's nest-building proclivities and his participation in subsequent parental care, i.e., if a male assists in building, he will also assist in feeding the young.

During nest-building the male customarily sings as vigorously as at any other time during the breeding cycle. Copulation occurs throughout nest-building, though the frequency is not known.

Length of Time Involved

The total length of time involved in nest-building is difficult to determine because seldom is the beginning observed. Nests, because they are constructed in protected sites, are usually discovered as a result of building activities already under way. Furthermore, the cessation is sometimes indeterminable. Nest-building may stop suddenly a day or more before egg-laying, or it may continue for some time after egg-laying has begun. The *number of days* is used as a measure of he time involved in the construction of nests. Due to the irregularities of nest-building in most species, no more accurate measure can be applied unless the observer has sufficient persistence and good fortune to follow daily activities from start to finish, in which case he can count the total *number of hours* involved.

Remarkably few records are available to show even the number of days involved in nest-building. Most passerine birds may require about six, three for constructing the outside of the nest and three more for finishing the interior and lining it. But judging by the few precise records at hand, the length of time is subject to wide inter specific and intra specific variation.

Inter specific variation is accounted for in part by two factors; (1) Type of Nest. Obviously, certain nests that are more elaborate than others require more time for construction. To build their long, pendulous nests, the Alta Mira or Lichtenstein's Oriole (*Icterus gularis*) in Mexico took "at least 18 and perhaps as many as 26 days" and the Wagler's Oropendola (*Zarhynchus wagleri*) in the Canal Zone, Panama, "about one month". (2) Climate. Species in the tropics take more time than closely allied species in the temperate regions. Derby of Kiskadee Flycatchers (*Pitangus sulphuratus*) in Mexico took 24 days whereas most tyrannid flycatchers farther north need only three to rarely more than 13 days for finishing their nests.

Two factors account in part for the intra specific variation in time required for nest-building: (1) Weather Condition. Cool or inclement weather may retard nest-building. On the other hand, rainfall may stimulate it, at least in the case of the European Wren (*Troglodytes troglodytes*) because it makes the nesting material more flexible and easier to manipulate. (2) Renesting. An individual, or pair, building a second or third nest in the same season takes less time.

Though the length of time taken in nest-building is measured in number of days, nest-building does not proceed steadily during those days. Usually only the early parts of the day are involved, at which time there are periods of building (i.e., attentive periods) and periods of no building (i.e., inattentive periods). As nest–building progresses, the attentive periods lengthen, and inattentive periods shorten. At the height of nest-building the average length of attentive and inattentive periods often quite closely corresponds to the average length of such periods during incubation.

A few birds modify or repair their nests after they have laid eggs and begun incubation, or even later. Bald Eagles (*Haliaeetus leucocephalus*) frequently add materials to their large aeries throughout the nesting season. The marsh-dwelling rails, gallinules, and coots are noted for their ability to build up their nests, whenever the water rises, to keep the flood from their eggs. The king Rail (*Rallus elegans*) has been reported to elevate its eggs in this way by as much as a foot.

Nest-building in Young Birds

The nest-building abilities of young birds need investigation. From meager evidence it appears that younger birds may build their nests as quickly and expertly as older birds do (Nice. 1943). This is perhaps not true among species that build elaborate nests.

In rare instances, young from a nest built early in a season assist adults in building a second nest later in the season. Independent attempts nest nest-building have been seen among young birds-of-the-year.

False Nests

Certain species of birds build structures called "nests" which are not true nests, because they are not constructed for the purpose of containing eggs.

Cock "Nests"

Sometimes called "dummy nests." These are nest constructed by males, notably wrens, in the vicinity of the regular nest. They resemble the regular nest, but they are not completely constructed and are without lining. Herrick (1935) suggests that such nests "are simply the work of male birds in response to an inherited predisposition to this sort of activity while singing and waiting upon their chosen territory for a suitable mate, or even after, when such a mate is engaged in the prosaic task of incubation."

Refuge "Nests"

A few birds, especially burrow and hole-nesters, often create "nests," similar to their regular nests, for roosting and for shelter during unfavorable weather. They may be constructed at any time of year but particularly in the fall. Some birds merely use cavities already available.

Re-use of Nests

A few species of birds, notably some of the large hawks, the Golden and Bald Eagles, and the Osprey, use the same nest year after year. When the same nest is used, it is generally either "repaired" and enlarged each year, or a completely new nest is built on top of the old.

Certain species appropriate the nests used by other species in the preceding season. Sometimes the new tenants alter the nests to suit their own requirements. A Mourning Dove (Zenaidura macroura), for instance, may build its platform nest on top of the cupped nest of a Robin (Turdus migratorius). Or the new tenants may accept the nest as it is. This is true of a Great Horned Owl (*Bubo virginaus*) taking over a hawk or eagle nest, and of the Solitary Sandpiper (*Tringa solitaria*) using the cupped tree nest of a Robin, Rusty Blackbird (Euphayus carolinus), or Eastern Kingbird (*Tyrannus tyrannus*).

Among passerine birds, as well as among many other groups, it is exceptional for pairs to renest in the same nest or at the same nesting site.

Protection of Nests

Nests are often concealed through the choice of nesting sites. If they are not concealed, then they may be inaccessible as when nests are placed on small islands, on the shelves of cliffs, and near the ends of slender branches. They theory that snake skins, sometimes found in nests, serve as a means of frightening enemies has been discredited.

Nests may gain protection against predators when they are placed close together in a colony, or singly in a colony of other species, because there are many more occupant birds on the alert to attack and take deterrent action. Nests in tropical regions also gain protection when they are placed - deliberately by some species - in the immediate vicinity of aggressive social ants and wasps, or near the nests of larger, more aggressive bird species. In Mexico, Pettingill (1942) found one nest each of the Social or Vermilion-crowned Flycatcher (*Myiozetetes similes*) and the larger kiskadee Flycatcher in a bull's horn acacia, a shrub about 12 feet high, that was tenanted by countless thousands of small ants ready to bite and sting. Although the ants did not annoy the nest occupants, they viciously swarmed over anyone touching the shrub. Both nests obviously benefited from the protection afforded by the bellicose ants while the nest of the Social Flycatcher benefited further by being close to the nest of the larger and more belligerent Kiskadee Flycatcher.

Nest Fauna

Nests of birds, particularly those in burrows, cavities, or bird boxes, or under the eaves of buildings, are snug havens for many small invertebrates, especially arthropods - mites, spiders, insects, etc. Most of hem are visitors stopping only for shelter. (For an idea of the variety of insects that may be found in birds' nests, see the check-list by Hicks, 1959). But along with the visitors are a number of flies and fleas that pass their life cycles specifically in birds nests and at one stage are parasitic on the birds. They are to be distinguished from the obligate ectoparasites on the feathers and/or other parts of birds.

Chief among the nest parasites, from the viewpoint of their serious effect on the occupant birds, are blowflies (*Protocalliphora* spp.) whose larvae live in the nest cup and, beginning at twilight, attach themselves on the nestlings for a blood meal during the night. If the nest infestation by larvae is heavy, the nestlings can be greatly weakened, even killed, from loss of blood.

THE STUDY OF NESTS AND NEST-BUILDING

Select a species for he study of nests and nest-building and concentrate on the activities of one pair.

Procedure

When possible, begin the study soon after the territory is established, thus observing the selection of the nesting site and the start of nest-building. Follow all activities of the pair through the searching and nest-building periods. Practice extreme caution in watching the birds during he nest-building, since they are not strongly attached to the nesting site and will abandon it on the slightest provocation. Therefore, use a binocular or telescope and remain as far away as possible.

Take full notes on the activities of both sexes; time their respective attentive and inattentive periods, and the periods when the male is singing. When the birds are not present at the nest, take photographs of stages in construction, and make careful notes on the appearance of the nest at each stage.

Obtain full data on the completed nest, particularly the following:

Measurements

If ground nest without structure: inside diameter (i.e., diameter of the depression); inside depth (i.e., depth of the depression). If cavity nest without structure: length, or depth, of cavity from lower edge of entrance to floor; diameter (or diameters) of the entrance and diameter (or diameters) of the part of cavity containing the eggs. If elevated nest or re-adapted ground nest: outside diameter, outside depth; inside diameter, inside depth. (If nests are domed, consider outside depth to be from top of arch, and inside depth to be from the under surface of arch). If re-adapted cavity nest: use the measurements indicated above for directly adaptive cavity nests, plus the measurements of the inside diameter of the nest structure built in the cavity.

Location

Take notes on the location and support of the nest, naming all dominant vegetation in the vicinity and observing the condition and protective value. If the nest is in a tree, measure its height is too great fro accurate determination by measuring stick or tape. Also measure or estimate the distance of the nest from the lower edge of the entrance to the ground and note the direction (using a compass) that the entrance faces. Photograph the environments, with the nest occupying a central position in the background.

Description of the Nest

Take two or three photographs that show the support of the nest and the appearance of the nest close up. Disturb the surrounding cover as little as possible. When taking the close-up pictures of an open nest, place the camera partly above and partly to the side so that the far inside wall and near outside wall will show.

After the young have left, the nest should be taken to the laboratory and studied in detail. Its structure should be analyzed, all materials identified, and the relative quantity of each material determined. If, after a nest is completed, it is for some reason destroyed or deserted, attempt to follow the same pair and gather evidence of renesting. The following data are particularly desirable: (1) Time involved between the destruction or desertion of the first nest and the completion of the second. (2) variation among individuals in the selection of the second nesting site and manner of building. (3) Proximity of the first and second nests. Compare the data with the findings of other investigators.

Presentation of Results

Draw up a report on the nests and nest-building of the pair observed. Prepare the manuscript as directed in Appendix BG. An outline of suggested topics is given below.

Territory

Type; brief description; date discovered.

Selection of Nesting Site

Time and date of searching period; relation to weather conditions; role and activities of sexes, including number and duration of singing periods of male and number of copulatory acts per day; number and description of nest-building attempts; time of day and total length of time (in hours) devoted to searching; searching as related to territory and territorial behavior.

Beginning and Duration of Nest-building

Brief statements of date of beginning and the duration. Discussion of weather, vegetation, and other ecological conditions as related to the time of beginning.

Nest-building

Detailed chronological account of the stages and mechanics of nest-building. This should include the role of the sexes, number of copulations per day, times when the male sings, manner of gathering materials, sources of materials, and number of material gathering trips. Discuss the length of time involved separately. Include the hours of the day when nest-building takes place, a table summarizing attentive and inattentive periods, and statements indicating the total amount of time spent in building the nest. (In deriving the table, follow the directions for the table on incubation periods given in the next section of this book). Use photographs to illustrate stages in construction.

Description of Nest

Location: Describe in full, and give height from ground, if elevated. Use photographs to illustrate the nesting site. Discuss protective factors, etc. Measurements: include the measurements as directed above. Structure: Note any details of structure not determined when watching the nest under construction. Analyze the materials, using a wheel diagram to show relative quantities of different materials. Include photographs of the nest.

Miscellaneous Observations

Present any observations made on renesting, building ability of younger birds, presence of false nests, protection afforded the nest, and nest fauna.

10

Group Movement in Fishes

Among the various interesting peculiarities of fishes, migration is perhaps the most remarkable and fascinating. The word "Migration" has been derived from the Latin word (*Migrare*, meaning to go from one place to another) and is used to represent long journeys undertaken particularly by animals such as fishes and birds. The term "migration" is frequently employed to express three different kinds of movements. In the early stages of life the eggs and fry may be carried by the currents some distance form the place where they were born. This helpless drift is at first involuntary, but later the young fry are quite able to make their way in and out of currents or remain stationary. Then the young fishes and the adults spend most of their time seeking food; they scatter about within their area; but good feeding grounds or incoming currents rich in plankton food bring may together in a concentration or shoal which persists for some time. These may be called voluntary movements on the part of the fish. Not all fishes are migratory.

In fact, a majority of them restrict their movements to small territorial limits and do not go out of these during any part of their lives. Few species, however, travel over long distances for feeding and spawning either in the same type of water to which they belong or in search of water naturally different than their own, e.g., from freshwater to sea water or *vice versa*. Rightly considered, the true migration is something very different from the drift and wanderings. Here the fish is urged to leave its ordinary hunting grounds and even to stop feeding, in order to undertake sometimes long and perilous journeys. The call of the blood, a deeper necessity common to all living things, takes possession of the fish, and the thousands and millions in like case forgather year by year almost at that same place and time—a circumstance of great importance both in the past and in the present. Most of the great fisheries are based on the spawning migrations of fishes, and at the present day we can observe how fishing population move from place to place and even from one country to another in pursuit of their calling.

History tells us also, how the New Englanders would hardly have surmounted the hardships they encountered in the first years of their endeavours to settle in America but for the rich fisheries of the coast; and again, how the Dutch fishing population spread form one country to another in the seventeenth century, and how England and Holland went to war on a small matter of the fishing industry. In ancient times also the Greek and Phoenician colonists chose their settlements where the fishing was of importance, and the nature of the fishing is sufficiently indicated by the representation of the Tunny on the old medals of Carthage and Cadiz. These under-currents in the life of nations are not apparent in the surface histories, yet *Edward Forbes* was justified in his quaint comment, "the existence of Alexander may have been determined by the migration of a shell-fish." It is a question whether the leading or the feeding of peoples is of the greater importance. *Heape* (1931) has defined migration as "a class of movement which impels migrants to return on the region from which they have migrated."

Migration can be said to typed as following:

(i) *Climatic Migration.* For securing suitable climatic condition.

(ii) *Alimental Migration.* For the need of food and water.

(iii) *Osmoregulatory Migration.* For better watery conditions to live as far osmoregulation is concerned.

(iv) *Gametic Migration.* For reproduction and development.

A fish can make migratory movements by several methods.

A. By Drifting

The fishes are carried passively by water currents. This may result in 'directional movement' if the overall water movement is in one direction.

B. Random Locomotory Movements

Random movements in any direction may lead to distribution or aggregation. If fishes move uniformly from one point to all direction, it is called *dispersal*, which results in distribution.

C. Oriented Swimming Movements

The fishes swim in a particular direction (I) either towards or away from the source of stimulation or (ii) at some angle to an imaginary line between them ad point of stimulation. The migration pattern is believed to be related to the currents. The young ones drift with the water current to the nursery grounds. The spawning migration is against with the current. Two terms are used to describe the movements of fishes in relation to water currents. Swimming or migration along with the direction of current is called *denatant* and swimming or migrating against the water current, and the migration of the adult fishes towards the spawning area is called *contranatant*.

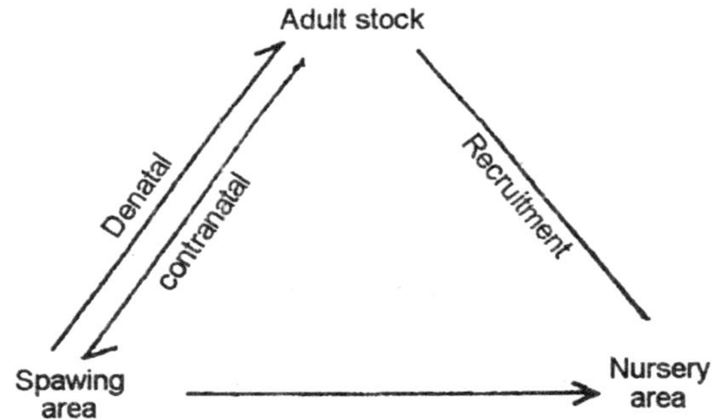

Fig. 10.1. Pattern of fish migration.

Types of Migratory Fishes

(i) *Anadromous*. Those fishes which spend a major part of their lives in the sea but migrate to fresh water during breeding season for spawning. *Salmon* and *Hilsa* have been found to travel several thousand miles in the sea and then several hundred miles inland to reach the spawning grounds.

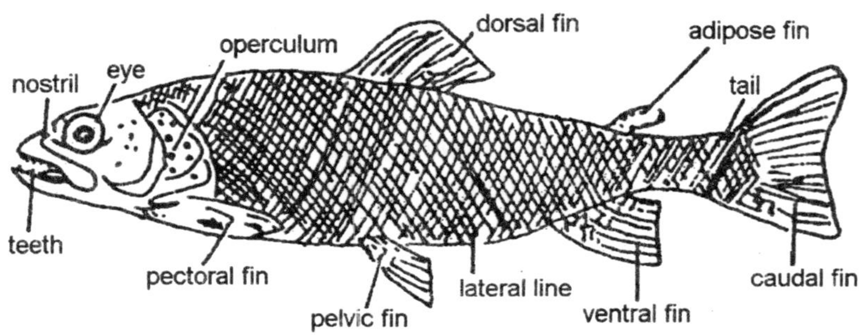

Fig. 10.2. Salmon.

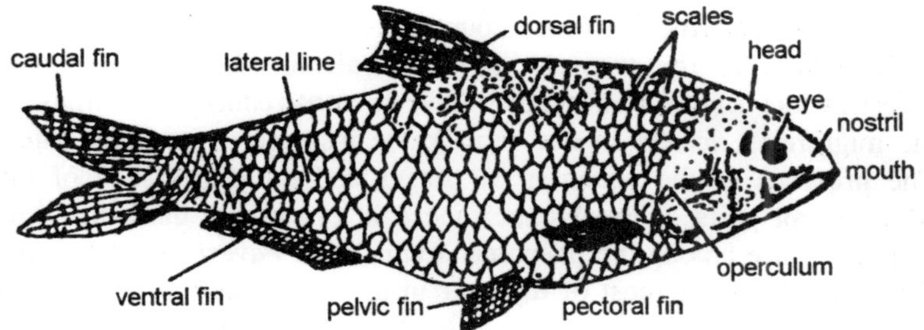

Fig. 10.3. Hilsa.

(ii) *Catadromous*. Those fishes which spend a major part of their lives in freshwater but migrate to the sea for breeding purpose. *Anguilla* travels several thousand miles starting from the rivers and reaching the spawning grounds in sea.

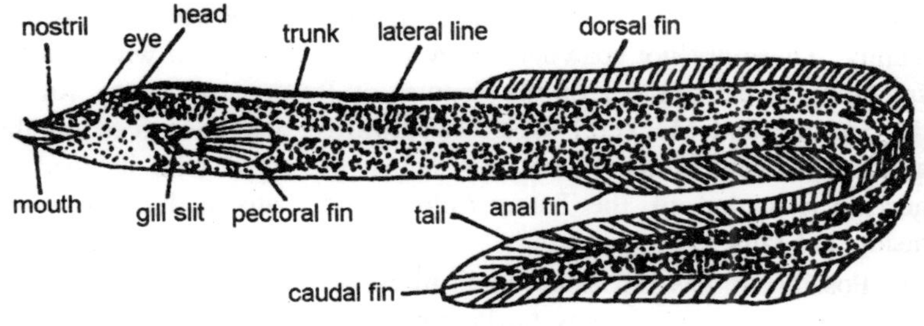

Fig. 10.4. Anguilla.

(iii) *Amphidromous*. Those fishes which migrate from fresh-water to the sea, or vice versa, not for breeding but regularly at some other stage of the life cycle.

Above three types are together called *Diadromous fishes*.

(iv) *Potamodromous*. Truly migratory fishes whose migration is confined to freshwater only e.g. *Clupea*, *Scomber*.

(v) *Oceanodromous*. Truly migratory fishes which live and migrate in the sea, e.g. Clupea, Scomber.

Migration of Different Fishes

Aristotle and the ancient writers presumably echoed the common belief of those days in saying, that the Tunny migrated from the Atlantic into the Mediterranean, following the current along the north coast of Africa of Egypt and Palestine and thence into the Black Sea. It was supposed to spawn in the Sea of Azov, thence returning along the northern coasts of the Mediterranean out into the Atlantic. Such wide-reaching ideas with regard to the migrations of fishes are the common property of fishermen of every country, for other fishes as well; it is their tribute to the mystery of their calling. But the scientific investigations of the past century have greatly reduced the limits of the migrations, except in one remarkable instance to be noted later. The presence of the young in all stages and the appearance of the adults in all parts of the Mediterranean at practically the same time (summer), are among the phenomena which led *Pavesi* to believe that the Tunny lives for most of the year in the deep water of the coasts.

As the spawning time approaches, the fish first all make a vertical migration, from the depths to the surface. It is said that new arrivals in the surface waters can be distinguished from the earlier by their darker colour. Then they move towards the shallower waters to spawn. It is the latter, the horizontal movement, which has given rise to the idea of immigration from the Atlantic and which accounts for the elaborate preparations made by the fishermen for the reception of the Tunny. These are not spawners; they are the larger fish which, after spawning in the summer on the Spanish or Portuguese coasts, follow the Sardine and Mackerel shoals northwards, and later feed on the autumn Herring of the North Sea. When the trawl with its load of herring is hauled up, the Tunny follows and snaps at the protruding heads and tails.

Possibly it was just as common in earlier years; there is a record by *Schonevelde* that it was fairly common near Eckernforde in the Western Baltic as far back as the beginning of he seventeenth century, also in connection with the presence of Mackerel. The Porpoise may often have been blamed for its depredations, but until the trawl was used for the capture of Herrings in large quantities, its occurrence was regarded as a rarity. A curious thing about the Herring, not known in any other species, is its ability to spawn at all seasons of the year ad in fresh as well as salt water, thus under the most diverse conditions as to temperature and salinity. Its variability in this respect suggests that from the beginning it has been a shallow-water or brackish-water form; it can readily adapt itself to any conditions from 0 down to 100 fathoms.

There are winter Herrings which spawn in the Firth of Forth and in the west; spring Herring in Norway, in the Schley, and north of Scotland; summer Herring off the north-east coast; autumn Herring spawning in the southern North Sea, and so on. But the Herring keeps to the temperate zone; in the tropical and subtropical regions the variability has changed it into other species. Another interesting problem in distribution is furnished by its relative, the Anchovy. This is a southern, subtropical form, with its main spawning places in the Mediterranean and adjacent waters. Yet a northern branch appears regularly year by year, and that in great quantities, in the Zuidersee and neighbouring brackish waters. In southern regions the species spawns in water of comparatively high temperature and salinity; here the water is brackish and cold. There is a wide gap between Spain and Zuidersee; few specimens have been taken in the Channel or Bay of Biscay. Are we to suppose that it migrates this distance yearly, or that it remains somewhere in the Channel, where it eludes capture?

In either case the origin of this branch remains a mystery; possibly it represents the original form of Anchovy, as we can more readily understand a migration to the pleasanter conditions of southern waters than the reverse. On the other hand, this particular branch may have suffered greatly from its enemies, Mackerel, Tunny, and others, and found a safe retreat from their pursuit in the Zuidersee. But no definite solution can be given to such a problem; of greater interest would be the investigation of the structural and developmental differences between forms of the same species growing up under such wide differences. The principal thing about the species so far mentioned is, that the migrations are from deeper to shallower water in the sea.

Some Clupeids, like the Shads, (*C. finta*, *alosa*, *sapidissima*, etc.), go further and ascend rivers to spawn in fresh water. Some are permanent inhabitants of the inland seas (Caspian, North Italian Lakes, and Great Lakes of America). The same phenomena are found among the nearly allied Salmonoids. Whilst some forms, like Argentina, spawn in deep water in the open sea, and others, like Mallotus (the Capelan), come close inshore to spawn, thus constituting a most important source of food, for example, for the Eskimos of Greenland, others like the Smelt enter brackish and fresh water, and the principal species of the group ascend rivers right to their sources, whilst whole families belong only to the fresh water. The *Salmon*, it is now believed, does not move far from the mouth of the river to which it belongs; yet like the Anchovy, it is rarely found in the sea except where it assembles to feed, as in the straits of the Pacific and in the Southern or Southeastern Baltic.

The latter case is of near interest; the Salmon are taken on hooks, but some escape and are caught later, with the hooks still attached, in the lakes of Finland. The Salmon is thus able to spend its whole life in fresh or nearly fresh water. In the Rhine again, when it was a clean river, the winter salmon used to remain nearly a whole year in the middle reaches below Basel before proceeding to spawn. In sounds almost incredible, therefore, that the Salmon does not feed after it has entered fresh water. Yet the evidence from all countries supports this view, and further, the digestive organs undergo a certain amount of degeneration. How then can we explain that the *Salmon* rises to the fly, and his even been taken with a Trout in its stomach? For one thing, it is certain that the *Salmon* of different regions, even different rivers, vary greatly in their powers and habits; for another, we have only to take a wider perspective to understand what happens.

In the larger Norwegian rivers the *Salmon* rise to the fly in the lower reaches, but provide no sport in the upper. In the still longer American waters the Salmon provide no sport at all. They feed, therefore, or endeavour to feed only when fresh from the sea. All the time they are in fresh water, it may be a whole year, they live on the materials stored up in their body. This has been considered a wonderful provision or adaptation of nature. Regarded from the Salmon's point of view. It seems more like a tragedy; the *Salmon* whishes to feed but cannot! Certainly its body must be an accumulator of energy of very good quality and enormous power. This fish ascends rivers against strongly running water, jumps numerous falls (probably not more than 10 ft. at a single leap however); if a male it has to fight other males and keep off enemies, then help to prepare a bed in which the female lays its eggs.

It is not to be wondered at, when the spawning is over, that the fish is completely exhausted and allows itself to drift tail-first down the steam. The Atlantic salmons spwan mainly during November and December. They ascend the rivers as the breeding period approaches, travelling several thousand miles in sea and then inland. On entering freshwater they stop feeding and lose weight. Their bright silvery colour changes to a dull reddish-brown shade and the skin becomes thick and spongy. The body of the male salmon becomes spotted with red, orange and large black spots. After selecting suitable spawning grounds the fish segregate into pairs.

Shallow saucer-like depressions are prepared in the river bed by the females where spawning takes place. After spawning, the return journey to the sea begins but few males survive to breed a second time. However, many females are able to reach the sea and start feeding. They soon recover their normal condition and silvery colour. The salmon does not usually spawn more than three times in its life-span of eight or nine years. From the time it reaches freshwater till spawning, the salmon lives on its reserve food accumulated in its beop in the form of fat. The king *salmon* of the Pacific coast of North America travels about 2250 miles from the sea in order to reach the rivers to spawn. However, all of them die after spawning. No male or female survives to undertake the return journey. Studies have shown that young salmon hatched in freshwater cannot return to the sea until their salt-secreting cells have developed.

Studies indicate that migratory behaviour in both, adults and the juveniles is not due to any special hormone regulation other than that which influences their general metabolism. Changes in their body states and behaviour may expose them to different environmental conditions to which they react in specific ways, such as travelling up freshwater streams by adults and downstream by juveniles. Several theories have been proposed to account for the unique way salmons are able to return to the very place where they were hatched. Some authorities believe that the fish have an extremely sensitive chemical sense that enable them to perform this feat. Another theory links the accelerated metabolism at spawning time with the need for more oxygen which increases as they ascend the hard waters of a stream. Still another idea stresses the importance of a carbon dioxide gradient in the water as being the determining factor.

Recent researches strongly favour the theory that sense of smell in fishes is the determining factor in locating home grounds, for different streams have different odours. The *Hilsa* of the Indian Ocean also migrates to rivers for spawning. This species is known to ascend up the Ganges and go up to Varanasi and Allahabad. During the monsoon months *Hilsa* also migrates to many rivers of southern India. However, the details of its migratory habit are not known. On the whole, it may be said that the run of Salmon up the river is connected with spawning, yet some important exceptions have to be made.

Many specimens will go some distance up a river only to turn back (and this may happen several times), or even remain months in definitely fresh water before proceeding to spawn. There may be a simple explanation of these eccentricities: it must be remembered that the Salmon is not far from being a freshwater fist. According to Paton and Newbigin (1900) the factor determining migration from sea to river is not the "*nisus generativus*, but the state of nutrition." The spawning comes later. With the Salmon we may finish this account of the one stream of migratory fishes, those which journey from deeper to shallower and even fresh water in order to fulfil their life purpose.

This is by far the most common form of migration; a very large number of marine fishes, even the Sharks, Skates, and Rays, follow in the main the same course.

As suggested by Meek, the term "anadromous," formerly used only for those entering fresh water, may well be applied to all those fishes which follow the same general rule in their migration, namely, an ascent from deeper to shallower waters. To a very large extent, this at once indicates the nature of the problem to be solved and even suggests the solution. But there are two groups of apparent exceptions. In the one the fishes exhibit but little tendency to undertake any spawning migrations. They spawn, as it were, where they find themselves. This applies naturally to the stationary fishes, though even among these a certain amount of migration can be noted. Thus, among the Blennies the Butterfish or Gunnel has a short, anadromous migration of a few miles, whilst the Angler or Frog-fish shows a longer migration in the reverse direction.

Among the Gadoids, again, most species prefer the deep water for spawning; but the Cod, if it has any preference, comes from the deep water for spawning; but the Cod, if it has any preference, comes from the deep water to shallower banks. These groups, in which the migratory movements are of little account or in which different tendencies are shown by different species, may be called neutral for the moment. The probability is, indeed, that they are not so much affected by the spawning process. The seeming migrations may be either a return to the natural home place, following upon the drift of the young away from these, or they may be mixed up with the movements in search of food.

The Cod, for example, is a rover with a circumpolar distribution like the Herring, and it shows a decided preference for the latter as food. It pursues the later at all seasons of the year and at the proper time it is able to spawn whenever it goes—within certain limits, of course but they are very wide and correspond closely to those of the Herring. From the biological standpoint these neutral groups are of great importance in any theory regarding the causes of spawning migrations. We may take it, that the species composing them are so well-adapted to their surrounds that they are not urged to make any great change when the need of spawn comes upon them. Viewing their life history as a whole, they show a better balance between the inner and outer conditions. The second exception is represented by the Eel group.

With the Common Flounder the Eel constitutes the old group of katadromous fishes, those which migrate from the fresh water to spawn in the sea. The Flounder is certainly katadromous like the Plaice seeks water of greater density than its ordinary feeding grounds for the purpose of spawning; but other Flatfishes like the Soles show the reverse procedure, and it seems better to regard the group as neutral. There is in any case no comparison with the life-history and migrations of the Eel. The Eel seems to belong to quite different world from other fishes; it would be difficult to imagine a greater contrast. Whilst other fishes are content with a reasonable area of distribution, thus admitting the limits of the piscine nature, the Eel range between the most extreme conditions on the globe, from mountain stream to the depths of the Atlantic. It passes through two transformations in the course of its life, when it enters and when it leaves fresh water, other fishes have only one or none at all.

The Eel takes some ten to fourteen years (in fresh water, plus three more in salt water) to become mature, other fishes everywhere only one to three. Other Eels, Conger, Muraena, etc., do not show such a great contrast, so far as we yet know their life-histories. All seem to be katadromous fishes, migrating out from shallow to deeper water, but only the *Anguilla* genus enters fresh water. Eels have two distinct phases in their life-history, each morphologically different from the other. Yellow eels represents the feeding and spawning phase while silver eels represent breeding phase. Yellow eels are found both in salt and freshwater and vary in length from a few inches to five or six feet.

During autumn a number of yellow eels become silvery and prepare to leave for spawning grounds. They stop feeding, their eyes become large, lips thinner, and the snouts sharper. Their colour becomes silvery along the sides with a blackish back. The reproductive organs develop while the alimentary canal shrinks. The eels migrate down the rivers to reach the sea. They travel about three to four thousand miles in the Atlantic to reach their breeding grounds in the western Atlantic, south of Bermuda. It is believed that the eels spawn at a depth of about 400 metres below the surface and the parents die soon after. The young larvae that hatch out are called leptocephali. They are flat, leaf like, tiny creatures provided with long needle-shaped teeth for feeding. They begin their homeward journey in the easterly direction. The grow rapidly during the first few months and when reach the coast of western European they are about three inches long and two year old. At this point they undergo metamorphosis and stop feeding.

The needle-like teeth are lost and the transparent body assumes a cylindrical shape. They are now called elvers, or glass eels. The elvers, when about three years old, start ascending the rivers in large numbers. They are able to cross all obstacles and reach their suitable resting places where they feed and grow for some years to develop into yellow eels. Finally on attaining maturity they change into silver eels and start moving towards their breeding site in the sea like their parents. The American species, *Anguilla rostrata* also has similar breeding habits. In the western Atlantic larvae of both species are found living together. The interesting and mysterious aspect of this strange migratory phenomenon are, how the tiny elvers without any guidance are able to find their way across the see and why the two species which practically hatch in the same region do not get mixed.

The explanation lies in the fact their the American eel grows more rapidly (only in one year against three years in the case of the European species). Thus, if the larva of the European eel travels westward instead of an easterly direction, it will reach the coast of America long before it is ready to change into an elver. And conversely, if the larva of the American eel migrates in an easterly direction it will undergo metamorphosis in the middle of the Atlantic. In other words, the larval life in each case in geared to the distance it has to travel. A recent theory, however, suggests that both American and European eels belong to the same species and that the adult European forms die before reaching the spawning grounds. All eggs are laid by American eels but, when hatched, the larvae in the northern part of the spawning grounds are carried away by currents to Europe. Those in the southern part are carried towards the American coast. By a series of brilliant observations and deductions *Johs*.

Schmidt has at length traced the Eel to its spawning places (1921), in an indirect yet quite convincing manner. The post-larval Eels (Leptocephali) were found in ever increasing quantities the further he proceeded westwards, by they became gradually smaller. Successive line or zones could thus be marked off, according to the size, until at length he found the smallest larva with yolk-sac still attached and event the eggs our over the deepest parts of the Atlantic near the West Indies and nearer the American than the European coasts. These were taken in the middle and upper layers, and there is therefore some doubt whether the adults actually spawn at the bottom. The depths on the "spawning grounds" were about 3000 fathoms. Two more of Schmidt's discoveries are of interest. At the same place where he found the larvae of the European Eel he also took those of the American Eel.

The spawning grounds of two forms, living thousands of miles apart, are practically the same, and the young of both species were occasionally taken in the same haul of the net. The second discovery is even more extraordinary. The two streams of post larvae spread out and move northwards towards the Bermudas; there they separate, one going east to Europe, the other west to America. The latter form spends only one year in the Atlantic and is then ready to enter fresh water; the European form takes three years to cross the Atlantic before it is ready to change its constitution and enter the rivers. How these tiny fishes can settle which direction to take, even allowing for a difference in constitution, is a profound mystery. In illustration of the immense quantities of these young Leptocephali it may be mentioned that Johs. Schmidt took as many as eight hundred specimens in one haul out in mid-Atlantic.

Factor Affecting Migration

Many efforts have been made to connect the migrations of fishes with the external conditions, temperature, salinity, light, etc., and it is probable that each has some influence, but the problem has another side. In the first place, it is essential to distinguish the food concentrations and wanderings from the spawning migrations, a matter not so easy as at first appeared, It would seem, indeed, that a feeding concentration, in many cases if not at all, precedes to actual spawning, and it may be that the storing of foodstuffs leads directly to the internal changes in metabolism which give the stimulus to change of grounds and conditions (Paton, 1900). Physiologically the feeding and the migratory impulses are very different. In the one the fish is seeking to obtain something, in the other to get rid of something. The latter may be called a morbid, even pathological condition, affacing every part and sufficiently powerful to make the fish ignore terrible injuries and even to over-come the impulse to feed.

The digestive organs partially or wholly degenerate, and the fish in then urged to seek relief by unusual or extraordinary means, that is, to find if possible external conditions in better agreement with its changing or changed constitution. One cannot say definitely what happens in any particular case, owing mainly to the protean nature of fishes. A Dog-fish, for example, is more in harmony with the surrounding medium than a Salmon, and Stickleback will survive under conditions that kill most fishes. These phenomena depend apparently on their ability to alter the blood-pressure, or concentration of salts, as they pass from one set of conditions to another. And this ability, again, depends upon how far the gill-membrane and the skin permit of the passage of water outwards and inwards, a question that has not yet been sufficiently investigated. It is possible, therefore, that the internal stimulus to migration comes from some internal change in the blood; probably in the blood-pressure. This is always much less than the pressure of the surrounding water, but it varies.

In the Salmon, *Greene* found that the concentration of salts in the blood was far below that of the salt water in which the fishes was feeding voraciously, and gradually altered as the fish entered fresh water. It may be that a slight reduction in the pressure within the blood is a source of discomfort to the fish and urges it to enter the rivers. In the case of Eel, on the other hand, there would seem to be a great increase of the internal pressure accompanied by an increase in the amount of carbonic acid at the time when the fishes are beginning to migrate from fresh to salt water. If they are detained on their journey in any way, the body thickens and swells and even increase greatly in weight, though the fish are apparently not feeding.

Petersen has shown that the digestive tract is in process of degeneration at this time and even the air-bladder decreases in size. The average Silver Eel in its breeding dress is about 70 cm. (27 to 28 ins.) in length and weighs about 1 ½ lb. The large specimens referred to were from 110 cm., and more, and weighed from 4 to over 8 lb. Most remarkable of all is the great increase in the size of the eyes. These become more than four times their former size and weigh twice as much. Without being able to say positively that such is the case, one may conclude that these phenomena point to a great increase in the internal concentration of salts and absorption of water. And this increased pressure with seems to be the cause of the Eel leaving the rivers to seek the greater pressures of the salt water. When the impulse to migrate comes upon them, the Eels are restless and ill at ease, moving up and down for a time in a manner quite unlike their usual sluggishness, then definitely turning down the rivers.

At every stage of their descent they come under increasing pressures without. And it may be recalled here, that the Eels are well-known to be extremely sensitive to the differences of pressure produced by thundery weather. In the case of marine fishes like the Plaice or God, we know that when the spawning fish are brought up in the trawl, the spawn continues to run even when the fishes are still alive. This indicates that the internal pressure persists for some time at the level of the deep water whence the fishes came. Since the low pressure inshore would help in getting rid of the spawn, the migration of the Plaice from shallow to deep water would be unnecessary for that purpose. Hence, we may conclude, that an increase of pressure within has occurred with the maturation of the sexual organs, and the resulting discomfort has urged the fish to seek deeper water and greater pressures. Similarly in the case of other fishes. Each species is urged to seek the external conditions, whether of lower or greater pressure, which corresponds best to the change in its own constitution.

Migration in Fishes and Birds

Though the migrations of fishes may appear the same superficially, they re widely different in important matters. They are not seasonal migrations in the same sense and the young take no part in them. Fishes have nothing to fear from the winter, not those we are dealing with at any rate; both the Salmon and the Herring spawn during the winter months. Further, the migration brings an advantage or benefit to the birds. On the one hand, they are urged on by the constitutional fever within; on the other, they are enticed forwards by what lies at the end of journey, this is wanting in the fishes; the urge is there but not the inducement or reward. If an Eel could foresee what lay before it in the depths of the Atlantic, would it undertake the journey? We may say, then, that the migrations of fishes cannot be regarded as adaptations, as we apply that term to birds. They are on a lower level, more primitive and crude. In all probability they are an inheritance from remote ancestors, which the fishes from lack of sufficient intelligence perhaps have not been able to overcome. Again, it should be noted that the birds do not show contradictory phenomena; the migrations are always to avoid the severity of the external conditions.

We might say this of the summer spawners among fishes; but all sorts and conditions are to be found, and the extreme cases, Salmon and Eel, seem wilfully to be seeking the most dangerous and most deadly conditions to be found on the globe. How far the fish in its migrations is guided in any way by its own sense organs, we cannot say. The external conditions—temperature, salinity, dissolved gasse, etc.—all make for differences in pressure in diverse ways. Possibly, as Hofer has suggested, the extreme sensibility of the lateral line system to slight variations in the strength of currents, may enable the Salmon to detect the entrance of a river or side-stream. But on the whole, one may doubt whether the senses play much part in the directing of the spawning migrations.

INDEX

I

J

K

L

M

N